Proceedings
of the
Steklov Institute of Mathematics

edited by
S. M. Nikol'skiĭ

Number 138 (1975)

Approximation of functions and operators

edited by

S. B. Stečkin

Providence, Rhode Island
AMERICAN MATHEMATICAL SOCIETY
1977

Академия Наук
Союза Советских Социалистических Республик

ТРУДЫ
ордена Ленина
МАТЕМАТИЧЕСКОГО ИНСТИТУТА
имени В. А. СТЕКЛОВА
CXXXVIII

Ответственный редактор
академик С. М. НИКОЛЬСКИЙ

ПРИБЛИЖЕНИЕ ФУНКЦИЙ
И ОПЕРАТОРОВ

Сборник статей
под редакцией
С. Б. СТЕЧКИНА

Издательство "Наука"
Москва 1975

AMS (MOS) subject classifications (1970). Primary 26A84, 41A05, 41A15, 41A35, 41A50, 41A65, 46E30; Secondary 15A42, 41A15, 41A30, 41A35, 34B05, 47E05.

ABSTRACT. This collection consists of original papers devoted to the investigation of approximations of functions and operators, and related problems. Approximations of functions by splines (N. L. Zmatrakov, N. I. Černyh) are studied, as well as interpolation in the mean by splines (Ju. N. Subbotin). The cycle of papers by V. V. Arestov and N. P. Kupcov is devoted to the approximation of operators, to Kolmogorov's inequalities in $L_2[0, \infty)$ and to the dual problem of approximating a class by a class. The collection will be of interest to workers in the area of approximation theory and also to specialists in computational mathematics.

Library of Congress Cataloging in Publication Data CIP

Main entry under title:

Approximations of functions and operators.

(Proceedings of the Steklov Institute of Mathematics ; 138 (1975))
Translation of Pribilizhenie funktsiĭ i operatorov.
Includes bibliographies.
1. Approximation theory--Addresses, essays, lectures.
I. Stechkin, S. B. II. Series: Akademiia nauk SSSR. Mathematicheskiĭ institut. Proceedings ; no 138.
QA1.A413 no. 138 [QA221] 511'.4 77-8940
ISBN 0-8218-3038-4

PROCEEDINGS OF THE STEKLOV INSTITUTE OF MATHEMATICS
IN THE ACADEMY OF SCIENCES OF THE USSR

(Труды математического института им. В. А. Стеклова, т. CXXXVIII, 1975)

TABLE OF CONTENTS

Trudy Mat. Inst. Steklov.
138 (1975)

Proc. Steklov Inst. Math.
138 (1975)

ON SOME EXTREMAL PROBLEMS FOR DIFFERENTIABLE FUNCTIONS OF ONE VARIABLE

UDC 517.5

V. V. ARESTOV

ABSTRACT. The duality of the problem concerning the least constant in the Hadamard-Kolmogorov inequality with the problem concerning the best approximation of a class of differentiable functions by another similar class of smoother functions as well as the duality of the Stečkin problem concerning the best approximation of a differentiation operator by bounded linear operators with the problem concerning the linear approximation of a class by a class are proved in the L_p-spaces on the real line and on the half-line $[0, \infty)$.

Bibliography: 28 titles.

Introduction

In the present paper we state dual problems for sharp Hadamard-Kolmogorov inequalities involving the norms of functions and their derivatives and for the Stečkin problem concerning the best approximation of differentiation operators by bounded operators.

In the Introduction the problems are posed, a survey of the known results is given and some simple properties of the quantities under consideration are cited.

In the sequel k and n are integers such that $0 \leqslant k < n$, the letters p, q and r denote parameters satisfying the condition $1 \leqslant p, q, r \leqslant \infty$, and I is either the real line or the half-line $[0, \infty)$. The notation $\|x\|_p$ will be used for the norm of a function x in the space $L_p = L_p(I)$, i.e. we put

$$\|x\|_p = \|x\|_{L_p(I)} = \begin{cases} \left\{\int_I |x(t)|^p \, dt\right\}^{1/p}, & 1 \leqslant p < \infty, \\ \operatorname*{ess\,sup}_{t \in I} |x(t)|, & p = \infty. \end{cases}$$

Let W_p^n for $1 < p \leqslant \infty$ be the set of functions x whose derivative $x^{(n-1)}$ of order $n-1$ is locally absolutely continuous on I and whose nth derivative $x^{(n)} \in L_p$, and let W_1^n be the set of functions x of the form

$$x(t) = \sum_{i=0}^{n-1} \frac{a_i}{i!} t^i + \frac{1}{(n-1)!} \int_0^t (t-\tau)^{n-1} \, d\eta(\tau),$$

where the a_i are real numbers, η is a function of bounded variation and $\eta(0) = -a_{n-1}$ (and hence $x^{(n-1)} = \eta$); when $x \in W_1^n$, we will write $\|x^{(n)}\|_1$ in place of $\bigvee x^{(n-1)}$.

AMS (MOS) subject classifications (1970). Primary 41A50, 41A65, 26A84; Secondary 41A35, 41A30.

Finally, we denote by $W^n_{p,r}$ the intersection of the spaces L_r and W^n_p, and by $Q = Q^n_{p,r}$ the set of functions $x \in W^n_{p,r}$ having the property $\|x^{(n)}\|_p \leqslant 1$.

We will consider the following quantity for an arbitrary pair of numbers $\mu_0 > 0$, $\mu_n > 0$:

$$\mu_k = \sup \| x^{(k)} \|_q,$$

where the supremum is taken over all functions $x \in W^n_{p,r}$ satisfying the conditions $\|x\|_r \leqslant \mu_0$, $\|x^{(n)}\|_p \leqslant \mu_n$. By making use of the homogeneity of the norm in $L_s(I)$ under the substitution of $cx(ht)$ for $x(t)$, one can obtain the dependence of μ_k on μ_0 and μ_n; namely, the following formula holds (see [1]):

$$\mu_k = K \mu_0^\alpha \mu_n^\beta ,$$

in which

$$\alpha = \frac{n - k - p^{-1} + q^{-1}}{n - p^{-1} + r^{-1}}, \qquad \beta = 1 - \alpha,$$

and K is a constant not depending on μ_0 or μ_n.

It follows that the functions $x \in W^n_{p,r}$ satisfy the inequality

$$\| x^{(k)} \|_q \leqslant K \| x \|_r^\alpha \| x^{(n)} \|_p^\beta . \tag{0.1}$$

Such inequalities were first considered by Hardy and Littlewood [2], Landau [3] and Hadamard [4]. Important results in the problem of investigating inequality (0.1) have been obtained by Kolmogorov [5] and Sz.-Nagy [1]. At the present time the constant K and the set of extremal functions on which (0.1) becomes an equality are known for many values of the parameters k, n, p, q, r and I (see the bibliographies in [6]–[8]). Definitive conditions for the finiteness of the constant K in (0.1) have been obtained by Gabušin [6]. He proved that K is finite if and only if

$$\frac{n-k}{r} + \frac{k}{p} \geqslant \frac{n}{q} . \tag{0.2}$$

We will also consider the Stečkin problem [9], [10] concerning the best approximation

$$E(N) = \inf_{\|T\|_{L_r}^{L_q} \leqslant N} \ \sup_{x \in Q^n_{p,r}} \| x^{(k)} - Tx \|_q \tag{0.3}$$

of the differentiation operator of order k by bounded linear operators T from L_r into L_q on the class of n-times differentiable functions. The investigation of the quantity (0.3) includes a study of the existence, uniqueness and other properties of an extremal operator, i.e. an operator T satisfying the conditions $\|T\| = \|T\|_{L_r}^{L_q} \leqslant N$ and $E(N) = U(T)$, where

$$U(T) = \sup_{x \in Q} \| x^{(k)} - Tx \|_q .$$

If $x \in W^n_{p,r}$, then $x^{(k)}$ is a continuous function unless $k = n - 1$ and $p = 1$. It is assumed in this connection that if q or r is infinite, then L_q or respectively L_r is the space C of continuous functions (unless $q = \infty$, $k = n - 1$, $p = 1$).

Stečkin [10] noted a connection between the quantity $E(N)$ and the least constant K in (0.1). This connection is expressed by the inequality

$$E(N) \geqslant \Omega(N), \tag{0.4}$$

where

$$\Omega(N) = \sup_{x \in Q} (\| x^{(k)} \|_q - N \| x \|_r) = \beta \alpha^{\alpha/\beta} K^{1/\beta} N^{-\alpha/\beta}. \tag{0.5}$$

The following formula holds for $E(N)$:

$$E(N) = N^{-\gamma} E(1), \qquad \gamma = \frac{\alpha}{\beta} = \frac{n - k - p^{-1} + q^{-1}}{k + r^{-1} - q^{-1}};$$

it is proved in [10] for $p = q = r$, while the general case, which is obtainable by the same method, is cited in [11]. A solution of problem (0.3) for concrete values of the parameters is given in [7]–[10] and [12]–[18]. In some of these cases the corresponding sharp inequality (0.1) was previously known, while in others (see [14], [15], [17] and [18]) the solution permitted one to find the least constant K in (0.1). We note that a solution of problem (0.3) can be obtained in the case $p = q = r = \infty$, $I = (-\infty, \infty)$ with the use of a result of Domar [19] (see [10] for $n = 2, 3$, and [12] for $n = 4, 5$).

The quantity $E(N)$ is not always finite; as Gabušin has shown [20], it is finite if and only if

$$q \geqslant p, \qquad q \geqslant r. \tag{0.6}$$

For $q = \infty$ it is natural to consider in conjunction with $E(N)$ the quantity

$$e(N) = \inf_{\|T\|_{L_r^*} \leqslant N} \ \sup_{x \in Q^n_{p,r}} \{x^{(k)}(0) - Tx\} \tag{0.7}$$

of best approximation of the value of a differentiation operator at $t = 0$ by bounded linear functionals T (see [9]–[12], [14]–[16] and others). Gabušin [16] proved that the equality

$$e(N) = \sup_{x \in Q} \{x^{(k)}(0) - N \| x \|_r\} = \beta \alpha^{\alpha/\beta} K^{1/\beta} N^{-\alpha/\beta} \tag{0.8}$$

holds (here $q = \infty$). But $e(N)$ coincides with $E(N)$ for $q = \infty$ (see [11] as well as [9], [10], [12], [14]–[16], [18] and Corollary 2 below); therefore inequality (0.4) becomes an equality when $q = \infty$. The equality $E(N) = \Omega(N)$ also holds in some other cases, e.g. (see [13]) when $p = q = r = 2$ and $I = (-\infty, \infty)$.

The constant K in (0.1) and the quantity (0.3) are connected with the approximation of one class of functions by another, i.e. with a quantity of the form

$$\sup_{\psi \in \Psi} \inf_{\varphi \in \Phi} \| \psi - \varphi \|_{L_s}.$$

The latter problem arose as a subsidiary problem in, for example, [21] and [22]. It was studied as an independent problem by Taĭkov [23], who indicated a connection between it and a problem of form (0.3). It will be discussed here for two pairs of classes of differentiable functions.

Let $1 \leqslant p', q', r' \leqslant \infty$ and let m, n be integers such that $0 < m \leqslant n$, let $B^n_{r'}(N) = B^n_{r'}(N, I)$ be the set of functions $\phi \in W^n_{r'}$ for which $\|\phi^{(n)}\|_{r'} \leqslant N$, let $B^m_{q'} = B^m_{q'}(1, I)$ and let $\mathfrak{M}(N) = \mathfrak{M}(N, I)$ be the set of linear (homogeneous and additive) operators S from $W^m_{q'}$ into $W^n_{r'}$ such that

$$\|(S\psi)^{(n)}\|_{r'} \leqslant N\|\psi^{(m)}\|_{q'} \tag{0.9}$$

for $\psi \in W^m_{q'}$. We put

$$F(\psi, N) = F(\psi, N, I) = \inf_{\varphi \in B^n_{r'}(N)} \|\psi - \varphi\|_{p'}, \qquad \psi \in W^m_{q'},$$

$$J(S) = J(S, I) = \sup_{\psi \in B^m_{q'}} \|\psi - S\psi\|_{p'}, \qquad S \in \mathfrak{M}(N).$$

Then the quantity

$$F(N) = F(N, I) = \sup_{\psi \in B^m_{q'}} F(\psi, N, I) \tag{0.10}$$

is the approximation in the metric of $L_{p'}$ of the class $B^m_{q'}$ by the class $B^n_{r'}(N)$ while

$$G(N) = G(N, I) = \inf_{S \in \mathfrak{M}(N)} J(S) \tag{0.11}$$

is the corresponding linear approximation of a class by a class. Clearly,

$$F(N) \leqslant G(N). \tag{0.12}$$

The quantities $F(N)$ and $G(N)$ have been studied in [24], [13], [25] and [26]. In particular, it is known [25] that

$$F(N) = N^{-\gamma}F(1), \qquad G(N) = N^{-\gamma}G(1),$$
$$\gamma = \left(m + \frac{1}{p'} - \frac{1}{q'}\right) \Big/ \left(n - m + \frac{1}{q'} - \frac{1}{r'}\right), \tag{0.13}$$

and that the quantity $F(N)$ is finite if [26] and only if [25]

$$\frac{m}{r'} + \frac{n-m}{p'} \leqslant \frac{n}{q'} \tag{0.14}$$

We cite values of the parameters p', q', r' and n for which the quantity $F(N)$ is known (Table 1).

TABLE 1

n	q'	r'	p'	Authors
arbitrary	2	2	2	Taĭkov, Subbotin [13]
$2 \leqslant n \leqslant 5$	1	1	1	Subbotin [24]
arbitrary	∞	∞	∞	Subbotin [24], Arestov, Gabušin [25]

We will consider similar problems for another pair of classes. Let $w^n_{r'}$ and $b^n_{r'}(N) = b^n_{r'}(N, I)$ be respectively the sets of functions $x \in W^n_{r'}$ and $x \in B^n_{r'}(N)$ satisfying the condition $x^{(i)}(0) = 0$ for $i = 0, 1, \ldots, n-1$; in exactly the same way, let $w^m_{q'}$ and

$b^m_{q'}$ be the sets of functions $x \in W^m_{q'}$ and $x \in B^m_{q'}$ with the property $x^{(i)}(0) = 0$ for $i = 0, 1, \ldots, m-1$. We put

$$f(\psi, N) = f(\psi, N, I) = \inf_{\varphi \in b^n_{r'}(N)} \|\psi - \varphi\|_{p'},$$
$$f(N) = f(N, I) = \sup_{\psi \in b^m_{q'}} f(\psi, N, I). \tag{0.15}$$

In addition, we denote by $\mathfrak{M}'(N) = \mathfrak{M}'(N, I)$ the set of linear operators S from $w^m_{q'}$ into $w^n_{r'}$ satisfying (0.9), and put

$$j(S) = j(S, I) = \sup_{\psi \in b^m_{q'}} \|\psi - S\psi\|_{p'},$$
$$g(N) = g(N, I) = \inf_{S \in \mathfrak{M}'(N)} j(S). \tag{0.16}$$

It is not difficult to show that formulas (0.13) with F and G replaced by f and g hold for (0.15), (0.16) (see the proof of (0.13) in [25]). By virtue of the equality

$$F(N) = \sup_{\psi \in b^m_{q'}} \inf_{\varphi \in B^n_{r'}(N)} \|\psi - \varphi\|_{p'}$$

and the imbedding $b^n_{r'}(N) \subset B^n_{r'}(N)$ we have

$$F(N) \leqslant f(N). \tag{0.17}$$

Also

$$G(N) \leqslant g(N), \tag{0.18}$$

since the formula

$$S\psi = \psi - \psi_0 + S'\psi_0,$$

where

$$\psi_0(t) = \psi(t) - \sum_{i=0}^{n-1} \psi^{(i)}(0) \frac{t^i}{i!}, \qquad \psi \in W^m_{q'},$$

associates with each operator $S' \in \mathfrak{M}'(N)$ an operator $S \in \mathfrak{M}(N)$ for which $j(S') = J(S)$.

Sometimes $G(N)$ (and hence $F(N)$) is bounded from above by $E(N)$ (see [23], [24], [13] and [25]). More precisely, suppose

$$m = n - k, \qquad p = r = q', \qquad p' = r' = q, \tag{0.19}$$

problem (0.3) has an extremal operator T that is defined, linear and permutable with the differentiation operator of order n on W^n_p and (cf. (0.3))

$$\sup_{x \in B^n_p} \|x^{(k)} - Tx\|_q = E(N). \tag{0.20}$$

Then

$$G(N) \leqslant E(N), \tag{0.21}$$

and hence

$$F(N) \leqslant E(N). \tag{0.22}$$

In fact, with each function $\psi \in W_p^{n-k}$ we associate a function $x \in W_p^n$ such that $x^{(k)} = \psi$, and define an operator S by the formula

$$S\psi = Tx, \quad \psi \in W_p^{n-k}.$$

We have

$$(S\psi)^{(n)} = (Tx)^{(n)} = Tx^{(n)} = T\psi^{(n-k)},$$

so that condition (0.9) is satisfied for S and, consequently, $S \in \mathfrak{M}(N)$. Further, by virtue of (0.20)

$$\| \psi - S\psi \|_q = \| x^{(k)} - Tx \|_q \leqslant E(N) \| x^{(n)} \|_p = E(N) \| \psi^{(n-k)} \|_p,$$

which implies (0.21).

In the present paper we establish a connection between problem (0.1) and problems (0.10), (0.15) and between problem (0.3) and problems (0.11), (0.16) under the following relations between the parameters:

$$m = n - k, \quad \frac{1}{p} + \frac{1}{p'} = 1, \quad \frac{1}{q} + \frac{1}{q'} = 1, \quad \frac{1}{r} + \frac{1}{r'} = 1. \tag{0.23}$$

It is shown in particular (Theorem 2) that the following equalities hold under the fulfillment of conditions (0.23):

$$\Omega(N) = \beta\alpha^{\alpha/\beta} K^{1/\beta} N^{-\alpha/\beta} = \begin{cases} F(N), & \text{if} \quad I = (-\infty, \infty), \\ f(N), & \text{if} \quad I = [0, \infty), \end{cases} \tag{0.24}$$

i.e. problems (0.10) and (0.15) are dual to (0.1) when $I = (-\infty, \infty)$ and $I = [0, \infty)$ respectively. In addition, we prove (Theorem 4) that the quantities $F(N)$ and $f(N)$ do not depend on the set I. These two facts permit us to state the values of $F(N)$ and $f(N)$ in those cases when a sharp inequality (0.1) is known. Many of these cases for $F(N)$ are collected in Table 2; in the last column we indicate the authors of inequality (0.1) for the corresponding values of the parameters (when $I = (-\infty, \infty)$).

TABLE 2

n	q'	r'	p'	Literature
arbitrary	1	1	1	Kolmogorov [5]
1	$\leqslant r'$	arbitrary	arbitrary	Sz.-Nagy [1]
arbitrary	1	2	2	Taĭkov [14]
2	2	arbitrary	$\frac{r'}{r'-1}$	Hardy, Littlewood, Pólya [27]

We note that the inequality

$$F(N) \leqslant E(N) \tag{0.25}$$

follows from (0.4), (0.24) and (0.17) under conditions (0.23), which differ from conditions (0.19) in (0.22).

In Theorem 3 we assert that the equalities

$$E(N)=\begin{cases}G(N), & \text{if} \quad I=(-\infty,\infty),\\ g(N), & \text{if} \quad I=[0,\infty);\end{cases} \tag{0.26}$$

are valid for $1<q\leqslant\infty$, $1<p\leqslant\infty$; here, as before, the parameters are subject to conditions (0.23).

Theorems 2 and 3 are concretizations of results of the note [28].

If $k=0$ and $q=r$, then, clearly, $K=1$ while $E(N)=0$ for $N\geqslant 1$ and $E(N)=\infty$ for $N<1$. In exactly the same way, if $m=n$ and $q'=r'$, then $F(N)=G(N)=0$ for $N\geqslant 1$ and $F(N)=G(N)=\infty$ for $N<1$; an analogous remark can be made for $f(N)$, $g(N)$.

Thus, when $k=0$ and $q=r$, i.e. (see (0.23)) when $m=n$ and $q'=r'$, equalities (0.26) hold while equalities (0.24) are degenerate ($\beta=0$). Therefore the conditions

$$\frac{n-k}{r}+\frac{k}{p}\geqslant\frac{n}{q}, \quad \text{with} \quad q>r, \quad \text{if} \quad k=0, \tag{0.27}$$

will sometimes be used in place of (0.2).

§1. Auxiliary assertions

Let $\mathcal{W}_s^l$ denote the space W_s^l if $I=(-\infty,\infty)$, and the space w_s^l if $I=[0,\infty)$; the classes $\mathcal{B}_s^l(N)$, $\mathcal{B}_s^l$ and quantities $\mathcal{F}(\psi,N)$, $\mathcal{F}(N)$, $\mathcal{J}(S)$, $\mathcal{G}(N)$ are defined analogously. Finally, we associate with an integer $i\in[0,n]$ and a function $\chi\in\mathcal{W}_s^{n-i}$ the functional

$$M_\chi^i x=\begin{cases}(-1)^{n-i}\int x^{(i)}(t)\,\chi^{(n-i)}(t)\,dt, & 1<s\leqslant\infty,\\ (-1)^{n-i}\int x^{(i)}(t)\,d\chi^{(n-i-1)}(t), & s=1;\end{cases} \tag{1.1}$$

we put $M_\chi^0=M_\chi$ when $i=0$. Here and below the symbol I is omitted in integrals over I.

LEMMA 1. *Suppose conditions* (0.27) *and* (0.23) *are satisfied and the difference* $\omega=\psi-\phi$ *of a pair of functions* $\psi\in\mathcal{W}_{q'}^{n-k}$, $\varphi\in\mathcal{W}_{r'}^{n}$ *belongs to* $L_{p'}$, *the function* ψ *being continuous outside some finite interval if* $p=1$, $q=\infty$ *and* $k=n-1$.

Then for any function $x\in W_{p,r}^n$

$$M_\psi^k x-M_\varphi x=\begin{cases}\int x^{(n)}(t)\,\omega(t)\,dt, & 1<p\leqslant\infty,\\ \int\omega(t)\,dx^{(n-1)}(t), & p=1.\end{cases} \tag{1.2}$$

PROOF. Let κ be an infinitely differentiable function on I such that $0\leqslant\kappa\leqslant 1$, $\kappa(t)=1$ for $|t|\leqslant 1$ and $\kappa(t)=0$ for $|t|\geqslant 2$. We put $x_\delta(t)=\kappa(t\delta^{-1})x(t)$ ($\delta>0$). Clearly, the function x_δ has compact support and belongs to $W_{p,r}^n$ (if $x\in W_{p,r}^n$).

We obtain (1.2) for x_δ by integrating the integrals in its right sides by parts and taking into account the fact that $\varphi\in W_{r'}^n$ and $\psi\in W_{q'}^{n-k}$. We will show that its validity for x follows in the limit for $\delta\longrightarrow\infty$.

The relation $M_\varphi x_\delta\longrightarrow M_\varphi x$ is obvious.

In the case $1<p\leqslant\infty$ we have, according to Leibniz' formula,

$$x_\delta^{(n)}(t) = x^{(n)}(t)\varkappa(t\delta^{-1}) + \rho(t),$$

where

$$\rho(t) = \sum_{i=1}^{n} C_n^i x^{(n-i)}(t)\,\delta^{-i}\varkappa^{(i)}(t\delta^{-1}).$$

The derivatives $x^{(i)}$ for $i = 0, 1, \ldots, n-1$ are bounded for $q = \infty$ by virtue of the finiteness of the constant K in (0.1), so that

$$\|\rho\|_C \leqslant c_1\delta^{-1}.$$

And since ρ vanishes outside the interval $I = [-2\delta, 2\delta]$,

$$\|\rho\|_p \leqslant c_2\delta^{(1/p)-1}, \qquad 1 < p \leqslant \infty.$$

Hence

$$\int x_\delta^{(n)}(t)\,\omega(t)\,dt \to \int x^{(n)}(t)\,\omega(t)\,dt \quad (\delta \to \infty)$$

for $1 < p \leqslant \infty$.

Suppose $p = 1$. We note at once that the case $r = \infty$, $n = 1$ is excluded by condition (0.27), since then $k = 0$ and, according to (0.2), $q = \infty$, i.e. $k = 0$, $q = r$.

Inasmuch as $\bigvee x^{(n-1)} < \infty$, the derivative $x^{(n-1)}$ has limits for $t \to \pm\infty$, and since $x \in L_r$, these limits are equal to zero. As before,

$$x_\delta^{(n-1)}(t) = x^{(n-1)}(t)\,\varkappa(t\delta^{-1}) + \rho_1(t),$$

the function ρ_1 being absolutely continuous and such that

$$\bigvee \rho_1 = \|\rho_1'\|_L \leqslant c_3\|x^{(n-1)}\|_{L_\infty(I_\delta)} + c_4\delta^{-1},$$

where I_δ is the set of points $t \in I$ for which $|t| \geqslant \delta$. Hence $\bigvee \rho_1 \to 0$ as $\delta \to \infty$. Since the function $\omega = \psi - \phi$ is continuous (when $q = \infty$, $p = 1$ and $k = n-1$ on I_δ for sufficiently large δ), it follows by virtue of Helly's theorem that

$$\int \omega(t)\,dx_\delta^{(n-1)}(t) \to \int \omega(t)\,dx^{(n-1)}(t) \quad (\delta \to \infty).$$

The relation

$$M_\psi^k x_\delta \to M_\psi^k x \quad (\delta \to \infty)$$

is verified analogously. The lemma is proved.

Let us consider the quantity

$$u(\psi, T) = \sup_{x \in Q_{p,r}^n} \{M_\psi^k x - Tx\} \tag{1.3}$$

for a function $\psi \in W_{q'}^{n-k}$ and a linear functional $T \in L_r^*$.

When $1 \leqslant r < \infty$, the functional T has the form

$$Tx = \int x(t)\,\lambda(t)\,dt, \tag{1.4}$$

where $\lambda \in L_{r'}$ and $\|T\| = \|\lambda\|_{r'}$; if, on the other hand, $r = \infty$, then, at least on the set

C_0 of continuous functions having a zero limit at infinity, we have

$$Tx = \int x(t)\, d\Lambda(t), \tag{1.5}$$

where $\bigvee\Lambda = \|T\|_{C_0^*} \leqslant \|T\|_{L_\infty^*}$.

In the case $r = \infty$ we denote by T_0 the functional that is defined by (1.5) on the space C of continuous functions and linearly extended with preservation of norm onto the set L_∞; clearly,

$$\|T_0\|_{L_\infty^*} = \|T_0\|_{C^*} = \bigvee\Lambda \leqslant \|T\|_{L_\infty^*}.$$

When $1 \leqslant r < \infty$ we put $T_0 = T$.

LEMMA 2. *Suppose conditions* (0.23) *and* (0.27) *are satisfied. If a functional* $T \in L_r^*$ *and a function* $\psi \in W_{q'}^{n-k}$ *are such that*

$$u(\psi, T) < \infty,$$

the function ψ *being continuous outside some finite interval for* $p = 1$, $q = \infty$ *and* $k = n - 1$, *then there exists a unique function*

$$\varphi \in \mathcal{W}_{r'}^{n} \tag{1.6}$$

with the properties

$$(-1)^n \varphi^{(n)} = \lambda \qquad or \qquad (-1)^n \varphi^{(n-1)} = \Lambda, \tag{1.7}$$

$$\|\psi - \varphi\|_{p'} = u(\psi, T_0) \leqslant u(\psi, T), \tag{1.8}$$

and the representation

$$M_\psi^k x - T_0 x = \begin{cases} \int \{\psi(t) - \varphi(t)\}\, x^{(n)}(t)\, dt, & 1 < p \leqslant \infty, \\ \int \{\psi(t) - \varphi(t)\}\, dx^{(n-1)}(t), & p = 1 \end{cases} \tag{1.9}$$

holds for any function $x \in W_{p,r}^n$.

PROOF. Let H denote the set of infinitely differentiable functions with compact support, let H_p^n denote the set of functions $x \in H$ with $\|x^{(n)}\|_p \leqslant 1$ and let

$$v(T) = \sup_{x \in H_p^n} \{M_\psi^k x - Tx\};$$

clearly

$$v(T) = v(T_0) \leqslant u(\psi, T). \tag{1.10}$$

A function $x \in H$ is uniquely determined by its nth derivative, so that

$$M_\psi^k x - Tx = Sx^{(n)}, \tag{1.11}$$

where S is a functional on the set Y of nth derivatives $x^{(n)}$ of the functions $x \in H$. We will regard Y as a (normed) subspace of the space L_p for $1 \leqslant p < \infty$ and of the space C_0 for $p = \infty$. It is not difficult to verify that S is linear on Y and $\|S\|_{Y^*} = v(T)$.

Extending S linearly with preservation of norm from Y onto L_p or C_0, we get

$$Sy = \int y(t)\,\omega(t)\,dt, \qquad 1 \leqslant p < \infty,$$

$$Sy = \int y(t)\,d\nu(t), \qquad p = \infty,$$

where

$$v(T) = \|\omega\|_{p'} \qquad \text{or} \qquad v(T) = \bigvee \nu. \tag{1.12}$$

Integrating the integral in (1.4) or (1.5) by parts, we get

$$Tx = T_0 x = \int x^{(n)}(t)\,\overline{\varphi}(t)\,dt, \quad x \in H, \tag{1.13}$$

where $\overline{\varphi}$ is a function having properties (1.6) and (1.7). In exactly the same way, we find (on H) that

$$M_\psi^k x = \int x^{(n)}(t)\,\psi(t)\,dt.$$

When $p < \infty$, (1.11) takes the form

$$\int \{\psi(t) - \overline{\varphi}(t)\}\,x^{(n)}(t)\,dt = \int \omega(t)\,x^{(n)}(t)\,dt, \qquad x \in H.$$

It follows that $\omega = \psi - \overline{\varphi} - P_0$, where P_0 identically vanishes when $I = [0, \infty)$, and is a polynomial of degree $n-1$ when $I = (-\infty, \infty)$. The function $\varphi = \overline{\varphi} + P_0$ also yields the representation (1.13) of T and has properties (1.6) and (1.7).

For $p = \infty$ we have

$$\int \{\psi(t) - \overline{\varphi}(t)\}\,x^{(n)}(t)\,dt = \int x^{(n)}(t)\,d\nu(t), \qquad x \in H,$$

or

$$\int \{\delta(t) - \nu(t)\}\,x^{(n+1)}(t)\,dt = 0, \qquad x \in H,$$

where

$$\delta(t) = \int_0^t \{\psi(\tau) - \overline{\varphi}(\tau)\}\,d\tau + \nu(0).$$

And this means that $\nu = \delta - P_1$, where P_1 identically vanishes if $I = [0, \infty)$, and is a polynomial of degree n if $I = (-\infty, \infty)$. Consequently, the function ν is absolutely continuous, its derivative $\omega = \nu' = \psi - \overline{\varphi} - P_1'$ belongs to L_1, and $\|\omega\|_1 = \bigvee \nu$. In (1.13) the function $\varphi = \overline{\varphi} + P_1'$ can be taken in place of $\overline{\varphi}$.

If φ is actually chosen in this way, then, clearly,

$$\|\omega\|_{p'} = \|\psi - \varphi\|_{p'} = v(T_0) = v(T) \leqslant u(\psi, T) < \infty.$$

According to (1.7) and (1.1), $M_\varphi = T_0$. In addition, the conditions of Lemma 1 on the parameters and the functions ψ and φ are satisfied. Therefore (1.9) holds for any function $x \in W_{p,r}^n$. Hence

$$u(\psi, T_0) \leqslant \|\psi - \varphi\|_{p'} = v(T_0),$$

which implies (1.8) by virtue of (1.10).

If $I = [0, \infty)$, then the uniqueness of φ follows from (1.7) and (1.6) (in this case $\varphi^{(i)}(0) = 0$ for $i = 0, 1, \ldots, n-1$). Suppose $I = (-\infty, \infty)$ and $\bar{\varphi}$ also satisfies conditions (1.6)–(1.9). By virtue of (1.7), $\bar{\varphi}$ differs from φ by a polynomial P of degree $n-1$. According to (1.8), we have $\|P\|_{p'} = \|\bar{\varphi} - \varphi\|_{p'} < \infty$. It follows for $p' < \infty$ that $P \equiv 0$ and hence $\bar{\varphi} = \varphi$. If, on the other hand, $p' = \infty$, then $P = c =$ const. Clearly, the function $\bar{\varphi} = \varphi + c$ satisfies (1.6) and (1.7) for any constant c. It also satisfies (1.9), since $\lim_{t \to \pm\infty} x^{(n-1)}(t) = 0$ for $x \in W^n_{1,r}$; therefore

$$u(\psi, T_0) \leqslant \|\psi - \varphi - c\|_\infty.$$

The equality

$$\operatorname*{ess\,sup}_{t \in (-\infty, \infty)} \{\psi(t) - \varphi(t)\} = -\operatorname*{ess\,inf}_{t \in (-\infty, \infty)} \{\psi(t) - \varphi(t)\},$$

follows from (1.8), since we could otherwise choose the constant c so that

$$\|\psi - \varphi - c\|_\infty < \|\psi - \varphi\|_\infty = u(\psi, T_0).$$

Consequently, the function φ is also unique for $p' = \infty$. Lemma 2 is proved.

For a fixed function $\psi \in W^{n-k}_{q'}$ we put

$$\sigma(\psi, N) = \inf_{\|T\| \leqslant N} u(\psi, T), \tag{1.14}$$

where $\|T\|$ is the norm of a functional $T \in L^*_r$ and $u(\psi, T)$ is defined by (1.3) and (1.1).

COROLLARY 1. *Suppose conditions* (0.23) *and* (0.27) *are satisfied and* ψ *is a member of* $W^{n-k}_{q'}$ *that is continuous outside some finite interval when* $q = \infty$, $p = 1$ *and* $k = n - 1$. *Then*

$$\sigma(\psi, N) = \inf_{\varphi \in \mathcal{B}^n_{r'}(N)} \|\psi - \varphi\|_{p'}. \tag{1.15}$$

In fact, the inequality $\sigma(\psi, N) \leqslant F(\psi, N)$ follows from Lemma 1, since the functional T_φ def on $W^n_{p,r}$ by the formula $T_\varphi x = M_\varphi x$ has a natural extension onto L_r such that $\|T_\varphi\|_{L^*_r} = \|\varphi^{(n)}\|_{r'}$. The reverse inequality follows from Lemma 2.

REMARK 1. As can be seen from the proofs of Lemmas 1 and 2 and Corollary 1, in addition to (1.15) it is possible to assert that to each extremal functional of problem (1.14) there corresponds a function φ for which $\|\psi - \varphi\|_{p'} = F(\psi, N)$ and, conversely, each such function induces an extremal functional.

§2. Connection between problems in the case $q = \infty$

We first consider some properties of the quantity $e(N)$ and an extremal functional of problem (0.7).

Let θ be the function defined on I by the relation

$$\theta(t) = \frac{(-1)^{n-k}}{(n-k-1)!} t_+^{n-k-1}, \quad t_+ = \begin{cases} t & \text{for} \quad t > 0, \\ 0 & \text{for} \quad t \leqslant 0; \end{cases} \tag{2.1}$$

clearly $\theta \in b^{n-k}_1$, $\theta \in B^{n-k}_1$. We have $M^k_\theta x = x^{(k)}(0)$, so that

$$e(N) = \sigma(\theta, N),$$

and hence, by virtue of Corollary 1,

$$e(N) = \inf_{\varphi \in \mathcal{B}_{r'}^{n}(N)} \|\theta - \varphi\|_{p'}. \tag{2.2}$$

Here the case $k = 0$, $r = \infty$ is generally excluded. But it is not difficult to verify directly that (2.2) is also valid in this case.

There exists an extremal functional T in problem (0.7) (see [11] and [16]). By virtue of Remark 1, this means that there exists a function $\varphi \in \mathcal{B}_{r'}^{n}(N)$ on which the infimum in (2.2) is achieved. Suppose $1 < p, r < \infty$ ($q = \infty$). Then T has the form

$$Tx = \int x(t)\lambda(t)\,dt \tag{2.3}$$

and, by virtue of Lemma 2, for any function $x \in W_{p,r}^{n}$

$$x^{(k)}(0) - Tx = \int \omega(t)\,x^{(n)}(t)\,dt, \tag{2.4}$$

where

$$\|\omega\|_{p'} = e(N), \qquad \omega = \theta - \varphi, \quad \varphi \in \mathcal{B}_{r'}^{n}(N), \tag{2.5}$$

the function θ is defined in (2.1), and $\lambda = (-1)^n \varphi^{(n)}$.

According to the results of [16] (Theorems 1 and 2), there exists a function $y \in W_{p,r}^{n}$ such that $Ty = \|T\| \cdot \|y\|_r$, $y^{(k)}(0) - Ty = e(N)\|y^{(n)}\|_p$ and y is an extremal in (0.1) ($q = \infty$). Hence

$$y = c_1 |\lambda|^{1/(r-1)} \operatorname{sgn} \lambda, \tag{2.6}$$

$$y^{(n)} = c_2 |\omega|^{1/(p-1)} \operatorname{sgn} \omega, \tag{2.7}$$

where c_1 and c_2 are positive constants (relation (2.6) is given in [16]).

Relations (2.5)–(2.7) imply the following properties of ω:

1) The function ω has continuous derivatives up to order n on the half-lines $[0, \infty)$ and $(-\infty, 0]$ (if $I = (-\infty, \infty)$).

2) The conditions

$$\omega^{(i)}(0) = \begin{cases} (-1)^{n-k}, & \text{if} \quad i = n-k-1, \\ 0, & \text{if} \quad i \neq n-k-1, \end{cases} \qquad i = 0, 1, \ldots, n-1,$$

are satisfied at $t = 0$ when $I = [0, \infty)$, while the conditions

$$\omega^{(i)}(+0) - \omega^{(i)}(-0) = \begin{cases} (-1)^{n-k}, & \text{if} \ \ i = n-k-1, \\ 0, & \text{if} \ \ i \neq n-k-1, \quad i = 0, 1, \ldots, n-1; \end{cases}$$

hold at $t = 0$ when $I = (-\infty, \infty)$.

3) The function $|\omega^{(n)}|^{1/(r-1)} \operatorname{sgn} \omega^{(n)}$ has continuous derivatives up to order $n - 1$ on I.

4) The function ω satisfies the equation

$$\frac{d^n}{dt^n}\{|\omega^{(n)}|^{1/(r-1)}\operatorname{sgn}\omega^{(n)}\} = (-1)^{n-1}c|\omega|^{1/(p-1)}\operatorname{sgn}\omega, \tag{2.8}$$

where c is a positive constant; here $t \neq 0$ when $k = n-1$ and $I = (-\infty, \infty)$.

Indeed, we can assert,

In order for (2.3) *to be an extremal functional in problem* (0.7) *when* $N = \|\lambda\|_{r'}$ *and* $1 < p, r < \infty$ *it is necessary and sufficient that there exist a function* $\varphi \in B^n_{r'}(N)$ *such that* $\lambda = (-1)^n\varphi^{(n)}$ *while the difference* $\omega = \theta - \varphi \in L_{p'}$ *have properties* 1)–4); *in this connection,*

$$e(\|\varphi^{(n)}\|_{r'}) = \|\theta - \varphi\|_{p'}. \tag{2.9}$$

The necessity has just been proved. Let us prove the sufficiency. The functions φ and $\psi = \theta$ satisfy the conditions of Lemma 1. It follows by virtue of the equalities $x^{(k)}(0) = M^k_\theta x$ and $T = M_\varphi$ that (2.4) holds for any x, and this implies the estimate

$$e(\|\omega^{(n)}\|_{r'}) \leqslant u(\theta, T) \leqslant \|\omega\|_{p'}.$$

Let us estimate $e(N)$ from below. We put

$$y = (-1)^{n-1}c^{-1}\|\omega\|_{p'}^{-1/(p-1)}|\omega^{(n)}|^{1/(r-1)}\operatorname{sgn}\omega^{(n)}.$$

By virtue of properties 3) and 4), we have $y \in W^n_{p,r}$ and

$$y^{(n)} = \|\omega\|_{p'}^{-1/(p-1)}|\omega|^{1/(p-1)}\operatorname{sgn}\omega;$$

consequently $y \in Q^n_{p,r}$.

Further, it is not difficult to verify that $Ty = \|T\|\|y\|_r = \|\omega^{(n)}\|_{r'}\|y\|_r$. Hence (see [10])

$$e(\|\omega^{(n)}\|_{r'}) \geqslant y^{(k)}(0) - \|\omega^{(n)}\|_{r'}\|y\|_r = y^{(k)}(0) - Ty,$$

and since

$$y^{(k)}(0) - Ty = \int \omega(t)\,y^{(n)}(t)\,dt = \|\omega\|_{p'},$$

we have

$$e(\|\omega^{(n)}\|_{r'}) \geqslant \|\omega\|_{p'}.$$

The assertion is proved.

It is evident that (2.8) is Euler's equation for problem (2.2) (and hence for problem (0.7)).

We cite another corollary of Lemmas 1 and 2.

COROLLARY 2. *If* $q = \infty$, *then*

$$E(N) = e(N). \tag{2.10}$$

A proof is given in [11] for $1 \leqslant r < \infty$, and in [18] for $k = n-1, p = 1$. We will therefore prove it here for $r = \infty$, $1 \leqslant p \leqslant \infty$, $0 \leqslant k < n$, excluding the case $k = n-1, p = 1$.

The quantity $e(N)$ is finite [11], and there exists an extremal functional T [11], [16]. It can be assumed by virtue of Lemma 2 that it has the form (1.5) on the set C $(r = \infty)$. We define an operator T_C from C into C by the formula

$$(T_C x)(\tau) = Tx_\tau, \qquad x_\tau(t) = x(t+\tau).$$

Clearly, $\|T_C\|_C^C = \|T\|_{C^*} \leqslant N$. Moreover, for any function $x \in W^n_{p,r}$ we have

$$|x^{(k)}(\tau) - (T_C x)(\tau)| = |x_\tau^{(k)}(0) - Tx_\tau| \leqslant e(N)\|x^{(n)}\|_p.$$

Hence

$$E(N) \leqslant e(N).$$

The reverse inequality can be obtained in the same way as was done in [11] for $1 \leqslant r < \infty$. Relation (2.10) also follows from the last inequality, (0.4) and (0.8). The corollary is proved.

THEOREM 1. *If* $m = n - k$, $q = \infty$, $q' = 1$, $1 \leqslant p, r \leqslant \infty$, $p' = p/(p-1)$, $r' = r/(r-1)$, *then*

$$G(N) = F(N) = F(\theta, N) = e(N) \qquad \text{for } I = (-\infty, \infty),$$
$$g(N) = f(N) = f(\theta, N) = e(N) \qquad \text{for } I = [0, \infty).$$

PROOF. The assertion is valid for $r = \infty$, $k = 0$. Further, it suffices by virtue of (2.2) and the inequalities $F(\theta, N) \leqslant F(N) \leqslant G(N)$ to prove the inequality

$$G(N) \leqslant e(N). \tag{2.11}$$

We will first exclude the case $k = n - 1$ and $p = 1$.

Let T be an extremal functional of problem (0.7). If $1 \leqslant r < \infty$, it has the form (2.3). If, on the other hand, $r = \infty$, it can be assumed by virtue of Lemma 2 (with $\psi = \theta$) that it has the form (1.5) on C.

According to Lemma 2, there exists a function $\varphi_0 \in W^n_{r'}$ having properties (1.7) and such that for any function $x \in W^n_{p,r}$

$$x^{(k)}(0) - Tx = \int \rho(t)\, x^{(n)}(t)\, dt \tag{2.12}$$

(cf. (2.4)) when $1 < p \leqslant \infty$, while

$$x^{(k)}(0) - Tx = \int \rho(t)\, dx^{(n-1)}(t)$$

when $p = 1$, where

$$\rho = \theta - \varphi_0, \tag{2.13}$$

the function θ being defined by relation (2.1). In this connection (cf. (2.5)),

$$\|\rho\|_{p'} = \|\theta - \varphi_0\|_{p'} = e(N). \tag{2.14}$$

For an arbitrary function $\psi \in W_1^{n-k}$ we put $\mu = (-1)^{n-k}\psi^{(n-k-1)}$ and define a functional T_ψ on L_r (or C if $r = \infty$) by the relation

$$T_\psi x = \int T x_\tau \, d\mu(\tau), \quad x_\tau(t) = x(t+\tau); \tag{2.15}$$

clearly, $T_\psi \in L_r^*$ and

$$\|T_\psi\|_{L_r^*} \leqslant \|T\|_{L_r^*} \bigvee \mu \leqslant N \bigvee \psi^{(n-k-1)}. \tag{2.16}$$

On the basis of definition (1.1) we can write

$$\int \{x^{(k)}(\tau) - T x_\tau\} \, d\mu(\tau) = M_\psi^k x - T_\psi x. \tag{2.17}$$

Suppose now $1 < p \leqslant \infty$. By virtue of (2.12) we have

$$x^{(k)}(\tau) - T x_\tau = \int \rho(t) x^{(n)}(t+\tau) \, dt.$$

for $x \in W_{p,r}^n$ and any $\tau \in I$. Using this representation, we obtain from (2.17) the identity

$$M_\psi^k x - T_\psi x = \int x^{(n)}(t) R(t) \, dt, \tag{2.18}$$

in which

$$\begin{aligned} R(t) &= \int_{-\infty}^{\infty} \rho(t-\tau) \, d\mu(\tau), \quad \text{if} \quad I = (-\infty, \infty), \\ R(t) &= \int_0^t \rho(t-\tau) \, d\mu(\tau), \quad \text{if} \quad I = [0, \infty). \end{aligned} \tag{2.19}$$

Analogously, if $p = 1$, we have

$$M_\psi^k x - T_\psi x = \int R(t) \, dx^{(n-1)}(t), \qquad x \in W_{p,r}^n,$$

where the function R is also given by (2.19).

We recall that $\mu = (-1)^{n-k} \psi^{(n-k-1)}$. Let

$$S\psi = \psi - R; \tag{2.20}$$

clearly, S is a linear operator on the set $\mathfrak{W}_1^{n-k}$. We will promptly prove that S is an operator of class $\mathfrak{M}(N)$ or $\mathfrak{M}'(N)$ and that

$$\mathfrak{J}(S) \leqslant e(N). \tag{2.21}$$

This will imply (2.11) and, consequently, the equalities

$$e(N) = G(N) = \mathfrak{J}(S).$$

And this means, in particular, that the operator S determines a best linear method of approximation of a class by a class.

Applying Minkowski's inequality for integrals, we have

$$\|R\|_{p'} \leqslant \|\rho\|_{p'} \bigvee \mu. \tag{2.22}$$

It follows by virtue of (2.20), (2.19) and (2.14) that

$$\|\psi - S\psi\|_{p'} = \|R\|_{p'} \leqslant e(N) \bigvee \psi^{(n-k-1)},$$

and hence that (2.21) is valid.

Let us prove that if $\psi \in \mathcal{W}_1^{n-k}$, then $S\psi \in \mathcal{W}_{r'}^n$ and

$$\|(S\psi)^{(n)}\|_{r'} \leqslant N \bigvee \psi^{(n-k-1)}. \tag{2.23}$$

This assertion can be verified directly by differentiating the right side of (2.20). We proceed differently, however. From (2.18) and (2.22) we have for $u(\psi, T_\psi)$ (see (1.3)) the estimate

$$u(\psi, T_\psi) \leqslant e(N) \bigvee \mu < \infty.$$

According to Lemma 2, for a function ψ and a functional T_ψ there exists a function $\varphi \in \mathcal{W}_{r'}^n$ such that

$$\|\varphi^{(n)}\|_{r'} \leqslant \|T_\psi\|_{L_r^*} \tag{2.24}$$

and (1.9) holds on the class $W_{p,r}^n$. In the present case $T_0 = T_\psi$, so that from (2.24) and (1.9) we have for infinitely differentiable functions x with compact support the relation

$$\int [\psi(t) - \varphi(t) - R(t)]\, x^{(n)}(t)\, dt = 0.$$

It follows that

$$S\psi = \psi - R = \varphi + P,$$

where P identically vanishes if $I = [0, \infty)$, and is a polynomial of degree $n - 1$ if $I = (-\infty, \infty)$. But this, as is easily seen, implies $S\psi \in \mathcal{W}_{r'}^n$, and (2.23) holds by virtue of (2.16) and (2.24).

Thus, with the exception of the case $k = n - 1$, $p = 1$, Theorem 1 is proved.

Suppose now $p = 1$ and $k = n - 1$. If $I = (-\infty, \infty)$, then $\mathcal{W}_1^1 = W_1^1$ is a class of functions of bounded variation. Such functions have a limit for $t \longrightarrow \pm\infty$. We define an operator S by putting

$$S\psi = \frac{1}{2}\{\psi(+\infty) + \psi(-\infty)\}. \tag{2.25}$$

We have

$$\psi(t) - (S\psi)(t) = {}^1/_2\{\psi(t) - \psi(+\infty)\} + {}^1/_2\{\psi(t) - \psi(-\infty)\},$$

which implies $\|\psi - S\psi\|_{L_\infty} \leqslant ½ \bigvee \psi$. Consequently,

$$G(N) \leqslant J(S) \leqslant {}^1/_2. \tag{2.26}$$

When $I = [0, \infty)$ we put

$$S\psi \equiv 0, \quad \psi \in w_1^1. \tag{2.27}$$

Since $\psi(0) = 0$, it follows that $|\psi(t)| = |\psi(t) - \psi(0)| \leqslant \bigvee\psi$, whence $\|\psi - S\psi\|_{L_\infty} \leqslant 1$ and

$$g(N) \leqslant j(0) \leqslant 1. \tag{2.28}$$

It is known [18] that when $p = 1$ and $k = n - 1$ ($r < \infty$ if $k = 0$)

$$e(N) = \sup_{x \in Q} x^{(n-1)}(0) = u(\theta, 0) = \begin{cases} 1/2 & \text{for} \quad I = (-\infty, \infty), \\ 1 & \text{for} \quad I = [0, \infty). \end{cases} \tag{2.29}$$

Inequality (2.11) follows from (2.26), (2.28) and (2.29) for the case under consideration. Thus Theorem 1 is completely proved.

The operator S defined on $\mathcal{W}_1^{n-k}$ by (2.27), (2.25) and (2.20) yields a best linear method of approximation of the class $\mathcal{B}_1^{n-k}$ by the class $\mathcal{B}_{r'}^n(N)$. Let us make some remarks concerning (2.20).

Suppose first $I = [0, \infty)$. For $\psi \in w_1^{n-k}$, integrating by parts $n - k - 1$ times, we get

$$\int_0^t \theta(t - \tau)\, d\psi^{(n-k-1)}(\tau) = \psi(t).$$

Therefore (2.20) takes the form

$$(S\psi)(t) = (-1)^{n-k} \int_0^t \varphi_0(t - \tau)\, d\psi^{(n-k-1)}(\tau). \tag{2.30}$$

Here φ_0 is the function of class $b_{r'}^n(N)$ connected with an extremal operator T of problem (0.7) by relations (1.7), (1.4), (1.5), (2.13) and (2.14). If we now associate with a function $\psi \in w_1^{n-k}$ the function $X \in w_1^n$ such that $X^{(k)} = \psi$, then, by way of illustration, (2.30) can also be written in the form

$$S\psi = (-1)^k \int_0^t \lambda(t - \tau) X(\tau)\, d\tau \tag{2.31}$$

for $1 \leqslant r < \infty$.

If $I = (-\infty, \infty)$, (2.20) has the form

$$(S\psi)(t) = \psi(t) - \int_{-\infty}^{\infty} [\theta(t - \tau) - \varphi_0(t - \tau)]\, d\psi^{(n-k-1)}(\tau). \tag{2.32}$$

Suppose the function $\omega = \theta - \varphi_0$ and its derivatives up to order $n - 1$ tend sufficiently rapidly to zero at infinity; more precisely, suppose $\omega^{(i)}(t) = O(|t|^{-i})$ for $t \longrightarrow \pm\infty$ (when $i = 0, 1, \ldots, n - 1$). Then

$$(S\psi)(t) = TX_t, \tag{2.33}$$

where $X_t(\tau) = X(t + \tau)$ and the function X satisfies the condition $X^{(k)} = \psi$. In fact, since $X^{(n-1)}$ is a function of bounded variation, we have $X^{(n-1-i)}(t) = O(|t|^i)$ for $t \longrightarrow \pm\infty$ when $i = 0, 1, \ldots, n - 1$; and integrating the integrals

$$\int_t^{\infty} \omega(t - \tau)\, d\psi^{(n-k-1)}(\tau), \qquad \int_{-\infty}^{t} \omega(t - \tau)\, d\psi^{(n-k-1)}(\tau)$$

by parts n times, we get (2.33) from (2.32).

Taĭkov [14] solved problem (0.7) for $I = (-\infty, \infty)$ and $p = r = 2$. In the present case it has turned out that the function ω (and its derivatives) satisfy the relation $\omega(t) = O(e^{-\nu|t|})$ with $\nu > 0$ for $t \to \pm\infty$. Consequently, here the best linear method of approximation (of the class W_1^{n-k} by the class W_2^n in the metric of L_2) can be written in the form (2.33).

Theorem 1 implies

COROLLARY 3. *When $I = (-\infty, \infty)$, the inequality*

$$\|\theta - \varphi\|_{p'}^{\beta} \|\varphi^{(n)}\|_{r'}^{\alpha} \geqslant \beta^{\beta}\alpha^{\alpha} K, \tag{2.34}$$

holds for any function $\varphi \in W_{r'}^n$; when $I = [0, \infty)$, (2.34) holds for any function $\varphi \in w_{r'}^n$. Here the function θ is defined by (2.1), K is the least constant in (0.1), $q = \infty$ and conditions (0.23) are satisfied. Inequality (2.34) is sharp.

In fact, according to Theorem 1 and (0.4) we have

$$\mathcal{F}(\theta, N) = \inf_{\varphi \in \mathcal{B}_{r'}^n(N)} \|\theta - \varphi\|_{p'} \geqslant \Omega(N),$$

from which (2.34) readily follows. The sharpness of (2.34) is a consequence of (0.8).

As was noted by Stečkin, this inequality was essentially proved (in another way) for $p = r = 2$ and $I = (-\infty, \infty)$ by Taĭkov [14].

§3. Connection of inequality (0.1) with the approximation of a class by a class

In the present section we will essentially make use of Lemma 1 of [16], which we cite in the following form.

LEMMA 3 (V. N. Gabušin [16]). *Suppose X is a Banach space, A is a (not necessarily bounded) linear functional on X, and Q is a symmetric convex set belonging to the domain of definition of A. Then the equality*

$$\inf_{\|T\|_{X^*} \leqslant N} \sup_{x \in Q} (Ax - Tx) = \sup_{x \in Q} (Ax - N\|x\|_X)$$

holds for any $N > 0$, and there exists an extremal functional.

We call a function ψ maximal in problems (0.10), (0.15) if, respectively,

$$F(\psi, N) = F(N), \quad \psi \in B_{q'}^m,$$
$$f(\psi, N) = f(N), \quad \psi \in b_{q'}^m.$$

THEOREM 2. *Suppose $1 \leqslant p, q, r \leqslant \infty$,*

$$m = n - k, \quad \frac{1}{p} + \frac{1}{p'} = 1, \quad \frac{1}{q} + \frac{1}{q'} = 1, \quad \frac{1}{r} + \frac{1}{r'} = 1 \tag{3.1}$$

and condition (0.27) is satisfied. Then

$$\beta\alpha^{\alpha/\beta} K^{1/\beta} N^{-\alpha/\beta} = \begin{cases} F(N) & \text{for} \quad I = (-\infty, \infty), \\ f(N) & \text{for} \quad I = [0, \infty). \end{cases} \tag{3.2}$$

In addition, the function θ (see (2.1)) is maximal in problems (0.10), (0.15) for $q = \infty$.

If there exists an extremal function z in (0.1) *for* $1 \leqslant q < \infty$ *and it is normalized by the conditions*

$$\| z^{(n)} \|_p = 1, \qquad \| z \|_r = (\alpha K)^{1/\beta} N^{-1/\beta}, \tag{3.3}$$

then the function η *for which*

$$\eta^{(n-k)} = \| z^{(k)} \|_q^{1-q} \, | z^{(k)} |^{q-1} \operatorname{sgn} z^{(k)}$$

will be maximal.

We note that conditions (0.2) and (0.14) coincide, so that if (0.2) is violated, both of the quantities in (3.2) are infinite. If, on the other hand, $k = 0$ and $q = r$, the left side of (3.2) is degenerate ($\beta = 0$).

PROOF. The case $q = \infty$, $p = 1$, $k = n - 1$ is excluded from the proof presented below. When $q = \infty$, however, all of the assertions of the theorem follow from Theorem 1 and (0.8).

Consider the quantity

$$\sigma(N) = \sup_{\psi \in \mathscr{B}_{q'}^{n-k}} \sigma(\psi, N),$$

where $\sigma(\psi, N)$ is defined by (1.14), (1.3) and (1.1). By virtue of (1.15), we have

$$\sigma(N) = \sup_{\psi \in \mathscr{B}_{q'}^{n-k}} \mathscr{F}(\psi, N) = \mathscr{F}(N).$$

On the other hand, according to Lemma 3,

$$\sigma(\psi, N) = \sup_{x \in Q_{p,r}^n} \{ M_\psi^k x - N \| x \|_r \}, \tag{3.4}$$

which implies

$$\sigma(N) = \sup_{x \in Q} \sup_{\psi \in \mathscr{B}_{q'}^{n-k}} (M_\psi^k x - N \| x \|_r).$$

And since

$$\sup_{\psi \in \mathscr{B}_{q'}^{n-k}} M_\psi^k x = \| x^{(k)} \|_q, \tag{3.5}$$

we finally obtain

$$\sigma(N) = \sup_{x \in Q} (\| x^{(k)} \|_q - N \| x \|_r) = \Omega(N).$$

Relation (3.2) is thereby proved.

Let us prove the second part of the theorem. If $q = \infty$, we have

$$\begin{aligned} \Omega(N) &= \sup_{x \in Q} (\| x^{(k)} \|_C - N \| x \|_r) \\ &= \sup_{x \in Q} (x^{(k)}(0) - N \| x \|_r) = \sup_{x \in Q} (M_\theta^k x - N \| x \|_r), \end{aligned}$$

from which, with the use of (3.4) for $\psi = \theta$ and (1.15), we get

$$\Omega(N) = \sigma(\theta, N) = \mathscr{F}(\theta, N),$$

and since $\Omega(N) = F(N)$, the function θ is maximal.

Suppose $1 \leqslant q < \infty$. We will show that under the conditions of the theorem

$$\Omega(N) = \sigma(\eta, N) = \mathcal{F}(\eta, N). \tag{3.6}$$

Clearly $\|\eta^{(n-k)}\|_{p'} = 1$, and $M_\eta^k x \leqslant \|x^{(k)}\|_q$, so that

$$\sigma(\eta, N) \leqslant \Omega(N).$$

On the other hand,

$$\sigma(\eta, N) \geqslant M_\eta^k z - N\|z\|_r = \|z^{(k)}\|_q - N\|z\|_r.$$

The function z is an extremal in (0.1) and satisfies (3.3); consequently

$$\|z^{(k)}\|_q = K\|z\|_r^\alpha = \alpha^{\alpha/\beta}K^{1/\beta}N^{-\alpha/\beta},$$

from which we get

$$\sigma(\eta, N) \geqslant \beta\alpha^{\alpha/\beta}K^{1/\beta}N^{-\alpha/\beta} = \Omega(N).$$

Thus (3.6) holds. Theorem 2 is completely proved.

REMARK 2. If, at least when $1 < q' \leqslant \infty$ $(1 \leqslant q < \infty)$, a set $\mathcal{B}_0 \subset \mathcal{B}_{q'}^{n-k}$ is such that

$$\sup_{\psi \in \mathcal{B}_0} \int \psi^{(n-k)} \zeta \, dt = \|\zeta\|_q,$$

for any function $\zeta \in L_q$, then

$$\sup_{\psi \in \mathcal{B}_0} \mathcal{F}(\psi, N) = F(N) = \Omega(N).$$

This assertion readily follows from the proof of (3.2).

We note in addition that a maximal function is generally not unique. In fact, suppose $q = \infty$, let z be an extremal function in (0.1) and let t_i be the points at which

$$|z^{(k)}(t_i)| = \|z^{(k)}\|_C. \tag{3.7}$$

It will be assumed that z satisfies (3.3). We choose numbers a_i so that

$$\sum |a_i| = 1, \qquad a_i z^{(k)}(t_i) \geqslant 0,$$

and define a function η by the equality $\eta(t) = \Sigma a_i \theta(t - t_i)$. Then

$$\eta \in \mathcal{W}_1^{n-k}, \ \bigvee \eta^{(n-k-1)} = 1 \ \text{ and } \ \int z^{(k)}(t)\, d\eta^{(n-k-1)}(t) = \|z^{(k)}\|_C.$$

It follows in the same way as in the proof of Theorem 2 that η satisfies (3.6), i.e. η is maximal. It is evident that such a function is not unique if there exist several points t_i satisfying (3.7). From the results of [5] we see that this is the case, for example, when $p = q = r = \infty$ and $I = (-\infty, \infty)$.

§4. The approximation of a differentiation operator and the linear problem of approximation of a class by a class

A connection analogous to (3.2) exists between $E(N)$ and the quantities $G(N)$ and $g(N)$.

THEOREM 3. *If* $1 < q \leqslant \infty$, $1 < p \leqslant \infty$, $1 \leqslant r \leqslant \infty$, *and relations* (3.1) *are satisfied, then*

$$E(N) = G(N) \quad \text{for} \quad I = (-\infty, \infty), \tag{4.1}$$

$$E(N) = g(N) \quad \text{for} \quad I = [0, \infty). \tag{4.2}$$

PROOF. Since these assertions are contained in Theorem 1 when $q = \infty$, it will be assumed that $1 < q < \infty$.

Suppose $E(N) < \infty$ and suppose $U(T)$ is finite for some bounded linear operator T from L_r into L_q with $\|T\| \leqslant N$. With a function $\psi \in W_{q'}^{n-k}$ we associate a functional T_ψ by the formula

$$T_\psi x = \int \psi^{(n-k)} Tx\, dt;$$

we have

$$\|T_\psi\|_{L_r^*} \leqslant \|\psi^{(n-k)}\|_{q'} N,$$

$$\sup_{x \in Q} \left\{ \int \psi^{(n-k)} x^{(k)}\, dt - T_\psi x \right\} \leqslant U(T) \|\psi^{(n-k)}\|_{q'}.$$

According to Lemma 2, there exists a unique function $\varphi \in W_{r'}^n$ with the properties

$$\|\psi - \varphi\|_{p'} \leqslant U(T) \|\psi^{(n-k)}\|_{q'}, \qquad \|\varphi^{(n)}\|_{r'} \leqslant N \|\psi^{(n-k)}\|_{q'}. \tag{4.3}$$

Thus $\varphi = S\psi$, where S is a single-valued operator from $W_{q'}^{n-k}$ into $W_{r'}^n$; in this connection, S satisfies condition (0.9) and

$$J(S) \leqslant U(T). \tag{4.4}$$

Let us show that the operator S is linear. For any functions $\psi_1, \psi_2 \in W_{q'}^{n-k}$ we have $T_{\psi_1 + \psi_2} = T_{\psi_1} + T_{\psi_2}$. It follows by virtue of (1.7) that

$$(S(\psi_1 + \psi_2))^{(n)} = (S\psi_1)^{(n)} + (S\psi_2)^{(n)}$$

for $r < \infty$, while

$$(S(\psi_1 + \psi_2))^{(n-1)} = (S\psi_1)^{(n-1)} + (S\psi_2)^{(n-1)} + c$$

for $r = \infty$, where c is a certain constant. Hence

$$S(\psi_1 + \psi_2) = S\psi_1 + S\psi_2 + P,$$

where P is a certain polynomial of degree $n - 1$. According to the first of inequalities (4.3) we have $P \in L_{p'}$, and since $p' < \infty$, we get $P \equiv 0$, i.e. the operator S is linear.

It now follows from (4.4) that

$$G(N) \leqslant E(N). \tag{4.5}$$

Suppose $G(N) < \infty$, and suppose an operator S of class $\mathfrak{M}(N)$ or $\mathfrak{M}'(N)$ has the property $J(S) < \infty$. We will first consider the case $r' > 1$ $(r < \infty)$. It is not difficult to see that the value $(S\psi)^{(n)}$ is uniquely determined by the derivative $\psi^{(n-k)}$, so that

$(S\psi)^{(n)} = T^*\psi^{(n-k)}$. Clearly, T^* is a linear operator from $L_{q'}$ into $L_{r'}$ and $\|T^*\| \leqslant N$. Let T denote the adjoint of T^* when $1 < r' < \infty$, and the restriction to L_r of the adjoint of T^* when $r' = \infty$. Clearly, $\|T\|_{L_r}^{L_q} \leqslant N$; also, for any functions $\psi \in W_{q'}^{n-k}$ and $x \in L_r$

$$\int x\,(S\psi)^{(n)}\,dt = \int \psi^{(n-k)}\,Tx\,dt.$$

Applying Lemma 1 to the functions ψ and $\varphi = S\psi$, we obtain

$$\left|\int \psi^{(n-k)}\,\{x^{(k)} - Tx\}\,dt\right| = \left|\int \psi^{(n-k)}\,x^{(k)}\,dt - \int \psi^{(n-k)}\,Tx\,dt\right| \leqslant \|x^{(n)}\|_p \|\psi - S\psi\|_{p'}.$$

Hence $U(T) \leqslant J(S)$; therefore $E(N) \leqslant J(S)$, and consequently

$$E\,(N) \leqslant G\,(N). \tag{4.6}$$

When $r' = 1$, the value $(S\psi)^{(n-1)} - \lim_{t\to\infty}(S\psi)^{(n-1)}(t)$ is uniquely determined by the function $\psi^{(n-k)}$, and we arrive at an operator T^* from $L_{q'}$ into the space V of functions of bounded variation. By considering the restriction to C of the adjoint of T^*, we obtain (4.6) in the same way as when $r' > 1$.

Thus, if either $E(N)$ or $G(N)$ is finite, so the other is and they are equal. Theorem 3 is proved.

REMARK 3. In the proof of the latter theorem (at least when $1 < q < \infty$, $1 \leqslant r < \infty$) a one-to-one correspondence between the sets of operators S and T for which the corresponding deviations $J(S)$ and $U(T)$ are finite is established by the formula

$$(-1)^k\,(S\psi)^{(n)} = T^*\psi^{(n-k)}, \tag{4.7}$$

where T^* is the adjoint of the operator T; in this connection, $U(T) = J(S)$. It follows that if either S or T is an extremal operator, so is the other (in the appropriate problem).

§5. The independence of the approximation of a class by a class from the set I

Ju. N. Subbotin [24] obtained one and the same estimate from below of the quantities $F(N, [0, \infty))$ and $F(N, (-\infty, \infty))$ in the cases $p' = q' = r' = \infty$ and $p' = r' = q' = 1$, and also proved that this estimate is sharp for $n \leqslant 5$. In [25] there is obtained an upper estimate of these quantities for arbitrary $n \geqslant 2$ in the case $p' = q' = r' = \infty$ that coincides with an upper estimate in [24]. Consequently, in these cases $F(N)$ does not depend on I. This fact holds for arbitrary values of the parameters.

THEOREM 4. *The following equalities are valid*:

$$F\,(N,\,[0,\,\infty)) = F\,(N,\,(-\infty,\,\infty)), \tag{5.1}$$

$$f\,(N,\,[0,\,\infty)) = f\,(N,\,(-\infty,\,\infty)). \tag{5.2}$$

PROOF. Any function $y \in B_s^l(\mu, [0, \infty))$ can be extended by a polynomial of degree $l-1$ on the half-line $(-\infty, 0]$ to a function of class $B_s^l(\mu, (-\infty, \infty))$. Conversely, the restriction of an arbitrary function $y \in B_s^l(\mu, (-\infty, \infty))$ to the half-line $[0, \infty)$ is a function of class $B_s^l(\mu, [0, \infty))$. Therefore

$$F(N, [0, \infty)) = \sup_{\psi \in B_{q'}^m(1, (-\infty, \infty))} \inf_{\varphi \in B_{r'}^n(N, (-\infty, \infty))} \|\psi - \varphi\|_{L_{p'}[0, \infty)}$$

and consequently

$$F(N, [0, \infty)) \leqslant F(N, (-\infty, \infty)).$$

In exactly the same way,

$$f(N, [0, \infty)) \leqslant f(N, (-\infty, \infty)).$$

Let us prove the reverse inequalities. It will be assumed in this connection that condition (0.14) is satisfied, since otherwise $F(N) = \infty$ and hence $f(N) = \infty$ inasmuch as $f(N) \geqslant F(N)$.

We begin with the inequality

$$f(N, (-\infty, \infty)) \leqslant f(N, [0, \infty)). \tag{5.3}$$

Let $\psi \in b_{q'}^m(1, (-\infty, \infty))$, and let ψ_1 and ψ_2 denote the functions defined on $[0, \infty)$ by the formulas $\psi_1(t) = \psi(t)$ and $\psi_2(t) = \psi(-t)$. We put $\epsilon_i = \|\psi_i^{(m)}\|_{q'}$ $(i = 1, 2)$ and assume that $q'/r' = 0$ if $r' = \infty$ $(1 \leqslant q' < \infty)$. Then for the quantities

$$f(\psi, N) = f(\psi, N, (-\infty, \infty)) \quad \text{and} \quad f(\psi_i, N\varepsilon_i^{q'/r'}) = f(\psi_i, N\varepsilon_i^{q'/r'}, [0, \infty))$$

we have

$$f(\psi, N) \leqslant \Big\{\sum f^{p'}(\psi_i, N\varepsilon_i^{q'/r'})\Big\}^{1/p'} \tag{5.4}$$

when $1 \leqslant p' < \infty$, and

$$f(\psi, N) \leqslant \max f(\psi_i, N\varepsilon_i^{q'/r'}) \tag{5.5}$$

when $p' = \infty$. But

$$f(\psi_i, N\varepsilon_i^{q'/r'}) = \varepsilon_i f(\psi_i \varepsilon_i^{-1}, \quad N\varepsilon_i^{q'/r'-1}) \leqslant \varepsilon_i f(N\varepsilon_i^{q'/r'-1}, [0, \infty)),$$

which by virtue of the equality $f(N) = f(1)N^{-\nu}$ implies

$$f(\psi_i, N\varepsilon_i^{q'/r'}) \leqslant \varepsilon_i^{\gamma} f(N, [0, \infty)), \tag{5.6}$$

where

$$\nu = \left(m - \frac{1}{q'} + \frac{1}{p'}\right) \Big/ \left(n - m + \frac{1}{q'} - \frac{1}{r'}\right), \quad \gamma = 1 + \nu - \frac{\nu q'}{r'}.$$

Suppose $1 \leqslant p' < \infty$ Then $q' < \infty$, since, according to (0.14), $p' = r' = \infty$ if $q' = \infty$. From (5.4) and (5.6) we get

$$f(\psi, N) \leqslant f(N, [0, \infty)) \, (\varepsilon_1^{p'\gamma} + \varepsilon_2^{p'\gamma})^{1/p'}.$$

Since

$$\frac{p'\gamma}{q'}-1=\frac{1}{p'}\left(\frac{n}{q'}-\frac{n-m}{p'}-\frac{m}{r'}\right)\Big/\left(n-m+\frac{1}{q'}-\frac{1}{r'}\right),$$

we have $p'\gamma \geqslant q'$ by virtue of (0.14). Also, $0 \leqslant \epsilon_i \leqslant 1$. Therefore $\epsilon_2^{p'\gamma} + \epsilon_2^{p'\gamma} \leqslant \epsilon_1^{q'} + \epsilon_2^{q'} \leqslant 1$. Hence

$$f(\psi, N, (-\infty, \infty)) \leqslant f(N, [0, \infty)), \tag{5.7}$$

and this leads to (5.3).

If $p' = \infty$ and $1 \leqslant q' < \infty$, then

$$\gamma = q'\left(\frac{n}{q'}-\frac{m}{r'}\right)\Big/\left(n-m+\frac{1}{q'}-\frac{1}{r'}\right),$$

which implies $\gamma \geqslant 0$ according to (0.14); analogously, we also have $\gamma \geqslant 0$ if $p' = q' = \infty$ ($r' = \infty$). Inequality (5.7) (and hence (5.3)) for $p' = \infty$ now follows from (5.6) and (5.5).

Thus (5.2) is proved.

Let us establish the inequality

$$F(N, (-\infty, \infty)) \leqslant F(N, [0, \infty)). \tag{5.8}$$

It suffices to prove that if

$$\overline{N} > N, \qquad A > F(N, [0, \infty)), \tag{5.9}$$

then

$$F(\overline{N}, (-\infty, \infty)) \leqslant A, \tag{5.10}$$

since this implies $F(\overline{N}, (-\infty, \infty)) \leqslant F(N, [0, \infty))$, from which (5.8) follows by virtue of (0.13).

Let $\overline{B} = \overline{B}_{q'}^m(1, (-\infty, \infty))$ denote the set of functions $\psi \in B_{q'}^m(1, (-\infty, \infty))$ satisfying the condition $\psi(t) = 0$ for $t \leqslant 0$, and let us prove the equality

$$F(N, (-\infty, \infty)) = \sup_{\psi \in \overline{B}} F(\psi, N, (-\infty, \infty)), \tag{5.11}$$

which follows from Theorem 1 when $q' = 1$ since $\theta \in \overline{B}$.

Suppose $1 < q' \leqslant \infty$. For any function $\zeta \in L_q$ $(1/q + 1/q' = 1)$, we have

$$\|\zeta\|_q = \sup_{\|\xi\|_{q'} \leqslant 1} \int \xi\zeta\, dt.$$

Here it is possible to confine oneself to the set of functions with compact support, so that in (3.5) it is possible to confine oneself to the functions $\psi \in B_{q'}^m$ whose derivatives of order m vanish for $t \leqslant a$ (the number a depends on ψ). The class B_0 of such functions satisfies the conditions of Remark 2 to Theorem 2. But this implies (5.11).

Suppose $\psi \in \overline{B}$. For an arbitrary number $\delta \geqslant 1$ we put $\psi_\delta(t) = \psi(t-\delta)$. The restriction of ψ_δ to $[0, \infty)$ is a function of class $B_{q'}^m(1, [0, \infty))$. It will be assumed in the sequel that $\overline{N}$ and A are two fixed numbers satisfying (5.9). For each δ we choose a function $\varphi_\delta \in B_{r'}^n(N, [0, \infty))$ so that

$$F(\psi_\delta, N, [0, \infty)) \leqslant \|\psi_\delta - \varphi_\delta\|_{L_{p'}[0,\infty)} \leqslant A.$$

Let κ be an infinitely differentiable function on $(-\infty, \infty)$ such that $0 \leqslant \kappa \leqslant 1$, $\kappa(t) = 0$ for $t \leqslant 0$ and $\kappa(t) = 1$ for $t \geqslant 1$, and let $y_\delta(t) = \varphi_\delta(t)\kappa(t\delta^{-1})$. Since $\psi_\delta(t) = 0$ for $0 \leqslant t \leqslant \delta$, we have

$$\psi_\delta(t) - y_\delta(t) = \varkappa(t\delta^{-1})\{\psi_\delta(t) - \varphi_\delta(t)\}, \qquad t \geqslant 0,$$

and therefore

$$\|\psi_\delta - y_\delta\|_{p'} \leqslant \|\psi_\delta - \varphi_\delta\|_{p'} \leqslant A.$$

When $1 < r' \leqslant \infty$ we have, according to Leibniz' formula,

$$y_\delta^{(n)}(t) = \varphi_\delta^{(n)}(t)\varkappa(t\delta^{-1}) + \rho(t),$$

$$\rho(t) = \sum_{i=1}^{n} C_n^i \varphi_\delta^{(n-i)}(t)\,\delta^{-i}\varkappa^{(i)}(t\delta^{-1}).$$

On the intervals $[0, \delta]$ the functions φ_δ have the properties $\|\varphi_\delta\|_{p'} \leqslant A$ and $\|\varphi_\delta^{(n)}\|_{r'} \leqslant N$. It follows from Theorem 1 of [6] that when $\delta \geqslant 1$ there exists a constant c not depending on δ (and hence not on φ_δ) such that $\|\varphi_\delta^{(i)}\|_{C[0,\delta]} \leqslant c$ for $i = 0, 1, \ldots, n-1$. Therefore $\|\rho\|_{r'} \leqslant c_1\delta^{1/r'-1}$ (where c_1 does not depend on δ). Consequently, for sufficiently large δ

$$\|y_\delta^{(n)}\|_{L_{r'}[0,\infty)} = \|y_\delta^{(n)}\|_{L_{r'}(-\infty,\infty)} \leqslant \overline{N}.$$

Let y denote the function defined on $(-\infty, \infty)$ by the equality $y(t) = y_\delta(t + \delta)$. We have $\|y^{(n)}\|_{r'} \leqslant \overline{N}$ and

$$\|\psi - y\|_{L_{p'}(-\infty,\infty)} = \|\psi_\delta - y_\delta\|_{L_{p'}[0,\infty)} \leqslant A,$$

which implies

$$F(\psi, \overline{N}, (-\infty, \infty)) \leqslant A. \tag{5.12}$$

Inequality (5.10) follows from this result by virtue of (5.11). Thus (5.8) and hence (5.1) hold when $r' > 1$.

Suppose $r' = 1$. The case $n = 1$ ($m = n = 1$, $r' = 1$) is trivial (see the end of the Introduction); it will therefore be assumed that $n \geqslant 2$. Inasmuch as $\bigvee_0^\delta \varphi_\delta^{(n-1)} \leqslant N$, we have for some interval $I_\delta = [a, b] \subset [0, \delta]$

$$b - a = \sqrt{\delta}, \qquad \bigvee_a^b \varphi_\delta^{(n-1)} \leqslant 2N/\sqrt{\delta}. \tag{5.13}$$

Let us show that the function $\zeta = \varphi_\delta^{(n-1)}$ cannot be large on this interval, viz. that there exists a constant c depending only on A and N such that

$$|\varphi_\delta^{(n-1)}(t)| \leqslant c/\sqrt{\delta} \quad \text{for} \quad t \in I_\delta = [a, b]. \tag{5.14}$$

Suppose this is not true. Then, by virtue of (5.13), the function ζ will preserve its sign and satisfy the condition

$$|\zeta(t)| \geqslant \frac{c - 2N}{\sqrt{\delta}} = \chi(\delta)$$

on I_δ for $c > 2N$. But then $|\varphi_\delta^{(n-2)}(t)| \geqslant \chi(\delta)$ on either the interval $[a, (a+b)/2 - 1]$, or the interval $[(a+b)/2+1, b]$. An analogous result can be obtained for $\varphi^{(n-i)}$ by induction on $i \leqslant n-1$. In particular, there exists an interval $[\alpha, \beta] \subset [a, b]$ of length $2^{1-n}\sqrt{\delta} - 2$ on which $|\varphi_\delta(t)| \geqslant \chi(\delta)$. It readily follows that

$$\| \varphi_\delta \|_{L_{p'}(I_\delta)} \geqslant \chi(\delta) \left\{ \int_0^{\frac{|\beta-\alpha|}{2}} t^{p'}\, dt \right\}^{1/p'}.$$

Thus, if $\delta \geqslant 1$,

$$\| \varphi_\delta \|_{L_{p'}(I_\delta)} \geqslant c_1 \chi(\delta) \sqrt{\delta} = c_1 (c - 2N),$$

where c_1 depends only on p' and n. By assumption, the constant c can be taken arbitrarily large. But this contradicts the condition $\|\varphi_\delta\|_{L_{p'}[0,\delta]} \leqslant A$. Consequently a constant c satisfying (5.14) exists.

Let $y_\delta(t) = \varphi_\delta(t)\, \kappa((t-a)\delta^{-1/2})$. We then have

$$y_\delta^{(n-1)}(t) = \varphi_\delta^{(n-1)}(t)\, \varkappa((t-a)\,\delta^{-1/2}) + \rho(t),$$

$$\rho(t) = \sum_{i=1}^{n-1} C_{n-1}^{i} \varphi_\delta^{(n-1-i)}(t)\, \delta^{-\frac{i}{2}} \varkappa^{(i)}((t-a)\,\delta^{-1/2}),$$

which implies

$$\bigvee_{-\infty}^{\infty} y_\delta^{(n-1)} = \bigvee_{a}^{\infty} y_\delta^{(n-1)} \leqslant \bigvee_{-\infty}^{\infty} \varkappa \sup_{a \leqslant t \leqslant b} |\varphi_\delta^{(n-1)}(t)| + \bigvee_{0}^{\infty} \varphi_\delta^{(n-1)} + \| \rho' \|_{L(a,\, b)}. \tag{5.15}$$

As before, it follows from Theorem 1 of [6] by virtue of (5.2) and the inequality $\|\varphi_\delta\|_{L_{p'}(a,b)} \leqslant A$ that there exists a constant c_2 for which $\|\varphi_\delta^{(i)}\|_{C[a,b]} \leqslant c_2$, $i = 0, 1, \ldots, n-2$. Therefore

$$\| \rho' \|_{L(a,\, b)} \leqslant c_3 \{ \| \varphi_\delta^{(n-1)} \|_{L_\infty(a,\, b)} + \delta^{-1/2} \}. \tag{5.16}$$

According to (5.14)–(5.16), for sufficiently large δ

$$\bigvee_{-\infty}^{\infty} y_\delta^{(n-1)} \leqslant \overline{N}.$$

In addition,

$$\| \psi_\delta - y_\delta \|_{L_{p'}(-\infty,\, \infty)} \leqslant \| \psi_\delta - \varphi_\delta \|_{L_{p'}[0,\, \infty)} \leqslant A.$$

From these results, as in the case $r' > 1$, we obtain (5.12) and hence (5.8) and (5.1). Theorem 4 is proved.

Let us consider the dependence on I of the quantities $G(N)$ and $g(N)$.

The following inequality holds:

$$G(N, [0, \infty)) \leqslant G(N, (-\infty, \infty)). \tag{5.17}$$

In fact, let $S \in \mathfrak{M}(N, (-\infty, \infty))$, and let $\overline{\psi} \in B_{q'}^{m}(1, (-\infty, \infty))$ denote the function coinciding on $[0, \infty)$ with a given function $\psi \in B_{q'}^{m}(1, [0, \infty))$ and being a polynomial on the half-line $(-\infty, 0]$. We define an operator S^+ by letting $S^+\psi$ denote the restriction of the function $S\overline{\psi}$ to the half-line $[0, \infty)$. It is not difficult to verify that

$S^+ \in \mathfrak{M}(N, [0, \infty))$ and

$$\sup_{\psi} \| \psi - S^+\psi \|_{L_{p'}[0, \infty)} \leqslant \sup_{\bar\psi} \| \bar\psi - S\bar\psi \|_{L_{p'}(-\infty, \infty)},$$

where $\psi \in B_{q'}^m(1, [0, \infty))$ and $\bar\psi \in B_{q'}^m(1, (-\infty, \infty))$. But this implies (5.17).

The inequality

$$g(N, [0, \infty)) \leqslant g(N, (-\infty, \infty)) \tag{5.18}$$

can be obtained analogously to (5.17).

It will now be shown that the reverse inequality holds, and hence *The quantity $g(N)$ does not depend on I, i.e.*

$$g(N, [0, \infty)) = g(N, (-\infty, \infty)). \tag{5.19}$$

By virtue of Theorem 3, Remark 3 to Theorem 3, condition (0.6) for the finiteness of $E(N)$ and inequality (5.18), it suffices to obtain the inequality

$$g(N, [0, \infty)) \geqslant g(N, (-\infty, \infty))$$

under the assumption $r' \geqslant q'$, $p' \geqslant q'$.

Let $\bar S \in \mathfrak{M}'(N, [0, \infty))$, and let S denote the operator defined on $w_{q'}^m$ for $I = (-\infty, \infty)$ by the formula

$$(S\psi)(t) = \begin{cases} (\bar S\psi_1)(t), & t \geqslant 0, \\ (\bar S\psi_2)(-t), & t \leqslant 0, \end{cases}$$

where $\psi_1(t) = \psi(t)$, $\psi_2(t) = \psi(-t)$ for $t \in [0, \infty)$.

Clearly, $S\psi$ has a sufficient number of derivatives and, together with its derivatives up to order $n-1$, satisfies the null boundary condition at $t = 0$. In addition,

$$\|(S\psi)^{(n)}\|_{L_{r'}(-\infty, \infty)} = \left\{\sum \|(\bar S\psi_i)^{(n)}\|_{L_{r'}[0, \infty)}^{r'}\right\}^{1/r'} \leqslant N(\|\psi_1^{(m)}\|_{q'}^{r'} + \|\psi_2^{(m)}\|_{q'}^{r'})^{1/r'}$$
$$= N\left\{\sum (\|\psi_i^{(m)}\|_{q'}^{q'})^{r'/q'}\right\}^{1/r'}.$$

Hence, if $r' \geqslant q'$,

$$\|(S\psi)^{(n)}\|_{L_{r'}(-\infty, \infty)} \leqslant N \|\psi^{(m)}\|_{L_{q'}(-\infty, \infty)},$$

i.e. $S \in \mathfrak{M}'(N, (-\infty, \infty))$.

Let us estimate the quantity

$$\|\psi - S\psi\|_{p'} = \|\psi - S\psi\|_{L_{p'}(-\infty, \infty)}.$$

Clearly,

$$\|\psi - S\psi\|_{p'} = \left\{\sum \|\psi_i - \bar S\psi_i\|_{L_{p'}[0, \infty)}^{p'}\right\}^{1/p'}.$$

And since

$$\|\bar\psi - \bar S\bar\psi\|_{p'} \leqslant j(\bar S)\|\bar\psi^{(m)}\|_{l'}$$

for $I = [0, \infty)$ and $\bar\psi \in w_{q'}^m$, we have, as before,

$$\|\psi - S\psi\|_{p'} \leqslant j(\bar{S})\left\{\sum(\|\psi_i^{(m)}\|_{q'}^{q'})^{p'/q'}\right\}^{1/p'},$$

which implies the inequality

$$\|\psi - S\psi\|_{p'} \leqslant j(\bar{S})\,\|\psi^{(m)}\|_{L_{q'}(-\infty,\,\infty)}$$

for $p' \geqslant q'$. Consequently

$$j(S,\ (-\infty,\ \infty)) \leqslant j(\bar{S},\ [0,\ \infty)).$$

The assertion is thereby proved.

REMARK 4. When $q = \infty$, it follows from Theorems 1–3 and relations (5.17) and (5.19) that the quantities $F(N)$, $G(N)$, $f(N)$ and $g(N)$ do not depend on I and satisfy the equalities $F(N) = G(N)$ and $f(N) = g(N)$.

BIBLIOGRAPHY

1. B. Sz.-Nagy, *Über Integralungleichungen zwischen einer Funktion und ihrer Ableitung*, Acta Sci. Math. (Szeged) **10** (1941/43), 64–74. MR **2**, 351.

2. G. H. Hardy and J. E. Littlewood, *Contributions to the arithmetic theory of series*, Proc. London Math. Soc. (2) **11** (1912/13), 411–478.

3. Edmund Landau, *Einige Ungleichungen für zweimal differentierbare Funktionen*, Proc. London Math. Soc. (2) **13** (1913/14), 43–49.

4. Jacques Hadamard, *Sur le module maximum d'une fonction et de ses dérivées*, C. R. des Séances Soc. Math. France **41** (1914), 68–72; reprinted in *Oeuvres de Jacques Hadamard*, Tome 1, Centre Nat. Recherche Sci., Paris, 1968, pp. 379–382.

5. A. N. Kolmogorov, *On inequalities between the upper bounds of the successive derivatives of an arbitrary function on an infinite interval*, Učen. Zap. Moskov. Gos. Univ. Mat. **30** (1939), 3–13; English transl., Amer. Math. Soc. Transl. (1) **2** (1962), 233–243. MR **1**, 298.

6. V. N. Gabušin, *Inequalities for norms of a function and its derivatives in L_p-metrics*, Mat. Zametki **1** (1967), 291–298 = Math. Notes **1** (1967), 194–198. MR **34** #6518.

7. ———, *Inequalities for norms of a function and its derivatives, and their applications*, Candidate's Dissertation, Moscow, 1970. (Russian)

8. V. V. Arestov, *On best approximation of differentiation operators in the uniform metric*, Candidate's Dissertation, Moscow, 1969. (Russian)

9. S. B. Stečkin, *Inequalities between norms of derivatives of arbitrary functions*, Acta Sci. Math. (Szeged) **26** (1965), 225–230. (Russian) MR **32** #2534.

10. ———, *Best approximation of linear operators*, Mat. Zametki **1** (1967), 137–148 = Math. Notes **1** (1967), 91–99. MR **35** #2051.

11. V. V. Arestov, *Best uniform approximation of differentiation operators*, Mat. Zametki **5** (1969), 273–284 = Math. Notes **5** (1969), 167–173. MR **40** #4651.

12. ———, *Best approximation of differentiation operators*, Mat. Zametki **1** (1967), 149–154 = Math. Notes **1** (1967), 100–103. MR **35** #2052.

13. Ju. N. Subbotin and L. V. Taĭkov, *Best approximation of a differentiation operator in the space L_2*, Mat. Zametki **3** (1968), 157–164 = Math. Notes **3** (1968), 100–105. MR **37** #4471.

14. L. V. Taĭkov, *Inequalities of Kolmogorov type and best formulas for numerical differentiation*, Mat. Zametki **4** (1968), 233–238 = Math. Notes **4** (1968), 631–634. MR **38** #1459.

15. V. N. Gabušin, *Best approximation of a differentiation operator on the half-line*, Mat. Zametki **6** (1969), 573–582 = Math. Notes **6** (1969), 804–810. MR **41** #4080.

16. ———, *Best approximation of functionals on certain sets*, Mat. Zametki **8** (1970), 551–562 = Math. Notes **8** (1970), 780–785. MR **43** #2407.

17. V. I. Berdyšev, *Best approximation in $L_{[0,\infty)}$ of the differentiation operator*, Mat. Zametki **9** (1971), 477–481 = Math. Notes **9** (1971), 275–277. MR **44** #3064.

18. V. V. Arestov, *Sharp inequalities between the norms of functions and their derivatives*, Acta Sci. Math. (Szeged) **33** (1972), 243–276. (Russian) MR **47** #9263.

19. Yngve Domar, *An extremal problem related to Kolmogoroff's inequality for bounded functions*, Ark. Mat. **7** (1968), 433–441. MR **38** #2534.

20. V. N. Gabušin, *Best approximation of the differentiation operator in the* L_p *metric*, Mat. Zametki **12** (1972), 531–538 = Math. Notes **12** (1972), 756–760. MR **48** #2629.

21. N. P. Korneĭčuk, *Best uniform approximation on some classes of continuous functions*, Dokl. Akad. Nauk SSSR **140** (1961), 748–751 = Soviet Math. Dokl. **2** (1961), 1254–1257. MR **24** #A2181.

22. A. F. Timan, *Deformation of metric spaces and certain related questions of the theory of functions*, Uspehi Mat. Nauk **20** (1965), no. 2 (122), 53–87 = Russian Math. Surveys **20** (1965), no. 2, 49–83. MR **32** #6414.

23. L. V. Taĭkov, *Best approximation in the mean of certain classes of analytic functions*, Mat. Zametki **1** (1967), 155–162 = Math. Notes **1** (1967), 104–109. MR **35** #2047.

24. Ju. N. Subbotin, *Best approximation of a class of functions by another class*, Mat. Zametki **2** (1967), 495–504 = Math. Notes **2** (1967), 792–797. MR **36** #3031.

25. V. V. Arestov and V. N. Gabušin, *Approximation of classes of differentiable functions*, Mat. Zametki **9** (1971), 105–112 = Math. Notes **9** (1971), 63–67. MR **43** #7834.

26. Ju. N. Subbotin, *The relation of spline approximations to the problem of approximating a class by a class*, Mat. Zametki **9** (1971), 501–510 = Math. Notes **9** (1971), 289–294. MR **44** #7192.

27. G. H. Hardy, J. E. Littlewood and G. Pólya, *Inequalities*, Cambridge Univ. Press, New York, 1934; 2nd ed., 1952. MR **13**, 727.

28. V. V. Arestov, *Approximation of linear operators and related extremal problems*, Trudy Mat. Inst. Steklov. **138** (1975), 29–42 = Proc. Steklov Inst. Math. **138** (1975), 31–44.

Translated by S. SMITH

Trudy Mat. Inst. Steklov.
138 (1975)

Proc. Steklov Inst. Math.
138 (1975)

APPROXIMATION OF LINEAR OPERATORS AND RELATED EXTREMAL PROBLEMS

UDC 517.5

V. V. ARESTOV

ABSTRACT. For the modulus of continuity

$$\omega(\delta) = \sup_{\substack{\|x\|_X \leqslant 1 \\ x \in Q}} \|Ax\|_Z$$

and best approximation

$$E(N) = \inf_{\|T\|_X^Z \leqslant N} \sup_{x \in Q} \|Ax - Tx\|_Z$$

of the operator A on the class $Q = \{x \in X: \|Bx\|_Y \leqslant 1\}$ by bounded operators T, related problems are considered in the dual spaces. Thus it is shown that in the case $A = H \circ B$

$$\sup_{\delta > 0} \{\omega(\delta) - N\delta\} = \sup_{\psi \in \Psi} \inf_{\varphi \in \Phi(N)} \|\psi - \varphi\|_{Y^*},$$

where

$$\Psi = \{\varkappa \circ H : \|\varkappa\|_{Z^*} \leqslant 1\}, \qquad \Phi(N) = \{\varphi \in Y^{\#} : \|\varphi \circ B\|_X \leqslant N\};$$

while if $\overline{HY} = Z$ and Z is a reflexive space, then $E(N)$ coincides with the linear approximation of Ψ by $\Phi(N)$.

Bibliography: 6 titles.

1. In this paper we consider dual quantities on conjugate spaces for the modulus of continuity of a linear operator on a set of elements and for the approximation of such an operator by bounded operators.

Let X and Z be normed linear spaces, let A be a linear (homogeneous and additive) operator from X to Z, let $\mathcal{T}(N)$ be the set of all bounded linear operators T from X to Z with norm $\|T\| = \|T\|_X^Z \leqslant N$, and let Q be a set of elements of X. Then

$$U(T) = \sup_{x \in Q} \|(A - T)x\|_Z$$

is the deviation of the operator T from A on the set Q, and

$$E(N) = \inf_{T \in \mathcal{T}(N)} U(T) \tag{1}$$

is the best approximation to the operator A by bounded operators T. An operator T for which the lower bound is achieved in (1) is said to be extremal.

The quantity $E(N)$ was first studied by S. B. Stečkin [1]. In particular, he showed that this quantity is related to the modulus of continuity

$$\omega(\delta) = \sup_{\substack{x \in Q \\ \|x\|_X \leqslant \delta}} \|Ax\|_Z \tag{2}$$

of the operator A on the set Q. Namely, we have

AMS (MOS) subject classifications (1970). Primary 41A65, 41A50.

$$\omega(\delta) \leqslant \inf_{N>0} \{E(N) + N\delta\},$$

and

$$E(N) \geqslant \Omega(N), \tag{3}$$

where

$$\Omega(N) = \sup_{\delta>0} \{\omega(\delta) - N\delta\} = \sup_{x \in Q} \{\|Ax\|_Z - N\|x\|_X\}. \tag{4}$$

In addition, if there exist an element $x \in Q$ and a bounded linear operator T such that

$$\begin{gathered} \|Tx\|_Z = \|T\| \, \|x\|_X, \\ U(T) = \|Ax - Tx\|_Z = \|Ax\|_Z - \|Tx\|_Z, \end{gathered} \tag{5}$$

then

$$\begin{gathered} E(\|T\|) = U(T) = \|Ax\|_Z - \|T\| \, \|x\|_X = \Omega(\|T\|), \\ \omega(\|x\|_X) = \|Ax\|_Z. \end{gathered} \tag{6}$$

The first of these inequalities signifies that T is an extremal operator for $N = \|T\|$.

If W is a normed linear space we denote, as usual, the set of linear functionals on W by $W^{\#}$ and the set of bounded linear functionals by W^*.

The following proposition is due to Gabušin [2].

Lemma 1. *If A is a linear functional on the normed linear space X and $Q \subset X$ is a symmetric convex set, then*

$$\inf_{\|T\|_{X^*} \leqslant N} \sup_{x \in Q} (A - T)x = \sup_{x \in Q} (Ax - N\|x\|_X).$$

Thus equality holds in (3) for functionals (under the conditions of the lemma).

2. In many concrete problems the set Q is defined by a linear operator in the following way (see [1]). Let Y be a normed linear space and B a linear operator from X onto Y. We put $Q = \{x \in X \colon \|Bx\|_Y \leqslant 1\}$. Then

$$\omega(\delta) = \sup_{\|Bx\|_Y \leqslant 1, \, \|x\|_X \leqslant \delta} \|Ax\|_Z, \tag{7}$$

$$E(N) = \inf_{\|T\| \leqslant N} \sup_{\|Bx\|_Y \leqslant 1} \|(A - T)x\|_Z. \tag{8}$$

In the sequel we shall consider the functions $\omega(\delta)$ and $E(N)$ only in this particular case.

We note that, by the definition of $\omega(\delta)$, for any element $x \in X$ with $Bx \neq 0$ we have

$$\|Ax\|_Z \leqslant \|Bx\|_Y \, \omega\left(\frac{\|x\|_X}{\|Bx\|_Y}\right). \tag{9}$$

The following generalization of a theorem of Sard [3] is useful in the study of the quantity $E(N)$.

Lemma 2. *If R is a linear operator from X to Z, B is a linear operator from X onto Y and*

$$U_R = \sup_{\|Bx\|_Y \leqslant 1} \|Rx\|_Z < \infty,$$

then $R \circ B^{-1}$ is a single-valued bounded linear operator from Y to Z, and

$$\| R \circ B^{-1} \|_Y^Z = U_R.$$

PROOF. If the operator $S = R \circ B^{-1}$ is not single-valued, there are elements x_1 and x_2 of X such that $Bx_1 = Bx_2$ and $Rx_1 \neq Rx_2$. Then the element $x = x_1 - x_2$ has the properties $Bx = 0$ and $Rx \neq 0$, which is impossible, since $U_R < \infty$.

Further, using the equation $BX = Y$ we obtain

$$\|S\|_Y^Z = \sup_{\|y\|_Y \leqslant 1} \|Sy\|_Z = \sup_{\|Bx\|_Y \leqslant 1} \|SBx\|_Z = U_R.$$

This proves our assertion.

We write $\mathfrak{N}(N)$ for the set of operators $T \in \mathcal{T}(N)$ such that the operator $(A - T) \circ B^{-1}$ is a single-valued bounded (linear) operator (from Y to Z); for $T \in \mathfrak{N}(N)$ we have

$$U(T) = \| (A - T) \circ B^{-1} \|_Y^Z. \tag{10}$$

The preceding lemma implies the following assertion.

COROLLARY. *We have*

$$E(N) = \inf_{T \in \mathfrak{N}(N)} \|(A - T) \circ B^{-1}\|_Y^Z; \tag{11}$$

in addition, $E(N) = \infty$ if the set $\mathfrak{N}(N)$ is empty.

Let $\mathcal{V}(N)$ be the set of all bounded linear operators V from Z^* to X^* such that $\|V\| \leqslant N$ and that for any functional $\kappa \in Z^*$ the functional $(\kappa \circ A - V\kappa) \circ B^{-1}$ is single-valued and bounded on Y. We put

$$J(V) = \sup_{\|\varkappa\|_{Z^*} \leqslant 1} \|(\varkappa \circ A - V\varkappa) \circ B^{-1}\|_{Y^*}, \tag{12}$$

$$G(N) = \inf_{V \in \mathcal{V}(N)} J(V); \tag{13}$$

if the set $\mathcal{V}(N)$ is empty we assume that $G(N) = \infty$.

THEOREM 1. *We have*

$$E(N) \geqslant G(N). \tag{14}$$

If Z is reflexive, then

$$E(N) = G(N), \tag{15}$$

and an operator T is extremal in (8) *if and only if the operator T^* is extremal in* (13).

PROOF. Let $E(N) < \infty$ and $T \in \mathfrak{N}(N)$. The norms of the operators T and T^* coincide. Besides, by (10) and (12), $U(T) = J(T^*)$. Hence (14) follows.

If Z is a reflexive space, then for any operator $V \in \mathcal{V}(N)$ the restriction T to X of the conjugate operator V^* on X^{**} is a bounded linear operator from X to Z such that $T^* = V$. Consequently $U(T) = J(V)$ and $\|T\| \leqslant N$. This gives the inequality $E(N) \leqslant G(N)$, and so (15).

REMARK. It is easy to verify that even if Z is not reflexive but (15) holds and T is an extremal operator in (8), then T^* is extremal in (13).

3. To an arbitrary linear functional κ of the unit ball K of the space Z^* and to a number $N > 0$ we associate the set $\Lambda(\kappa, N)$ of functionals $\lambda \in X^*$ with $\|\lambda\|_{X^*} \leqslant N$ such that $(\kappa \circ A - \lambda) \circ B^{-1}$ is a single-valued bounded linear functional on Y. We put

$$\begin{aligned} R(\varkappa, N) &= \inf_{\lambda \in \Lambda(\varkappa, N)} \|(\varkappa \circ A - \lambda) \circ B^{-1}\|_{Y^*}, \\ F(N) = \sup_{\varkappa \in K} R(\varkappa, N) &= \sup_{\|\varkappa\|_{Z^*} \leqslant 1} \inf_{\lambda \in \Lambda(\varkappa, N)} \|(\varkappa \circ A - \lambda) \circ B^{-1}\|_{Y^*}. \end{aligned} \tag{16}$$

A functional $\kappa \in Z^*$ satisfying the conditions

$$\|\varkappa\| = 1, \qquad F(N) = R(\varkappa, N) \tag{17}$$

is said to be maximal (in the problem (16)). Since for any operator $V \in \mathcal{V}(N)$ and functional $\kappa \in K$ we have $V\kappa \in \Lambda(\kappa, N)$, it follows that

$$G(N) \geqslant F(N). \tag{18}$$

The function $F(N)$ is connected in the following way with the modulus of continuity $\omega(\delta)$ of the operator A by means of the function $\Omega(N)$ defined in (4).

THEOREM 2. *We have*

$$F(N) = \Omega(N). \tag{19}$$

In addition, if there is an element $\bar{x} \in X$ with the property

$$\|B\bar{x}\|_Y = 1, \quad \Omega(N) = \|A\bar{x}\|_Z - N\|\bar{x}\|_X, \tag{20}$$

and the functional $\bar{\kappa} \in Z^*$ is defined by the conditions

$$\|\bar{\varkappa}\| = 1, \quad \bar{\varkappa}A\bar{x} = \|A\bar{x}\|_Z, \tag{21}$$

then the functional $\bar{\kappa}$ is maximal; that is, it satisfies (17).

PROOF. We put

$$\begin{aligned} \sigma(\varkappa, N) &= \inf_{\|\lambda\|_{X^*} \leqslant N} \sup_{\|Bx\|_Y \leqslant 1} (\varkappa Ax - \lambda x), \\ \sigma(N) &= \sup_{\|\varkappa\|_{Z^*} \leqslant 1} \sigma(\varkappa, N). \end{aligned}$$

The quantity $\sigma(\kappa, N)$ is a particular case of $E(N)$. According to (11),

$$\sigma(\varkappa, N) = \inf_{\lambda \in \Lambda(\varkappa, N)} \|(\varkappa \circ A - \lambda) \circ B^{-1}\|_{Y^*} = R(\varkappa, N), \tag{22}$$

and therefore $\sigma(N) = F(N)$.

On the other hand, by Lemma 1,

$$\sigma(\varkappa, N) = \sup_{\|Bx\|_Y \leqslant 1} (\varkappa Ax - N\|x\|_X). \tag{23}$$

Hence

$$\sigma(N) = \sup_{\|\varkappa\|_{Z^*}\leqslant 1} \sup_{\|Bx\|_Y\leqslant 1} (\varkappa Ax - N\|x\|_X) = \sup_{\|Bx\|_Y\leqslant 1} \{ \sup_{\|\varkappa\|_{Z^*}\leqslant 1} \varkappa Ax - N\|x\|_X\}$$

$$= \Omega(N), \quad (24)$$

since for any element $z \in Z$ we have

$$\|z\|_Z = \sup_{\|\varkappa\|_{Z^*}\leqslant 1} \varkappa z.$$

Thus (19) is correct.

We prove the second part of the proposition. Since $\bar{\kappa}Ax \leqslant \|Ax\|_Z$, we obtain $\sigma(\bar{\kappa}, N) \leqslant \Omega(N)$ from (23). Further, according to (20) and (21) we have

$$\Omega(N) = \bar{\varkappa}A\bar{x} - N\|\bar{x}\|_X = \inf_{\|\lambda\|_{X^*}\leqslant N} (\bar{\varkappa}A\bar{x} - \lambda\bar{x}) \leqslant \inf_{\|\lambda\|_{X^*}\leqslant N} \sup_{\|Bx\|_Y\leqslant 1} (\bar{\varkappa}\circ A - \lambda)x$$

$$= \sigma(\bar{\varkappa}, N);$$

that is, $\Omega(N) \leqslant \sigma(\bar{\kappa}, N)$. Therefore $\Omega(N) = \sigma(\bar{\kappa}, N)$. The maximality property (17) of $\bar{\kappa}$ now follows from (22) and (19).

This proves the theorem.

REMARK. If the set Σ of functionals $\kappa \in Z^*$ with $\|\kappa\| \leqslant 1$ is such that $\|z\|_Z = \sup_{\kappa\in\Sigma}\kappa z$ for any element $z \in AX$, then

$$\Omega(N) = F(N) = \sup_{\varkappa\in\Sigma} R(\varkappa, N).$$

This follows from the fact that we may restrict the functionals in (24) to the set Σ.

4. We consider the quantities (8), (13) and (16) in the particular case when the operator $A \circ B^{-1}$ is single-valued; that is,

$$A = H\circ B, \quad (25)$$

where $H = A \circ B^{-1}$ is a linear operator from Y to Z.

If we write $\mathcal{P}(N)$ for the set of linear operators P from Y to Z that satisfy the condition $\|P \circ B\|_X^Z \leqslant N$, it is then easy to verify that (11) may be written in the form

$$E(N) = \inf_{P\in\mathcal{P}(N)} \|H - P\|_Y^Z. \quad (26)$$

We transform $F(N)$. Let

$$\Phi(N) = \{\varphi\in N^{\#} : \|\varphi\circ B\|_{X^*} \leqslant N\},$$
$$\Psi = \{\varkappa\circ H : \|\varkappa\|_{Z^*} \leqslant 1\}. \quad (27)$$

If $\psi = \kappa \circ H$, then (see (26)) the quantity

$$F(\psi, N) = \inf_{\varphi\in\Phi(N)} \|\psi - \varphi\|_{Y^*}$$

coincides with $R(\kappa, N)$, and therefore

$$F(N) = \sup_{\|\varkappa\|_{Z^*}\leqslant 1} F(\varkappa\circ H, N) = \sup_{\psi\in\Psi} \inf_{\varphi\in\Phi(N)} \|\psi - \varphi\|_{Y^*}, \quad (28)$$

that is, $F(N)$ is an approximation of Ψ by the set $\Phi(N)$.

We show that in this case

$$G(N) = \inf_{m \in \mu(N)} \sup_{\|\varkappa\|_{Z^*} \leqslant 1} \|\varkappa \circ H - m\varkappa\|_{Y^*}, \tag{29}$$

where $\mu(N)$ is the set of linear operators m acting from Z^* to $Y^\#$ that satisfy the condition

$$\|m\varkappa \circ B\|_{X^*} \leqslant N \|\varkappa\|_{Z^*}.$$

Indeed, to each operator $V \in \mathcal{V}(N)$ we can associate an operator $m \in \mu(N)$ by the formula $m\kappa = V\kappa \circ B^{-1}$. Conversely, the formula $V\kappa = m\kappa \circ B$ associates an operator $V \in \mathcal{V}(N)$ to each operator $m \in \mu(N)$. Since the corresponding deviations are equal, we have (29).

Besides (25) we shall assume that $\overline{HY} = Z$; that is, any functional $\kappa \in Z^*$ is uniquely determined by the functional $\kappa \circ H$. Then, obviously, we may write

$$G(N) = \inf_{M \in \mathcal{M}(N)} \sup_{\|\varkappa\|_{Z^*} \leqslant 1} \|\varkappa \circ H - M(\varkappa \circ H)\|_{Y^*}, \tag{30}$$

where $M(N)$ is the set of linear operators M from the set $\{\kappa \circ H: \kappa \in Z^*\}$ to $Y^\#$ with the property

$$\|(M(\varkappa \circ H)) \circ B\|_{X^*} \leqslant N \|\varkappa\|_{Z^*}.$$

The preceding condition signifies that if $\psi \in \Psi$ then $\varphi = M\psi \in \Phi(N)$, and therefore $G(N)$ is here the linear approximation to the set Ψ by the set $\Phi(N)$.

We note that if (25) holds, $\overline{HY} = Z$, and Z is a reflexive space, then, by Theorem 1, $E(N) = G(N)$; and also $T \in \mathfrak{N}(N)$ is extremal for the problem (3) if and only if the operator $M \in M(N)$ connected with T by the relation

$$(M(\varkappa \circ H)) \circ B = T^*\varkappa, \quad \varkappa \in Z^*,$$

is extremal in (30); here T^* is the conjugate operator to T.

5. The quantities $F(N)$, $G(N)$ and $E(N)$ are not always finite.
Thus $F(N) = G(N) = E(N) = \infty$ *if*

$$N < \overline{N} = \sup_{Bx=0,\ x \neq 0} \frac{\|Ax\|_Z}{\|x\|_X}.$$

In view of (14) and (18) it is enough to prove this for $F(N)$. We assume that $F(N) < \infty$. If the quantity

$$\|(\varkappa \circ A - \lambda) \circ B^{-1}\|_{Y^*} = \sup_{\|Bx\|_Y \leqslant 1} (\varkappa \circ A - \lambda)\, x$$

is finite for the functionals $\kappa \in K$ and $\lambda \in \Lambda(\kappa, N)$, then $\lambda x = \kappa Ax$ on the set $X_0 = \{x: Bx = 0\}$. Hence

$$\sup_{x \in X_0} \frac{\varkappa Ax}{\|x\|_X} = \sup_{x \in X_0} \frac{\lambda x}{\|x\|_X} \leqslant \|\lambda\|_{X^*} \leqslant N.$$

Since $\kappa \in K$ is arbitrary, this gives the inequality $\overline{N} \leqslant N$.

We consider, as an example, the quantity

$$E_{n,k}(N) = \inf_{\|T\|_{L_r}^{L_q} \leqslant N} \sup_{\|x^{(n)}\|_{L_p} \leqslant 1} \|x^{(k)} - Tx\|_{L_q},$$

where $0 \leqslant k < n$ and $L_s = L_s[0, 1]$. Here $Z = L_q$, X is the set of all functions x

having absolutely continuous $(n-1)$th derivative $x^{(n-1)}$ and $x^{(n)} \in L_p$, $Y = \{x^{(n)}: x \in X\}$, $A = D^k$ and $B = D^n$. In this case X_0 is the set Π_{n-1} of polynomials of degree $n-1$.

In consequence,

$$E_{n,k}(N) < \infty \Rightarrow N \geqslant \overline{N} = \sup_{\pi \in \Pi_{n-1}} \frac{\|\pi^{(k)}\|_{L_q}}{\|\pi\|_{L_r}}. \tag{31}$$

6. We consider concrete cases of the problems (7), (8), (28) and (30).

Let S be a measurable set in the n-dimensional euclidean space R^n; let a and b be finite measurable functions on S; let the parameters α and β satisfy the conditions $1 \leqslant \alpha, \beta \leqslant \infty$.

We put

$$\omega(\delta) = \sup_{\substack{\|x\|_{L_\alpha} \leqslant \delta \\ \|bx\|_{L_\beta} \leqslant 1}} \|ax\|_L. \tag{32}$$

This is a particular case of (7); namely, if we write I for the identity operator, then here $A = aI$, $B = bI$, $Z = L$, $X = \{x \in L_\alpha: bx \in L_\beta\}$ and $Y = BX = \{bx: x \in L_\alpha, bx \in L_\beta\}$; in addition, we consider the norms of the spaces L_α and L_β on X and Y respectively.

For $1 \leqslant \alpha < \infty$ we have $\overline{X} = L_\alpha$ since b is finite, and therefore each bounded linear operator T from X to $Z = L$ can be extended without change of norm to all of L_α. Consequently, in this case $E(N)$ coincides with

$$\mathcal{E}(N) = \inf_{\|T\|_{L_\alpha}^{L} \leqslant N} \sup_{\substack{\|bx\|_{L_p} \leqslant 1 \\ x \in L_\alpha}} \|ax - Tx\|_L. \tag{33}$$

In this case $S = [0, \infty)$, $a(t) = t^\nu$ and $b(t) = t^\mu$ we have

$$\omega(\delta) = \omega(1)\,\delta^\gamma, \qquad \gamma = \frac{\mu - \nu + (1/\beta) - 1}{\mu + (1/\beta) - (1/\alpha)},$$

for the function (32), and then (9) takes the form

$$\|t^\nu x(t)\|_L \leqslant C\|x(t)\|_{L_\alpha}^{\gamma}\|t^\mu x(t)\|_{L_\beta}^{1-\gamma},$$

that is, the Carleson inequality. The least constant $C = \omega(1)$ in this inequality (and so the quantity (32) in the present case) was found by V. I. Levin under more general conditions on the parameters (see also the generalization of this result to many variables by Andrianov [5]). Taĭkov [6] actually found (32) and (33) in studying the problems (7) and (8) for the differentiation operator in the case $S = (-\infty, \infty)$, $a(t) = t^\nu$, $b(t) = t^\mu$ and $\alpha = \beta = 2$.

The quantities $E(N)$ and $\omega(\delta)$ for arbitrary α and β ($1 \leqslant \alpha, \beta \leqslant \infty$) are connected by the inequality

$$\mathcal{E}(N) \geqslant \Omega(N) = \sup_{\delta > 0}\{\omega(\delta) - N\delta\}; \tag{34}$$

in addition, if the operator T and the function x are such that

$$\| Tx \|_L = \| T \|^L_{L_\alpha} \| x \|_{L_\alpha}, \qquad \| bx \|_{L_\beta} = 1,$$
$$U(T) = \| Ax - Tx \|_L = \| Ax \|_L - \| Tx \|_L, \tag{35}$$

then we have

$$\mathcal{E}(\| T \|) = U(T) = \Omega(\| T \|), \qquad \omega(\| x \|_{L_\alpha}) = \| ax \|_L. \tag{36}$$

In the case $1 \leqslant \alpha < \infty$ this is a particular case of the relations (3)–(6); in the case $\alpha = \infty$ we must use the inequality $\mathcal{E}(N) \geqslant E(N)$.

We estimate $\mathcal{E}(N)$. Starting from the form of the operator A it is natural to consider the operator $T = gI$. For each such operator we have

$$\| T \|^L_{L_\alpha} = \| g \|_{L_{\alpha'}},$$
$$U(T) = \sup_{\substack{\|bx\|_{L_\beta} \leqslant 1 \\ x \in L_\alpha}} \int \left| \frac{a-g}{b} bx \right| dt = \left\| \frac{a-g}{b} \right\|_{L_{\beta'}}, \tag{37}$$

where $(1/\alpha) + (1/\alpha') = 1$ and $(1/\beta) + (1/\beta') = 1$. Hence

$$\mathcal{E}(N) \leqslant \inf_{\|g\|_{L_{\alpha'}} \leqslant N} \left\| \frac{a-g}{b} \right\|_{L_{\beta'}}. \tag{38}$$

For a linear functional $\Theta \in L^*_\alpha$ we put

$$u(\Theta) = \sup_{x \in Q} \left\{ \int ax\, dt - \Theta x \right\}$$

and consider the quantity

$$e(N) = \inf_{\|\Theta\|_{L^*_\alpha} \leqslant N} u(\Theta). \tag{39}$$

Each bounded linear operator T from L_α to L defines a bounded linear functional $\Theta \in L^*_\alpha$ by the formula

$$\Theta x = \int Tx\, dt, \qquad x \in L_\alpha.$$

In addition, $\|\Theta\| \leqslant \|T\|$ and $u(\Theta) \leqslant U(T)$. Hence

$$e(N) \leqslant \mathcal{E}(N). \tag{40}$$

In the case $1 \leqslant \alpha < \infty$ an arbitrary functional $\Theta \in L^*_\alpha$ has the form

$$\Theta x = \int xg\, dt,$$

where $g \in L_{\alpha'}$ and $\|\Theta\|_{L^*_\alpha} = \|g\|_{L_{\alpha'}}$. And then (see (38))

$$u(\Theta) = \left\| \frac{a-g}{b} \right\|_{L_{\beta'}},$$

and consequently

$$e(N) = \inf_{\|g\|_{L_{\alpha'}} \leqslant N} \left\| \frac{a-g}{b} \right\|_{L_{\beta'}}. \tag{41}$$

Further, by Lemma 1 of V. N. Gabušin, we have

$$e(N)=\sup_{x\in Q}\left\{\int ax\,dt-N\|x\|_{L_\alpha}\right\}=\Omega(N). \tag{42}$$

The relations (42), (41), (40) and (38) now imply the following assertion.

For $1\leqslant\alpha<\infty$ *and* $1\leqslant\beta\leqslant\infty$ *we have*

$$\mathcal{E}(N)=e(N)=\inf_{\|g\|_{L_{\alpha'}}\leqslant N}\left\|\frac{a-g}{b}\right\|_{L_{\beta'}}=\Omega(N). \tag{43}$$

We remark that by (34) and (37) for $1\leqslant\alpha,\beta\leqslant\infty$ and any function $g\in L_{\alpha'}$ we have

$$\left\|\frac{a-g}{b}\right\|_{L_{\beta'}}\geqslant\Omega(\|g\|_{L_{\alpha'}});$$

on putting $g=\varphi b$ we may write this in the form

$$\left\|\frac{a}{b}-\varphi\right\|_{L_{\beta'}}\geqslant\Omega(\|\varphi b\|_{L_{\beta'}}). \tag{44}$$

This inequality was in fact proved by Taĭkov [6] for the case $\alpha=\beta=2$, $a(t)=t^\nu$, $b(t)=t^\mu$ and $S=(-\infty,\infty)$.

We consider the quantity $F(N)$. For simplicity we shall assume that the functions a and b are nonzero almost everywhere on S. Then $A=H\circ B$, where $H=aI/b$. We may therefore use (28).

We elucidate the form of functionals in the set Ψ. A functional $\kappa\in Z^*=L^*$ is defined by some function $\xi\in L_\infty$ by

$$\varkappa z=\int_S \xi(t)\,z(t)\,dt,$$

and $\|\kappa\|_{Z^*}=\|\xi\|_{L_\infty}$. Thus

$$\overline{\psi}y=\varkappa Hy=\int\xi\frac{a}{b}y\,dt,\qquad y\in Y,$$

if $\overline{\psi}=\kappa\circ H\in\Psi$ (that is, $\|\kappa\|\leqslant 1$), where $\|\xi\|_{L_\infty}\leqslant 1$. Conversely, this formula defines a functional $\overline{\psi}$ in the set Ψ. In consequence, the set Ψ is formed by functionals $\overline{\psi}$ of the form

$$\overline{\psi}y=\int\psi(t)\,y(t)\,dt,$$

where $\|\psi b/a\|_{L_\infty}\leqslant 1$.

In the case $1\leqslant\alpha<\infty$ we can establish in a similar way that $\Phi(N)$ is the set of functionals $\overline{\varphi}$ of the form

$$\overline{\varphi}y=\int\varphi(t)\,y(t)\,dt,$$

where φ is an arbitrary function satisfying $\|\varphi b\|_{L_{\alpha'}}\leqslant N$. In addition, we have

$$\|\overline{\psi}-\overline{\varphi}\|_{Y^*}=\sup_{\substack{\|bx\|_{L_\beta}\leqslant 1\\ x\in L_\alpha}}\int(\psi-\varphi)\,bx\,dt=\|\psi-\varphi\|_{L_{\beta'}};$$

here $1/\beta+1/\beta'=1$. In consequence, for $1\leqslant\alpha<\infty$ the quantity $F(N)$ coincides with

$$\mathcal{F}(N)=\sup_{\left\|\psi\frac{b}{a}\right\|_{L_\infty}\leqslant 1}\ \inf_{\|\varphi b\|_{L_{\alpha'}}\leqslant N}\|\psi-\varphi\|_{L_{\beta'}}. \tag{45}$$

For $1 \leqslant \alpha, \beta < \infty$ we write $G(N)$ for the linear approximation of the set $\Psi = \{\psi: \|\psi b/a\|_{L_\infty} \leqslant 1\}$ by the set $\Phi(N) = \{\varphi: \|\varphi b\|_{L_{\alpha'}} \leqslant N\}$. In the present example, it is easy to check that $\overline{HY} = Z$. Hence, as for (45), the equation $G(N) = G(N)$ follows in the case $1 \leqslant \alpha < \infty$.

We put

$$\mathcal{F}(\psi, N) = \inf_{\|\varphi b\|_{L_{\alpha'}} \leqslant N} \|\psi - \varphi\|_{L_{\beta'}}$$

and prove the following assertion.

For $1 \leqslant \alpha, \beta < \infty$ *we have*

$$\mathcal{G}(N) = \mathcal{F}(N) = \mathcal{F}\left(\frac{a}{b}, N\right). \tag{46}$$

Since $a/b \in \Psi$, we have

$$\mathcal{F}\left(\frac{a}{b}, N\right) \leqslant \mathcal{F}(N) \leqslant \mathcal{G}(N). \tag{47}$$

Further, we associate the operator $M = b\rho I/a$ with the arbitrary function $\rho \in \Phi(N)$. We have

$$\|bM\psi\|_{L_{\alpha'}} = \left\|\frac{b}{a}\psi b\rho\right\|_{L_{\alpha'}} \leqslant \left\|\frac{b}{a}\psi\right\|_{L_\infty} \|b\rho\|_{L_{\alpha'}},$$

$$\|\psi - M\psi\|_{L_{\beta'}} = \left\|\frac{b}{a}\psi\left(\frac{a}{b} - \rho\right)\right\|_{L_{\beta'}} \leqslant \left\|\frac{b}{a}\psi\right\|_{L_\infty} \left\|\frac{a}{b} - \rho\right\|_{L_{\beta'}}.$$

In consequence, M defines a linear method of approximation of the set Ψ by the set $\Phi(N)$ and

$$\mathcal{G}(N) \leqslant \sup_{\psi} \|\psi - M\psi\|_{L_{\beta'}} = \left\|\frac{a}{b} - \rho\right\|_{L_{\beta'}}.$$

Hence $G(N) \leqslant F(a/b, N)$. This and (47) together imply (46).

It is clear from the proof of this assertion that if $T = gI$ is an extremal operator for the problem (33), then the formula $M = gI/a$ defines the best linear method of approximation of the set Ψ by the set $\Phi(N)$.

We make some remarks on the precise evaluation of the quantities (32) and (33), not assuming that $a \neq 0$ and $b \neq 0$ almost everywhere.

It is easy to see that replacing a and b by $|a|$ and $|b|$ does not change (32). We may therefore assume that $a \geqslant 0$ and $b \geqslant 0$. Then we may restrict ourselves in (43) to functions g satisfying $0 \leqslant g \leqslant a$. For functions $x \geqslant 0$ and operators $T = gI$ with $0 \leqslant g \leqslant a$ the conditions (35) take the form

$$\int gx\,dt = \|x\|_{L_\alpha} \|g\|_{L_{\alpha'}}, \tag{48}$$

$$\int \frac{a-g}{b} bx\,dt = \|bx\|_{L_\beta} \left\|\frac{a-g}{b}\right\|_{L_{\beta'}}, \tag{49}$$

$$\|bx\|_{L_\beta} = 1. \tag{50}$$

Let $1 < \alpha, \beta < \infty$. Then (48)–(50) signify that

$$x = \nu g^{\frac{1}{\alpha-1}}, \qquad \nu = \left\| b g^{\frac{1}{\alpha-1}} \right\|_{L_\beta}^{-1}, \tag{51}$$

and g is a solution of the equation

$$a = g + b^\beta (\varepsilon g)^{\frac{\beta-1}{\alpha-1}}, \tag{52}$$

where ϵ is a positive constant.

For any $c \geqslant 0$ and $\gamma \geqslant 0$ the function $\chi(u) = u + cu^\gamma$ is continuous and increases from 0 to ∞ on the interval $[0, \infty)$. The equation (52) therefore has a unique solution for any a, b and ϵ. If also

$$g \in L_{\alpha'}, \qquad b g^{\frac{1}{\alpha-1}} \in L_\beta, \tag{53}$$

and x is defined by (51), then, according to (36), (37) and (52),

$$\mathfrak{E}(\| g \|_{L_{\alpha'}}) = U(gI) = \left\| b (\varepsilon g)^{\frac{1}{\alpha-1}} \right\|_{L_\beta}^{\beta-1}, \tag{54}$$

$$\omega(\| x \|_{L_\alpha}) = \int axdt. \tag{55}$$

It can happen that $N_\epsilon = \|g\|_{L_{\alpha'}}$ does not take all values from 0 to ∞ when $\epsilon \in (0, \infty)$. In addition, one or both of the conditions (53) may be violated. We describe the solutions of the problem (33) in these cases; in addition, we assume that $b \not\equiv 0$.

We first consider some properties of the solution $g = g_\epsilon$ of the equation (52). It is clear that $0 \leqslant g_\epsilon \leqslant a$ and that

$$g_\varepsilon(t) = a(t) \quad \text{for} \quad t \in S(b) = \{t \in S : b(t) = 0\}. \tag{56}$$

Further, since for $t \notin S(b)$ (that is, $b(t) \neq 0$) the right-hand side of (52) is an increasing function of the variables ϵ and g, we have

$$g_\varepsilon(t) < g_{\varepsilon'}(t) \qquad \text{for} \qquad \varepsilon > \varepsilon', \qquad t \notin S(b). \tag{57}$$

On putting $\epsilon g_\epsilon = \rho$ and $\gamma = (\beta - 1)/(\alpha - 1)$, we transform (52) to the form

$$a = \frac{1}{\varepsilon}\rho + b^\beta \rho^\gamma.$$

The right-hand side of this equation increases with ρ and decreases with ϵ, so $\epsilon' g_{\epsilon'} < \epsilon g_\epsilon$ for $\epsilon > \epsilon'$. Thus, if $\epsilon > \epsilon'$, we have (on S) that

$$\frac{\varepsilon'}{\varepsilon} g_{\varepsilon'} < g_\varepsilon \leqslant g_{\varepsilon'}. \tag{58}$$

We assume that $N_\epsilon = \|g_\epsilon\|_{L_{\alpha'}}$ is finite for some $\epsilon > 0$. Then, by (58), it will be finite for any $\epsilon > 0$. Further, it follows from (56) and (57) that N_ϵ strictly decreases with respect to ϵ on $(0, \infty)$. And, since as is easy to verify, $g_\epsilon(t)$ is continuous in ϵ for any fixed $t \in S$, it is easy to prove, using Lebesgue's theorem on taking limits under the integral sign, that N_ϵ is also a continuous function of ϵ on the interval $(0, \infty)$. We put

$$N_0 = \lim_{\varepsilon \to +0} N_\varepsilon, \qquad N_\infty = \lim_{\varepsilon \to \infty} N_\varepsilon;$$

for any $\epsilon > 0$ we have

$$N_\infty < N_\varepsilon < N_0.$$

On analyzing the behavior of solutions of (52) as $\epsilon \to 0$ and $\epsilon \to \infty$ we obtain

$$N_0 = \|a\|_{L_{\alpha'}(S)}, \qquad N_\infty = \|a\|_{L_{\alpha'}(S(b))}.$$

It follows easily from (58) that if the second of the conditions (53) holds for some $\epsilon > 0$ it then holds for any $\epsilon > 0$. And then (54) gives the value of $E(N)$ for any $N \in (N_\infty, N_0)$.

We show that (54) also holds in the case when $bg^{1/(\alpha-1)} \notin L_\beta$; that is, we show that if $bg^{1/(\alpha-1)} \notin L_\beta$, then $E(N) = \infty$ for any $N \in (N_\infty, N_0)$. For fixed $\epsilon > 0$ and an arbitrary number $m > 0$ we write S_m for the set of points $t \in S$ that lie in the ball of radius m in the space R_n and satisfy the condition $b^\beta(t) g^{\beta/(\alpha-1)}(t) \leqslant m$. We put

$$y_m(t) = \begin{cases} g^{\frac{1}{\alpha-1}}(t) & \text{for} \quad t \in S_m, \\ 0 & \text{for} \quad t \notin S_m, \end{cases} \qquad x_m = \frac{y_m}{\|by_m\|_{L_\beta}}.$$

Then we have

$$\mathcal{E}(N_\varepsilon) \geqslant \int x_m a\, dt - N_\varepsilon \|x_m\|_{L_\alpha} = \int a x_m dt - \|g\|_{L_{\alpha'}(S_m)} \|x_m\|_{L_\alpha} + c_m,$$

where

$$c_m = \{\|g\|_{L_{\alpha'}}(S_m) - N_\varepsilon\} \|x_m\|_{L_{\alpha'}}.$$

On using (52) we may rewrite the preceding inequality in the form

$$\mathcal{E}(N_\varepsilon) \geqslant \left\| b(\varepsilon g)^{\frac{1}{\alpha-1}} \right\|^{\beta-1}_{L_\beta(S_m)} + c_m.$$

Since by definition

$$g \in L_{\alpha'}, \qquad bg^{\frac{1}{\alpha-1}} \notin L_\beta,$$

the first term on the right tends to ∞ as $m \to \infty$, and $c_m \to 0$. In consequence, $E(N_\epsilon) = \infty$.

We have thus studied the solution of the problem (33) in the case $1 < \alpha, \beta < \infty$ under the conditions $g \in L_{\alpha'}$ for $N \in (N_\infty, N_0)$. If $N_\infty > 0$, then for $N = N_\infty$, on putting

$$g_\infty(t) = \begin{cases} a(t) & \text{for} \quad b(t) = 0, \\ 0 & \text{for} \quad b(t) \neq 0, \end{cases}$$

we obtain

$$\mathcal{E}(N_\infty) = U(g_\infty I) = \left\{ \int_{b(t)\neq 0} \left(\frac{a}{b}\right)^{\beta'} dt \right\}^{1/\beta'}$$

irrespective of whether the last quantity is finite or infinite; for $N < N_\infty$ we have $E(N) = \infty$ (see §5 of this paper). If it happens that $N_0 < \infty$ (that is, $a \in L_{\alpha'}$), then for $N \geqslant N_0$ we shall have

$$\mathcal{E}(N) = \mathcal{E}(N_0) = U\,(aI) = 0.$$

Finally, if $g \notin L_{\alpha'}$ (for any or some $\epsilon > 0$), then we may show that $E(N) = \infty$ for any N.

We note that in the case $1 < \beta = \alpha < \infty$ the equation (52) has a particularly simple form; here $g = a/(1 + \epsilon b^{\alpha})$.

We consider in less detail solutions of the problems (33) and (32) for certain other values of the parameters α and β. In addition, we shall, in general, assume that all the integrals that occur are finite. Let $\alpha = 1$. If for $1 < \beta \leqslant \infty$ we put

$$\underline{S}_N = \{t \in S\colon\ a\,(t) \leqslant N\},\quad \overline{S}_N = \{t \in S\colon\ a\,(t) > N\},$$

$$g\,(t) = \begin{cases} a\,(t) & \text{for} \quad t \in \underline{S}_N, \\ N & \text{for} \quad t \in \overline{S}_N, \end{cases} \tag{59}$$

$x(t) = 0$ on the set $\underline{S}_N$ and

$$x\,(t) = \mu \frac{1}{b\,(t)} \left(\frac{a\,(t) - N}{b\,(t)}\right)^{\frac{1}{\beta - 1}}$$

on the set $\overline{S}_N$ (the constant μ to be chosen so that $\|bx\|_{L_\beta} = 1$), we arrive at the equation

$$\mathcal{E}(N) = U\,(gI) = \left\|\frac{a - N}{b}\right\|_{L_{\beta'}(\overline{S}_N)} \tag{60}$$

and at (55) with the chosen x and $\alpha = 1$.

(60) also holds when $\alpha = \beta = 1$ (with the function (59)). But here no function x with properties (48)–(50) exists, although it is not hard to construct a corresponding extremal sequence of functions. If in addition there is a point $t_N \in \overline{S}_N$ at which the functions a and b are continuous, and

$$b\,(t_N) \neq 0, \qquad \frac{a\,(t_N) - N}{b(t_N)} = \left\|\frac{a - N}{b}\right\|_{L_\infty(\overline{S}_N)},$$

then we have

$$\omega\left(\frac{1}{b\,(t_N)}\right) = \frac{a\,(t_N)}{b\,(t_N)}, \tag{61}$$

and by means of the δ-function an "ideal" extremal function may be written in the form

$$x\,(t) = \frac{1}{b\,(t_N)}\,\delta\,(t - t_N).$$

Finally, let $\alpha = \infty$. On analyzing the conditions (48) and (49) we see that the functions x and g must be chosen in the following manner. If $\beta = \infty$

$$g\,(t) = \begin{cases} a\,(t), & \text{if} \quad \varepsilon b\,(t) \leqslant 1, \\ 0, & \text{if} \quad \varepsilon b\,(t) > 1, \end{cases}$$

$$x\,(t) = \begin{cases} \varepsilon, & \text{if} \quad \varepsilon b\,(t) \leqslant 1, \\ \dfrac{\varepsilon}{b\,(t)}, & \text{if} \quad \varepsilon b\,(t) > 1; \end{cases}$$

for $1 \leqslant \beta < \infty$

$$g(t) = \max\{0,\ a(t) - (\varepsilon b(t))^{\beta}\},$$

for $1 < \beta < \infty$

$$x(t) = \frac{\mu}{b(t)} \left(\frac{a(t) - g(t)}{b(t)}\right)^{\frac{1}{\beta - 1}},$$

and for $\beta = 1$

$$x(t) = \begin{cases} \mu, & \text{if} \quad a(t) > \varepsilon b(t), \\ 0, & \text{if} \quad a(t) \leqslant \varepsilon b(t); \end{cases}$$

here ϵ is an arbitrary positive parameter and the constant $\mu > 0$ is such that $\|bx\|_{L_\beta} = 1$. Then we have

$$\mathcal{E}(\|g\|_L) = \left\|\frac{a-g}{b}\right\|_{L_{\beta'}} = \Omega(\|g\|_L), \qquad \omega(\|x\|_{L_\infty}) = \int ax\, dt,$$

and thus (43) holds for $N = \|g\|_L = N_\epsilon$ $(\alpha = \infty)$.

BIBLIOGRAPHY

1. S. B. Stečkin, *Best approximation of linear operators,* Mat. Zametki **1** (1967), 137–148 = Math. Notes **1** (1967), 91–99. MR **35** #2051.

2. V. N. Gabušin, *Best approximation of functionals on certain sets,* Mat. Zametki **8** (1970), 551–562 = Math. Notes **8** (1970), 780–785. MR **43** #2407.

3. Arthur Sard, *Integral representations of remainders,* Duke Math. J. **15** (1948), 333–345. MR **10**, 197.

4. V. I. Levin, *Exact constants in inequalities of the Carlson type,* Dokl. Akad. Nauk SSSR **59** (1948), 635–638. (Russian) MR **9**, 415.

5. F. I. Andrianov, *Multidimensional analogs of the Carlson inequality and of its generalization,* Izv. Vysš. Učebn. Zaved. Matematika **1967**, no. 1 (56), 3–7. (Russian) MR **35** #3019.

6. L. V. Taĭkov, *Inequalities of Kolmogorov type and best formulas for numerical differentiation,* Mat. Zametki **4** (1968), 233–238 = Math. Notes **4** (1968), 631–634. MR **38** #1459.

Translated by P. G. SPAIN

APPROXIMATION OF OPERATORS INVARIANT WITH RESPECT TO A SHIFT

UDC 517.5

V. V. ARESTOV

ABSTRACT. The quantity

$$E(N) = \inf_{\substack{L_s(R^m) \\ \|T\|_{L_r(R^m)} \leqslant N}} \sup_{x \in Q} \| Ax - Tx \|_{L_q(R^m)}$$

is investigated. It is porved that if the class Q and the operator A on Q are invariant with respect to an arbitrary shift, then it suffices in (*) to consider operators T invariant with respect to a shift (on R^m). With the help of this result and known properties of operators invariant with respect to a shift, the properties of the quantity $E(N)$ as a function of r, s, q and Q are deduced. An application of these results to the problem of approximating differential operators with constant coefficients on classes of smooth functions is given.

Bibliography: 22 titles.

§1. The invariance property of approximating operators

Let X be a Banach space, $\mathcal{Z}$ a linear space, A an operator from a set $Q \subset X$ into $\mathcal{Z}$, and Σ and Z two linear subspaces of $\mathcal{Z}$, in each of which a norm is introduced with respect to which it is a Banach space. For a bounded linear operator T from X to Z, we put

$$U(T) = \sup_{x \in Q} \| Ax - Tx \|_{\Sigma}; \tag{1.1}$$

if $z = Ax - Tx \notin \Sigma$, we consider that $\|z\|_{\Sigma} = \infty$. The quantity $U(T)$ is the deviation of the operator T from the operator A in the class Q. Then the quantity

$$E(N) = \inf_{\|T\|_X^Z \leqslant N} U(T) = \inf_{\|T\|_X^Z \leqslant N} \sup_{x \in Q} \| Ax - Tx \|_{\Sigma} \tag{1.2}$$

is the best approximation of the operator A on the set Q by bounded linear operators T from X into Z with norm $\|T\| = \|T\|_X^Z \leqslant N$. Investigation of $E(N)$ also includes the question of the existence, uniqueness, and other properties of an extremal operator T on which the lower bound in (1.2) is achieved.

The problem (1.2) was first studied by Stečkin [1]. In particular, he observed that in the case $\mathcal{Z} = \Sigma = Z$, the quantity $E(N)$ is related to the modulus of continuity

$$\omega(\delta) = \sup_{\substack{x \in Q \\ \|x\|_X \leqslant \delta}} \| Ax \|_Z \tag{1.3}$$

of the operator A on the class Q. Namely, we have the inequalities

$$\omega(\delta) \leqslant \inf_{N > 0} \{ E(N) + N\delta \}, \tag{1.4}$$

$$E(N) \geqslant \Omega(N), \tag{1.5}$$

AMS (MOS) subject classifications (1970). Primary 41A35, 41A50, 41A65; Secondary 41A30.

where

$$\Omega(N) = \sup_{x \in Q} (\|Ax\|_Z - N\|x\|_X) = \sup_{\delta > 0} \{\omega(\delta) - N\delta\}. \tag{1.6}$$

In this connection, Stečkin noted that if there exists an element $x \in Q$ and an operator T which satisfy the conditions

$$\|Tx\|_Z = \|T\| \|x\|_X, \quad U(T) = \|Ax\|_Z - \|Tx\|_Z, \tag{1.7}$$

then one has the inequalities

$$E(\|T\|) = U(T) = \Omega(\|T\|). \tag{1.8}$$

The papers of Stečkin [1], [2], Subbotin and Taĭkov [3], Taĭkov [4], Gabušin [5]–[8], Berdyšev [9], Arestov [10]–[12], and others examined particular cases of problem (1.2). Studied in the greatest detail was the approximation of the differentiation operator $A = d^k/dt^k$ on the class of n times $(0 \leqslant k < n)$ differentiable functions; in this case the value (1.3) is expressed in terms of the least constant in an Hadamard-Kolmogorov type inequality, and (1.4) gives an upper bound for this constant.

In the present paper, we consider the problem (1.2) in the function spaces L_r on m-dimensional Euclidean space R^m. We prove that if the operator A on Q is invariant with respect to a shift, then we may confine ourselves in (1.2) to operators T invariant with respect to a shift. With the aid of this assertion and of known properties of operators invariant with respect to a shift, we deduce certain properties of $E(N)$. We give an application of these results to the problem of approximating ordinary differential operators with constant coefficients on classes of smooth functions.

Below, p, q, r and s are parameters satisfying the condition $1 \leqslant p, q, r, s \leqslant \infty$, and p', q', r' and s' are the dual indices, i.e., for example, $1/p + 1/p' = 1$. For the norm of a function $x \in L_p = L_p(R^m)$ we use the notation $\|c\|_p$, so that

$$\|x\|_p = \begin{cases} \left\{\int |x(t)|^p\, dt\right\}^{1/p}, & 1 \leqslant p < \infty, \\ \operatorname{ess\,sup} |x(t)|, & p = \infty. \end{cases}$$

Let S be the space of functions which are measurable on R^m and locally integrable, Q some class of elements in L_r, and A an operator from Q into S. We put

$$E(N) = E(N; r, s) = \inf_{\|T\|_{L_r}^{L_s} \leqslant N} \sup_{x \in Q} \|(A - T)x\|_q. \tag{1.9}$$

It is clear that (1.9) is (1.2) in the case $\mathcal{Z} = S$, $\Sigma = L_q$, $Z = L_s$, $X = L_r$. Below, by L_r for $r = \infty$ we will always understand the space C_0 of continuous functions with limit 0 as $|t| \longrightarrow \infty$.

For $\tau \in R^m$, we define the shift operator σ_τ by the equation $(\sigma_\tau x)(t) = x(t + \tau)$. We will say that the set $W \subset S$ is invariant with respect to a shift if for arbitrary $\tau \in R^m$ the operator σ_τ carries W into itself, i.e. $\sigma_\tau W = W$. We will call an operator P from such a set W into S invariant with respect to a shift on W if $\sigma_\tau Px = P\sigma_\tau x$ for all $\tau \in R^m$ and $x \in W$.

We denote by $\mathcal{T} = \mathcal{T}_r^s$ the set of bounded linear operators T from L_r to L_s

which are invariant with respect to a shift on all of the space L_r.

LEMMA 1.1. *If the class Q and the operator A on Q are invariant with respect to a shift, then for any bounded linear operator T from L_r to L_s satisfying the condition*

$$U(T) = \sup_{x \in Q} \|(A - T)x\|_q < \infty, \tag{1.10}$$

there exists an operator

$$F \in \mathcal{T}_r^s \tag{1.11}$$

with the properties

$$\|F\|_{L_r}^{L_s} \leqslant \|T\|_{L_r}^{L_s}, \tag{1.12}$$

$$U(F) \leqslant U(T). \tag{1.13}$$

The proof consists in the following. We consider the operators

$$T_h = \frac{1}{|G_h|} \int_{G_h} \sigma_\tau T \sigma_{-\tau}\, d\tau,$$

where G_h is the ball in R^m of radius h with center at the point 0, $|G_h|$ in its volume, and the integral is understood in some weak sense. We prove that there exists a sequence of numbers $h_k \longrightarrow \infty$ for which the sequence of operators T_{h_k} has the "weak" limit F. We verify that the operator F has the needed properties.

Let Φ be the space of continuous functions with compact support (on R^m) and Φ_γ the space Φ with the norm of L_γ. Since $\Phi \subset L_\gamma$ for arbitrary γ, the expression

$$\int \varphi(t)(\sigma_\tau T \sigma_{-\tau} x)(t)\, dt = \eta(\tau)$$

makes sense for $\varphi \in \Phi$ and $x \in L_r$; here and below, in integrals over R^m, the domain of integration is not indicated. Clearly,

$$|\eta(\tau)| \leqslant \|\varphi\|_{s'} \|T\| \|x\|_r \tag{1.14}$$

and for any two points τ and τ_0 in R^m

$$\eta(\tau) - \eta(\tau_0) = \int \varphi \sigma_\tau T (\sigma_{-\tau} x - \sigma_{-\tau_0} x)\, dt + \int \zeta (\sigma_{-\tau}\varphi - \sigma_{-\tau_0}\varphi)\, dt,$$

where $\zeta = T\sigma_{-\tau_0} x \in L_s$. Hence

$$|\eta(\tau) - \eta(\tau_0)| \leqslant \|\varphi\|_{s'} \|T\| \|\sigma_{-\tau} x - \sigma_{-\tau_0} x\|_r + \|\zeta\|_s \|\sigma_{-\tau}\varphi - \sigma_{-\tau_0}\varphi\|_{s'}.$$

Both terms on the right side of this inequality approach zero as $\tau \longrightarrow \tau_0$. Consequently the function η is continuous on R^m.

For $h > 0$, we put

$$f_h(\varphi, x) = \frac{1}{|G_h|} \int_{G_h} \eta(\tau)\, d\tau = \frac{1}{|G_h|} \int_{G_h} \int_{R^m} \varphi \sigma_\tau T \sigma_{-\tau} x\, dt\, d\tau. \tag{1.15}$$

This is a bilinear form on the pair of spaces Φ and L_r, with, by virtue of (1.14),

$$|f_h(\varphi, x)| \leqslant \|\varphi\|_{s'} \|x\|_r \|T\|. \tag{1.16}$$

We denote by Ψ and Λ countable, everywhere dense subsets of the spaces $\Phi_{s'}$ and L_r, respectively; the product of the sets Ψ and Λ is countable. Using inequality (1.16), the Bolzano-Weierstrass theorem, and the Cantor diagonal process, it is easy to prove the existence of a sequence of (natural) numbers h_k, converging to ∞, such that for any functions $\varphi \in \Psi$ and $x \in \Lambda$ the sequence $\{f_{h_k}(\varphi, x)\}$ converges. After this we convince ourselves by means of the Cauchy criterion that the numbers $f_{h_k}(\varphi, x)$ form a convergent sequence for any functions $\varphi \in \Phi$ and $x \in L_r$.

On the pair of spaces Φ and L_r, we consider the bilinear form

$$f(\varphi, x) = \lim_{k\to\infty} f_{h_k}(\varphi, x); \tag{1.17}$$

by virtue of (1.16), we have

$$|f(\varphi, x)| \leqslant \|\varphi\|_{s'} \|x\|_r \|T\|_{L_r}^{L_s}. \tag{1.18}$$

Let $1 < s \leqslant \infty$, i.e. $1 \leqslant s' < \infty$. For a fixed element $x \in L_r$, the values of the form f define a bounded linear functional z on $\Phi_{s'}$, i.e. an element of the dual space $\Phi^*_{s'}$. Since $1 \leqslant s' < \infty$, we have $\Phi^*_{s'} = L_s$, and consequently $z \in L_s$. In addition, for $\varphi \in \Phi$

$$f(\varphi, x) = \int \varphi(t) z(t)\, dt. \tag{1.19}$$

For a fixed element $x \in L_r$ (and operator T), the element z is uniquely determined, i.e. $z = Fx$, where F is some operator from L_r to L_s. We examine the properties of this operator.

The representation (1.19) can be rewritten in the form

$$f(\varphi, x) = \int \varphi F x\, dt. \tag{1.20}$$

From (1.18) and (1.20), by virtue of the arbitrariness of $\varphi \in \Phi_{s'}$, we have $\|Fz\|_s \leqslant \|T\| \|x\|_r$, i.e.

$$\|F\|_{L_r}^{L_s} \leqslant \|T\|_{L_r}^{L_s}. \tag{1.21}$$

With the aid of (1.20), it is easy to verify the linearity of F. In fact, for any real numbers α and β, functions x_1 and x_2 in L_r, and any function $\varphi \in \Phi$ we obtain

$$\int \varphi \{F(\alpha x_1 + \beta x_2) - \alpha F x_1 - \beta F x_2\}\, dt = \int \varphi F(\alpha x_1 + \beta x_2)\, dt - \alpha \int \varphi F x_1 dt$$
$$- \beta \int \varphi F x_2 dt = f(\varphi, \alpha x_1 + \beta x_2) - \alpha f(\varphi, x_1) - \beta f(\varphi, x_2) = 0,$$

which (again by the arbitrariness of $\varphi \in \Phi_{s'}$) implies the equality $F(\alpha x_1 + \beta x_2) = \alpha F x_1 + \beta F x_2$, i.e. the linearity of F.

We will now denote by G_k the ball of radius h_k, and by $|G_k|$ its volume. Relations (1.20), (1.17) and (1.15) give the formula

$$\int \varphi F x\, dt = \lim_{k\to\infty} \frac{1}{|G_k|} \int\limits_{G_k} \int\limits_{R^m} \varphi \sigma_\tau T \sigma_{-\tau} x\, dt\, d\tau \tag{1.22}$$

for any $\varphi \in \Phi$, $x \in L_r$.

We check the invariance with respect to a shift of the operator F on L_r. Let $x \in L_r$, $\varphi \in \Phi$ and $\theta \in R^m$. By (1.22) we have

$$\int \varphi \{\sigma_\theta F \sigma_{-\theta} x - Fx\}\, dt = \lim_{k\to\infty} J_k, \tag{1.23}$$

where

$$J_k = \frac{1}{|G_k|}\left\{\int_{G_k}\int_{R^m} \varphi\sigma_{\tau+\theta} T\sigma_{-\tau-\theta} x\, dt\, d\tau - \int_{G_k}\int_{R^m} \varphi\sigma_\tau T\sigma_{-\tau} x\, dt\, d\tau\right\}.$$

In the first integral we replace the variable τ by $\tau' = \tau + \theta$, and denote by $G_k(\theta)$ the ball of radius h_k with center at θ. This leads to the relation

$$J_k = \frac{1}{|G_k|}\left(\int_{G_k(\theta)} - \int_{G_k}\right)\int_{R^m} \varphi\sigma_\tau T\sigma_{-\tau} x\, dt\, d\tau;$$

hence, by virtue of (1.14),

$$|J_k| \leqslant \frac{|G_k(\theta) - G_k|}{|G_k|}\|\varphi\|_{s'}\|x\|_r\|T\|,$$

where $|G_k(\theta) - G_k|$ is the volume of the solid $G_k(\theta) - G_k$. Since the radius h_k of the spheres approaches ∞ as $k \to \infty$, we have $J_k \to 0$ (here θ is fixed). Therefore (1.23) has the form

$$\int \varphi \{\sigma_\theta F\sigma_{-\theta} x - Fx\}\, dt = 0.$$

Consequently $\sigma_\theta F \sigma_{-\theta} = F$, and thus (1.11) holds.

We prove (1.13). By virtue of (1.22) and the invariance with respect to a shift of the operator A on the class Q, for $\varphi \in \Phi$ and $x \in Q$ we obtain the relation

$$\int \varphi(Ax - Fx)\, dt = \lim_{k\to\infty}\frac{1}{|G_k|}\int_{G_k}\int_{R^m} \varphi\sigma_\tau (A\sigma_{-\tau} x - T\sigma_{-\tau} x)\, dt\, d\tau.$$

If $x \in Q$, then $\sigma_{-\tau} x \in Q$, and therefore

$$\|A\sigma_{-\tau} x - T\sigma_{-\tau} x\|_q \leqslant U(T),$$

and consequently

$$\left|\int \varphi(Ax - Fx)\, dt\right| \leqslant \|\varphi\|_{q'} U(T).$$

Since φ is any function in the class $\Phi = \Phi_{q'}$, this implies the estimate $U(F) \leqslant U(T)$, i.e. (1.13).

All the assertions of the lemma have been proved for $1 < s \leqslant \infty$. We consider the case $s = 1$ $(s' = \infty)$.

The dual of Φ_∞ is the space $\mathcal{V}$ of finite Borel measures, i.e. countably additive set functions on the class of Borel subsets of the space R^m; we will denote by $V\zeta$ the total variation of the measure $\zeta \in \mathcal{V}$.

In the given case, in place of (1.19), we have, for fixed $x \in L_r$,

$$f(\varphi, x) = \int \varphi(t)\, d\zeta(t) \quad (\varphi \in \Phi),$$

where $\zeta \in \mathcal{V}$, and

$$V\zeta \leqslant \|T\|_{L_r}^{L_s} \|x\|_r.$$

Further, in the same way as for $s > 1$, we arrive at a linear operator $\mathcal{F}$ from L_r to $\mathcal{V}$ having the properties

$$\|\mathcal{F}\|_{L_r}^{\mathcal{V}} \leqslant \|T\|_{L_r}^{L_s}, \tag{1.24}$$

$$\sigma_\theta \mathcal{F} = \mathcal{F} \sigma_\theta \quad (\theta \in R^m), \tag{1.25}$$

$$f(\varphi, x) = \int \varphi d\mathcal{F}x \quad (\varphi \in \Phi,\ x \in L_r). \tag{1.26}$$

The first two relations mean that $\mathcal{F}$ is a bounded linear operator from L_r to $\mathcal{V}$, invariant with respect to a shift.

We will now prove that for any $x \in L_r$ the measure $\zeta = \mathcal{F}_x$ is absolutely continuous. We denote by K the space of functions of compact support and infinitely differentiable on R^m, and by K' the space of generalized functions (functionals) on K. For $x \in K$ we will regard the measure $\zeta = \mathcal{F}x$ as an element of the space K', i.e. as a functional of measure type, acting according to the formula

$$(\zeta, \varphi) = \int \varphi(t)\, d\zeta(t), \qquad \varphi \in K.$$

We will show that if $x \in K$, then in K' we have

$$D^l \mathcal{F} x = \mathcal{F} D^l x, \qquad l = (l_1, \ldots, l_m) \tag{1.27}$$

for any $l_1, \ldots, l_m$; here D^l is the differentiation operator. This fact is verified in a manner analogous to the proof of (1.15) in [13]. It is sufficient to consider a first order derivative. Take $\theta = (\tau, 0, \ldots, 0)$ and put $\varphi_\tau = \sigma_\theta \varphi$; in exactly the same way, we define x_τ. By virtue of the linearity and property (1.25) of the operator $\mathcal{F}$, for $x \in K$, $\varphi \in K$ and $\tau \neq 0$ we have

$$\int \frac{\varphi_\tau - \varphi}{\tau} d\mathcal{F}x = \int \varphi\, d\mathcal{F} \frac{x_{-\tau} - x}{-\tau}.$$

As $\tau \to 0$, the ratio $(\varphi_\tau - \varphi)/\tau$ converges uniformly to $\partial\varphi/\partial t_1$, and the ratio $(x_{-\tau} - x)/(-\tau)$ converges in the metric of L_r to $\partial x/\partial t_1$. Hence we obtain the equality

$$\int \frac{\partial \varphi}{\partial t_1} d\mathcal{F}x = -\int \varphi d\mathcal{F} \frac{\partial x}{\partial t_1},$$

i.e. the equality

$$\frac{\partial \mathcal{F}x}{\partial t_1} = \mathcal{F} \frac{\partial x}{\partial t_1}.$$

Consequently (1.27) is valid.

This formula means that any derivative of the generalized function ζ is a functional of measure type. Every such functional has order of singularity $\leqslant 1$ (see [14], §12, Problem 2). And from this it follows (see [14], §13) that the functional $\zeta = \mathcal{F}x$ (for $x \in K$) is an ordinary (even infinitely differentiable) function z. Thus, for $\varphi \in K$,

$$\int \varphi d\zeta = \int \varphi z dt;$$

it is clear that this equation also holds on the entire class Φ. The function z is uniquely determined by the function x, i.e. $z = Fx$. It is not difficult to show that F is a linear operator from $K \subset L_r$ to L_1, where

$$\| Fx \|_{L_1} \leqslant \| \mathcal{F} \|_{L_r}^{\mathcal{V}} \| x \|_r \tag{1.28}$$

and for $x \in K$ and $\varphi \in \Phi$ we have

$$\int \varphi F x dt = \int \varphi d\mathcal{F} x. \tag{1.29}$$

The space K is dense in L_r ($L_r = C_0$ if $r = \infty$); extending F from K to L_r by continuity, we arrive at an operator F from L_r to L_1 with the properties (1.28) and (1.29), this time for $x \in L_r$. Here the norm of the operator F does not increase, and therefore, by (1.28) and (1.24), we will have

$$\| F \|_{L_r}^{L_s} \leqslant \| T \|_{L_r}^{L_s}.$$

Moreover, (1.26) and (1.29) imply

$$f(\varphi, x) = \int \varphi F x dt, \quad \varphi \in \Phi, \ x \in L_r;$$

hence one easily obtains property (1.12) for the operator F. The invariance of the operator F with respect to a shift follows from (1.29) and (1.25).

Lemma 1.1 is fully proved.

We denote by $\mathcal{T}_r^s(N)$ the set of bounded linear operators T from L_r to L_s with norm $\|T\| \leqslant N$ which are invariant with respect to a shift on L_r.

THEOREM 1.1. *If the class Q and the operator A on Q are invariant with respect to a shift, then*

$$E(N) = \inf_{T \in \mathcal{T}_r^s(N)} U(T). \tag{1.30}$$

In fact, the definition (1.9) of $E(N)$ implies the inequality

$$E(N) \leqslant \inf_{T \in \mathcal{T}_r^s(N)} U(T).$$

On the other hand, by virtue of Lemma 1.1, the reverse inequality holds.

Everywhere below it is assumed that the operator A on Q (and thus also Q) is invariant with respect to a shift.

For functions $\eta \in L_{s'}$ and $x \in L_r$, we put

$$\nu(\eta, x) = \sup_{T \in \mathcal{T}_r^s(1)} \left| \int \eta T x \, dt \right|; \tag{1.31}$$

clearly

$$0 \leqslant \nu(\eta, x) \leqslant \| \eta \|_{s'} \| x \|_r. \tag{1.32}$$

We denote by $\Phi(\delta)$ the set of pairs (η, x) of functions $\eta \in L_{q'} \cap L_{s'}$, $x \in Q$ with the properties $\|\eta\|_{q'} \leqslant 1$ and $\nu(\eta, x) \leqslant \delta$. In analogy with (1.3) and (1.6), we introduce the quantities

$$\lambda(\sigma) = \sup_{(\eta, x) \in \Phi(\delta)} \int \eta A x \, dt, \tag{1.33}$$

$$\Lambda(N) = \sup_{\delta > 0} \{\lambda(\delta) - N\delta\} = \sup_{\substack{\|\eta\|_{q'} \leqslant 1,\ x \in Q \\ \eta \in L_{q'} \cap L_{s'}}} \left\{\int \eta A x \, dt - N\nu(\eta, x)\right\}. \tag{1.34}$$

We have the inequality

$$E(N) \geqslant \Lambda(N). \tag{1.35}$$

This can be written in the form

$$E(N) \geqslant \sup_{\substack{x \in Q \\ \|\eta\|_{q'} \leqslant 1 \\ \eta \in L_{q'} \cap L_{s'}}} \inf_{T \in \mathcal{T}_r^s(N)} \left\{\int \eta A x \, dt - \int \eta T x \, dt\right\},$$

and this inequality follows from (1.30).

If $s = q$ (and A maps from Q to L_q), then by virtue of (1.32)

$$\Lambda(N) \geqslant \Omega(N) = \sup_{x \in Q} (\|Ax\|_q - N\|x\|_r),$$

i.e., we arrive at (1.5).

Stečkin's relations (1.7) and (1.8) have the following analog.

If the operator T and the functions x and y satisfy the conditions

$$T \in \mathcal{T}_r^s, \ x \in Q, \quad y \in L_{s'} \cap L_{q'}, \quad \|y\|_{q'} = 1, \tag{1.36}$$

$$U(T) = \int y(Ax - Tx) \, dt, \tag{1.37}$$

$$\int yTx \, dt = \|T\|_{L_r}^{L_s} \nu(y, x), \tag{1.38}$$

then

$$E(N) = U(T) = \Lambda(N), \ \lambda(\nu(y, x)) = \int yAx dt \tag{1.39}$$

for $N = \|T\|_{L_r}^{L_s}$.

This assertion follows from the chain of relations

$$U(T) \geqslant E(N) \geqslant \Lambda(N) \geqslant \lambda(\nu(y, x)) - \|T\|\nu(y, x) \geqslant \int yAxdt - \int yTxdt = U(T).$$

In Theorem 1.1 of [13] it is asserted that if $s < r$ and $T \in \mathcal{T}_r^s$ then $T \equiv 0$. Therefore we have the following proposition.

COROLLARY 1.1. *If $s < r$, then, for any $N > 0$,*

$$E(N) = \sup_{x \in Q} \|Ax\|_q = E(0). \tag{1.40}$$

The operator classes $\mathcal{T}_r^s$ and $\mathcal{T}_{s'}^{r'}$ coincide, together with equality of the operator norms (see, for example, [13], Theorems 1.2 and 1.3). Hence, we obtain the following assertion.

COROLLARY 1.2. *If $Q \subset L_r \cap L_{s'}$, then*

$$E(N; r, s) = E(N; s', r'). \tag{1.41}$$

§2. The linear problem

The problem (1.9) is of interest in the case when the operator A is linear and the class Q is defined by some linear operator. More precisely, let A and B be two linear operators with the same domain of definition $\mathcal{W} \subset S$, which map $\mathcal{W}$ into S. As the class Q we take the set

$$Q_{p,r} = Q_{p,r}(\mathcal{W}, B) = \{x \in L_r \cap \mathcal{W}: \|Bx\|_p \leqslant 1\}; \tag{2.1}$$

here (just as for L_r), when $p = \infty$, we understand by L_p the space C_0. Then (1.9) assumes the form

$$E(N) = E(\mathcal{W}; N) = E(N; r, s; p, q) = \inf_{\|T\|_{L_r}^{L_s} \leqslant N} U(T), \tag{2.2}$$

where

$$U(T) = U(T; p, q) = \sup_{x \in Q_{p,r}} \|(A - T)x\|_q. \tag{2.3}$$

In what follows, it will be assumed that the set $\mathcal{W}$ and the operators A, B on $\mathcal{W}$ are invariant with respect to a shift. In this case the class $Q = Q_{p,r}$ will also be invariant with respect to a shift, and therefore the assertions in §1 hold for (2.2).

We put

$$W_r = W_{p,r} = \{x \in L_r \cap \mathcal{W} : Bx \in L_p\},$$
$$Y_p = BW_r = \{y = Bx : x \in L_r \cap \mathcal{W},\ Bx \in L_p\}.$$

These sets are linear; we will assume that they are provided with the norms of the spaces L_r and L_p, respectively.

LEMMA 2.1. *For any operator $T \in \mathfrak{T}_r^s$ satisfying the condition $U(T) < \infty$ we have on W_r the representation*

$$A - T = F \circ B, \tag{2.4}$$

where F is a bounded linear operator from Y_p to L_q with norm

$$\|F\|_{Y_p}^{L_q} = U(T), \tag{2.5}$$

invariant with respect to a shift on Y_p.

The representation (2.4), with a linear operator F having the property (2.5), was obtained in [15] (Lemma 1) by means of a certain modification of Sard's proof of Theorem 1 in [16]. We will prove the invariance of F with respect to a shift. For $y = Bx \in Y_p$ and $\tau \in R^m$, we have

$$F\sigma_\tau y = F\sigma_\tau Bx = FB\sigma_\tau x = (A - T)\sigma_\tau x = \sigma_\tau (A - T)x = \sigma_\tau FBx = \sigma_\tau Fy,$$

and thus $F\sigma_\tau = \sigma_\tau F$.

By Theorem 1.1 of [13], for $q < p$ the space $\mathfrak{T}_p^q$ contains only the operator $F \equiv 0$. From the proof of this theorem it is evident that, in fact, a stronger assertion is true. Namely, if $q < p$ and F is a bounded linear operator from a (not necessarily closed) linear subspace $Y_p \subset L_p$ into L_q, invariant with respect to a shift, then $F \equiv 0$. From this one easily obtains the following assertion.

LEMMA 2.2. *If $q < p$ and $E(N) < \infty$ for some N, then $E(N) = 0$ and the operator A can be extended from W_r to all of L_r, as an operator $T_A \in \mathfrak{T}_r^s(N)$.*

In fact, let $E(N) < \infty$ and $U(T) < \infty$ for $T \in \mathfrak{T}_r^s(N)$. Then by virtue of Lemma 2.1 we have the representation (2.4). Moreover, $F \equiv 0$ on Y_p; hence $A = T$ on W_r and $E(N) = U(T) = 0$.

The following assertion follows easily from the last lemma; it is useful in finding necessary conditions for the finiteness of the quantity $E(N)$.

COROLLARY 2.1. *If $q < p$ and*

$$\| A \|_{W_r}^{L_s} = \sup_{x \in W_r} \frac{\| Ax \|_s}{\| x \|_r} = \infty,$$

then $E(N) = \infty$ for any $N \geqslant 0$.

For the pair of parameters p, q, we have an analog of Corollary 1.2.

LEMMA 2.3. *Let $W_{p,r} = W_{q',r}$, and let the set $Y = BW_r$ be dense in the spaces L_p and $L_{q'}$. Then, for any N,*

$$E(N; r, s; p, q) = E(N; r, s; q', p'). \tag{2.6}$$

By virtue of Theorem 1.1 it is sufficient to prove that for any operator $T \in \mathfrak{T}_r^s$

$$U(T; p, q) = U(T; q', p'), \tag{2.7}$$

where the quantity $U(T; p, q)$ is defined in (2.3). We may assume that at least one of the quantities in (2.7), for example $U(T) = U(T; p, q)$, is finite. By virtue of Lemma 2.1 we have the representation $A - T = F \circ B$, in which the operator F is linear and invariant with respect to a shift on Y_p, and $\|F\|_{Y_p}^{L_q} = U(T) < \infty$. Since $\overline{Y}_p = L_p$, the operator F can be extended by continuity to L_p (or C_0, if $p = \infty$). Clearly this operator will be invariant with respect to a shift on L_r. Thus, on W_r,

$$A - T = F \circ B,$$

where $F \in \mathfrak{T}_p^q$ and $\|F\|_{L_p}^{L_q} = U(T)$.

But since $Y_p = Y_{q'}$ and $\overline{Y}_{q'} = L_{q'}$, we have $U(T; q', p') = \|F\|_{Y_{q'}}^{L_{p'}} = \|F\|_{L_{q'}}^{L_{p'}}$. But $\|F\|_{L_{q'}}^{L_{p'}} = \|F\|_{L_p}^{L_q}$, and thus (2.7) holds. The lemma is proved.

§3. Approximation of differential operators. Duality with respect to the parameters

Below, we consider problem (2.2) on the real axis ($m = 1$) for differential operators with constant coefficients. Thus, let

$$Ax = \sum_{\nu=0}^{k} a_\nu x^{(\nu)}, \qquad Bx = \sum_{\nu=0}^{n} b_\nu x^{(\nu)}, \tag{3.1}$$

where a_ν and b_ν are real numbers, and $a_k \neq 0$ and $b_n \neq 0$, where k and n are integers and $0 \leqslant k < n$. As the domain of definition of the operators A and B we will take the set W of functions whose derivatives of order $n-1$ are locally absolutely continuous on $R^1 = (-\infty, \infty)$. In this case $Q_{p,r} = Q_{p,r}(B)$ is the class of functions $x \in L_r$ for which the derivative $x^{(n-1)}$ is locally absolutely continuous and $\|Bx\|_p \leqslant 1$. We consider the quantity

$$E(N) = E(N; r, s; p, q) = \inf_{\|T\|_{L_r}^{L_s} \leqslant N} \sup_{x \in Q_{p,r}} \|(A-T)x\|_q. \tag{3.2}$$

Here, when $r = \infty$, L_r will mean C_0, but if $p = \infty$, then L_p is L_∞. As before, we denote the linear hull of the class $Q_{p,r}$ by $W_{p,r}$; thus

$$W_{p,r} = W_{p,r}(B) = \{x \in L_r : Bx \in L_p\}.$$

LEMMA 3.1. *For any function $x \in W_{p,r}$ and any $\epsilon > 0$, there exists a function $y \in K$ with the properties*

$$\|x-y\|_r \leqslant \varepsilon, \tag{3.3}$$

$$\|x^{(k)} - y^{(k)}\|_C \leqslant \varepsilon \quad \text{for} \quad k = 0, \quad 1, \ldots, \quad n-1, \tag{3.4}$$

$$\|By\|_p \leqslant \|Bx\|_p + \varepsilon. \tag{3.5}$$

PROOF. As follows from Gabušin [17], on the real line $J = (-\infty, \infty)$ and the half line $J = [0, \infty)$ the inequality

$$\|x^{(k)}\|_{C(J)} \leqslant c\,(\delta^{-k-r^{-1}} \|x\|_{L_r(J)} + \delta^{n-k-p^{-1}} \|Bx\|_{L_p(J)}) \tag{3.6}$$

holds for all $\delta < \delta_0$, where c and δ_0 do not depend on x.

Hence, if $x \in W_{p,r}(B)$, then $x^{(k)} \in C$ for $k = 0, 1, \ldots, n-1$. Writing down (3.6) for the half line $I = [0, \infty)$ for the functions $x_\tau(t) = x(t+\tau)$, and making $\tau \to \infty$, we verify that $x^{(k)}(t) \to 0$ as $t \to \infty$. Analogously, applying (3.6) for the adjoint operator B^*, we see that $x^{(k)}(t) \to 0$ as $t \to -\infty$. Thus, if $x \in W_{p,r}(B)$, then $x^{(k)} \in C_0$, $k = 0, 1, \ldots, n-1$.

Now let $x \in W_{p,r}(B)$, and let h, δ and σ be positive numbers with $\sigma < \delta$. We set

$$\eta(t) = \begin{cases} (Bx)(t), & \text{if} \quad t \in [-2h, 2h], \\ 0, & \text{if} \quad t \notin [-2h, 2h]. \end{cases}$$

For $t \in (-2h, 2h)$ we have

$$x(t) = \int \varkappa(\tau - t)\eta(\tau)\,d\tau + \varphi(t),$$

where κ is a fundamental solution for the operator B^* and the function φ satisfies the equation $B\varphi = 0$.

For $p < \infty$ we choose a polynomial ζ such that

$$\|\eta - \zeta\|_{L_p(-2h, 2h)} < \sigma. \tag{3.7}$$

In the case $p = \infty$, by virtue of the theorems of Luzin ([18], Chapter IV, §4, Theorem 4) and Weierstrass ([18], Chapter IV, §5, Theorem 2) there exists a polynomial

ζ such that on the set $M \subset [-2h, 2h]$ with measure meas $M \geqslant 4h - \sigma$ we have the inequality

$$\|\eta - \zeta\|_{L_\infty(M)} < \sigma,$$

and on the whole interval $[-2h, 2h]$

$$\|\zeta\|_{C[-2h, 2h]} \leqslant \|\eta\|_\infty + \sigma. \tag{3.8}$$

For every $\delta > 0$, the number $\sigma < \delta$ may be chosen so that the function

$$z(t) = \int_{-2h}^{2h} \varkappa(\tau - t)\zeta(\tau)\, d\tau + \varphi(t) \tag{3.9}$$

satisfies the conditions

$$\|x^{(k)} - z^{(k)}\|_{C[-2h, 2h]} < \delta \quad \text{for } k = 0, 1, \ldots, n-1. \tag{3.10}$$

If $t \in (-2h, 2h)$, then $(Bz)(t) = \zeta(t)$. Therefore the inequalities (3.7), (3.8), and $\sigma < \delta$ give the estimate

$$\|Bz\|_{L_p(-2h, 2h)} \leqslant \|Bx\|_p + \delta. \tag{3.11}$$

We denote by μ a function of class K with the properties

$$0 \leqslant \mu \leqslant 1, \qquad \mu(t) = \begin{cases} 1, & \text{if } |t| \leqslant 1, \\ 0, & \text{if } |t| \geqslant 2 \end{cases}$$

and put $\mu_h(t) = \mu(h^{-1}t)$. Consider the function $y = \mu_h z$. It is clear that $y \in K$. We will show that for sufficiently large h and small δ the function y also satisfies conditions (3.3)–(3.5).

We have

$$\|x - y\|_r \leqslant \|x - \mu_h x\|_r + \|\mu_h(x - z)\|_r;$$

hence by virtue of the choice of μ_h and property (3.10) (for $k = 0$) we obtain

$$\|x - y\|_r \leqslant 2\|x\|_{L_r(R \setminus [-h, h])} + \delta(4h)^{1/r}. \tag{3.12}$$

We now estimate $\|By\|_p$. By means of the Leibniz formula we find that (on $[-2h, 2h]$)

$$By = \sum_{\nu=0}^{n} b_\nu (\mu_h z)^{(\nu)} = \sum_{\nu=0}^{n} b_\nu \sum_{l=0}^{\nu} C_\nu^l z^{(\nu-l)} \mu_h^{(l)} = \mu_h Bz + \rho_n, \tag{3.13}$$

where

$$\rho_n(t) = \sum_{1 \leqslant l \leqslant \nu \leqslant n} b_\nu C_\nu^l z^{(\nu-l)} h^{-l} \mu^{(l)}(h^{-1}t).$$

Let $I_h = [-2h, -h] \cup [h, 2h]$ and

$$u(h) = \max_{0 \leqslant k \leqslant n-1} \|x^{(k)}\|_{C(I_h)};$$

as we have already shown, $u(h) \to 0$ as $h \to \infty$. By virtue of (3.10) and the choice of μ, for $h \geqslant 1$ we have the estimate

$$\|\rho_n\|_p = \|\rho_n\|_{L_p(I_h)} \leqslant ch^{-1+\frac{1}{p}} \{u(h) + \delta\}, \tag{3.14}$$

in which the constant c does not depend on h ($h \geqslant 1$) or on x. Relations (3.11), (3.13) and (3.14) lead to the inequality

$$\|By\|_p \leqslant \|Bx\|_p + \delta + ch^{\frac{1}{p}-1}\{u(h) + \delta\}. \tag{3.15}$$

Similarly, we obtain

$$y^{(k)} = \mu_h z^{(k)} + \rho_k,$$

where for $h \geqslant 1$ the function ρ_k satisfies the condition

$$\|\rho_k\|_C \leqslant c_k h^{-1}\{u(h) + \delta\},$$

and c_k is some constant. Hence

$$\begin{aligned} \|y^{(k)} - x^{(k)}\|_C &\leqslant \|x^{(k)} - \mu_h x^{(k)}\|_C + \|\mu_h(z^{(k)} - x^{(k)})\|_C + \|\rho_k\|_C \\ &\leqslant 2\|x^{(k)}\|_{L_\infty(R\setminus[-h,h])} + \delta + c_k h^{-1}\{u(h) + \delta\}. \end{aligned} \tag{3.16}$$

It is clear from (3.12), (3.15) and (3.16) that, choosing first a sufficiently large h and then a sufficiently small δ, one can obtain a function $y \in K$ with properties (3.3)–(3.5). The lemma is proved.

REMARK. When $1 \leqslant p < \infty$, for any function $x \in W_{p,r}(B)$ and any $\epsilon > 0$ there exists a function $y \in K$ satisfying (3.3), (3.4) and the condition

$$\|Bx - By\|_p \leqslant \varepsilon.$$

This fact follows easily from the proof of the lemma (see (3.7), (3.13) and (3.14)).

Along with (3.2), we consider the quantity

$$E(K, N) = \inf_{\|T\|_{L_r}^{L_s} \leqslant N} \sup_{x \in \mathcal{K}_p(B)} \|(A - T)x\|_q, \tag{3.17}$$

where $\mathcal{K}_p(B) = \{x \in K\colon \|Bx\|_p \leqslant 1\}$. Since $\mathcal{K}_p(B) \subset Q_{p,r}(B)$, we have $E(K, N) \leqslant E(N)$.

LEMMA 3.2. $E(N) = E(K, N)$.

PROOF. It is enough to show that if the operator T has the properties

$$\|T\|_{L_r}^{L_s} < \infty, \quad U(K, T) = \sup_{x \in \mathcal{K}_p(B)} \|Ax - Tx\|_q < \infty,$$

then

$$U(T) = U(K, T). \tag{3.18}$$

We denote by $Z = Z_{q'}$ the set of continuous functions z of compact support satisfying the condition $\|z\|_{q'} \leqslant 1$. We have

$$U(K, T) = \sup_{z \in Z} \sup_{x \in \mathcal{K}_p(B)} \int zAx\,dt - \int zTx\,dt.$$

By virtue of Lemma 3.1, for any function $x \in Q_{p,r}(B)$ there exists a sequence of functions $y_m \in \mathcal{K}_p(B)$ such that $\|x^{(\nu)} - y_m^{(\nu)}\|_C \longrightarrow 0$ ($\nu = 0, 1, \ldots, k$) and $\|x - y_m\|_r \longrightarrow 0$ as $m \longrightarrow \infty$. But then

$$\int zAy_m \, dt \to \int zAx \, dt,$$

and since the functional $\int zTx\,dt$ is bounded on L_r, we also have

$$\int zTy_m \, dt \to \int zTx \, dt.$$

Hence

$$\sup_{x \in Q_{p,r}(B)} \int z(Ax - Tx)\,dt = \sup_{x \in \mathcal{K}_p(B)} \int z(Ax - Tx)\,dt,$$

and this implies (1.38). The lemma is proved.

LEMMA 3.3. *The set $BK = \{y = Bx \colon x \in K\}$ is dense in C_0 and in L_p for $1 < p < \infty$. In order that the set BK be dense in L_1, it is necessary and sufficient that the polynomial $b(t) = \Sigma_0^n b_\nu t^\nu$ have no roots of the form $i\alpha$, $\alpha \in (-\infty, \infty)$.*

PROOF. It follows from the Hahn-Banach theorem on the extension of a linear functional that a (not necessarily closed) subspace X_0 of the Banach space X is dense in X if and only if, for every functional $f \in X^*$,

$$f(X_0) = \{0\} \Rightarrow f \equiv 0.$$

For $1 \leqslant p < \infty$ a functional $f \in L_p^*$ has the form $f(x) = \int \varphi x\,dt$, where $\varphi \in L_{p'}$. But if

$$f(Bx) = \int \varphi Bx \, dt = 0, \qquad \forall x \in K, \tag{3.19}$$

then φ is a (classical) solution of the equation $B^*\varphi = 0$. Hence φ is a linear combination of the functions $\varphi_1, \ldots, \varphi_n$ of the form

$$t^\nu \cos \alpha t \, e^{\beta t}, \qquad t^\nu \sin \alpha t \, e^{\beta t},$$

which form a linearly independent system of solutions of the equation $B^*y = 0$. From this easily follows the whole assertion of the lemma for $1 \leqslant p < \infty$.

But if $f \in C_0^*$, then $f(x) = \int x\,d\mu$, where $V\mu = \|f\| < \infty$. In place of (3.19), we have

$$\int Bx \, d\mu = -\int \mu Bx' \, dt = 0, \qquad \forall x \in K.$$

Hence $B^*\mu' = 0$, and thus $\mu = \text{const}$. The lemma is proved.

THEOREM 3.1. *For any values of the parameters, for* (3.2), *we have*

$$E(N; r, s; p, q) = E(N; s', r'; p, q). \tag{3.20}$$

In the case $p > 1$, $q < \infty$, we also have

$$E(N; r, s; p, q) = E(N; r, s; q', p'); \tag{3.21}$$

and if the polynomial $b(t) = \Sigma_0^n b_\nu t^\nu$ *has no roots of the form* $i\alpha$, *then the latter formula holds for* $p \geqslant 1$ *and* $q \leqslant \infty$, *i.e. for any values of the parameters.*

PROOF. By virtue of Lemma 3.2, it is sufficient to prove the analog of (3.20) and (3.21) for the quantity (3.17). For $E(K, N)$ the conditions of Corollary 1.2 hold, whence (3.20) follows. By virtue of Lemma 3.3, under the conditions of the second part of the theorem the set BK is dense in L_p and $L_{q'}$. Applying Lemma 2.3 to $E(K, N)$, we obtain (3.21). Everything is proved.

§4. Approximation of differential operators. Reduction to the problem of the approximation of a particular function by the class of multipliers

Hörmander [13] proved that if T is a bounded linear operator from L_α to L_β, invariant with respect to a shift (i.e., in our notation, $T \in \mathcal{T}_\alpha^\beta$), then there exists a (unique) function $\theta \in K'$ such that

$$\tilde{T}x = \theta * x \quad \text{for} \quad x \in K, \tag{4.1}$$

where $(\theta * x)(t) = (\theta, \sigma_t x)$ and $(\sigma_t x)(\tau) = x(t + \tau)$.

We denote the set of generalized functions $\theta \in K'$ which define, by (4.1), a bounded operator from K (with the L_α norm) to L_β by L_α^β, and we put

$$\|\theta\|_{L_\alpha^\beta} = \sup_{x \in K} \frac{\|\theta * x\|_\beta}{\|x\|_\alpha}.$$

Since K is dense in L_α (in C_0, if $\alpha = \infty$), every operator of the form (4.1) with $\theta \in L_\alpha^\beta$ can be extended, by continuity, from K to L_α (or C_0), and hence the spaces L_α^β and $\mathcal{T}_\alpha^\beta$ are isometric. Here we will use many of the properties of the spaces L_α^β which are set forth in [13].

We prove now that (3.2) coincides with the quantity

$$\varepsilon(N) = \inf_{\substack{\varphi \in K' \\ \|B^*\varphi\|_{L_r^s} \leqslant N}} \|A^*\varkappa - \varphi\|_{L_p^q} = \varepsilon(N;\ r, s;\ p, q), \tag{4.2}$$

where κ is a fundamental solution for the operator B^*.

THEOREM 4.1. *For* $p > 1$,

$$E(N) = \varepsilon(N), \tag{4.3}$$

and if the polynomial $b(t) = \Sigma_0^n b_\nu t^\nu$ *has no roots of the form* $i\alpha$, *then this equality also holds for* $p = 1$.

By virtue of Lemma 3.2, it is sufficient to show that $E(K, N) = \epsilon(N)$. We first prove the inequality $E(K, N) \leqslant \epsilon(N)$; here we may assume that $\epsilon(N) < \infty$. Let the function $\varphi \in K'$ be such that

$$\|B^*\varphi\| \leqslant N, \quad \|A^*\varkappa - \varphi\| < \infty.$$

We define the operator T_φ by the relation

$$T_\varphi x = (B^*\varphi) * x, \quad x \in K.$$

On K we have $A^*\kappa * Bx = Ax$ and $\varphi * Bx = T_\varphi x$, and consequently

$$U(K, T_\varphi) = \sup_{\|Bx\|_p \leqslant 1} \|(A^*\varkappa - \varphi) * Bx\|_q \leqslant \|A^*x - \varphi\|_{L_p^q}.$$

Hence $E(K, N) \leqslant \epsilon(N)$.

By virtue of Theorem 1.1,

$$E(K, N) = \inf_{T \in \mathcal{T}_r^s(N)} U(K, T).$$

Assume that $E(K, N) < \infty$ and that for the operator $T \in \mathcal{T}_r^s(N)$ we have $U(K, T) < \infty$. Then by Lemma 2.1 we have $A - T = F \circ B$ on K, where F is a bounded linear operator from BK to L_q, invariant with respect to a shift on BK. Under our assumptions the set BK is dense in L_p, and therefore F can be extended by continuity from BK to L_p for $p < \infty$, and to C_0 for $p = \infty$. This extension will be an operator invariant with respect to a shift, and therefore

$$Fy = f * y, \ y \in K,$$

where $f \in L_p^q$ and $\|f\|_{L_p^q} = U(K, T)$. The operator T has a representation (4.1) with $\|\theta\|_{L_r^s} \leqslant N$. Hence on K we have

$$(Ax)(0) - (\theta, x) = (f, Bx). \tag{4.4}$$

Since $(Ax)(0) = (A^*\kappa, Bx)$, taking $\overline{\varphi} \in K'$ such that $B^*\overline{\varphi} = \theta$, we can write (4.4) in the form

$$(A^*\varkappa - \overline{\varphi} - f, Bx) = 0, \ x \in K.$$

Consequently the functional $\eta = A^*\kappa - \overline{\varphi} - f$ satisfies the equation $B^*\eta = 0$, and is therefore an ordinary function. We put $\varphi = \overline{\varphi} + \eta$. Then we have $f = A^*\kappa - \varphi$ and $B^*\varphi = \theta$. Hence

$$U(K, T) = \|f\|_{L_p^q} = \|A^*\varkappa - \varphi\|_{L_p^q}, \qquad \|B^*\varphi\|_{L_r^s} = \|T\|_{L_r}^{L_s} \leqslant N,$$

and thus $\epsilon(N) \leqslant E(K, N)$. The theorem is proved.

We make several remarks.

1) From the proof of the theorem it is clear that the inequality

$$E(N) \leqslant \varepsilon(N) \tag{4.5}$$

holds for any values of the parameters.

2) If $q < \infty$ and $p = 1$ $(1 \leqslant r, s < \infty)$, then by virtue of (4.5) and (4.3)

$$E(N; r, s; 1, q) \leqslant E(N; r, s; q', \infty). \tag{4.6}$$

3) Relations (4.5) and (4.2) allow us to assert that if $\varphi \in K'$ and $\|B^*\varphi\|_{L_r^s} < \infty$, then

$$\|A^*\varkappa - \varphi\|_{L_p^q} \geqslant E(\|B^*\varphi\|_{L_r^s}). \tag{4.7}$$

4) We have (see [13]) the equalities $L_\alpha^\beta = L_{\beta'}^{\alpha'}$ and $\|\theta\|_{L_\alpha^\beta} = \|\theta\|_{L_{\beta'}^{\alpha'}}$. Therefore it follows from (4.3) that for $p > 1$ and $q < \infty$

$$\begin{aligned} E(N; r, s; p, q) &= E(N; s', r'; p, q), \\ E(N; r, s; q', p') &= E(N; s', r'; q', p'). \end{aligned} \tag{4.8}$$

The structure of the spaces L^β_α in the cases $\alpha = 1$, $1 \leqslant \beta \leqslant \infty$ ($1 \leqslant \alpha \leqslant \infty$, $\beta = \infty$) and $\alpha = \beta = 2$ is known (see [13], Theorems 1.4 and 1.5). If the pairs r, s and p, q satisfy these conditions, then (4.2) represents some extremal problem for ordinary functions φ. We will now examine certain of these cases. Thus, we have the equalities

$$\begin{aligned} L^\infty_\alpha &= L^{\alpha'}_1 = L_{\alpha'}, \qquad \text{if} \qquad 1 \leqslant \alpha < \infty, \\ L^\infty_\infty &= L^1_1 = \mathcal{V}, \end{aligned} \tag{4.9}$$

where $\mathcal{V}$ is the space of finite Borel measures on $(-\infty, \infty)$; here the corresponding norms are also equal. Therefore, for $q = s = \infty$, using Theorem 4.1, we obtain for (3.2) the following representation by means of (ordinary) functions φ whose (classical) derivatives of order $n - 1$ are locally absolutely continuous.

For $1 < p \leqslant \infty$, in the case $1 \leqslant r < \infty$,

$$E(N; r, \infty; p, \infty) = \inf_{\|B^*\varphi\|_{L_{r'}} \leqslant N} \|A^*\varkappa - \varphi\|_{L_{p'}} \tag{4.10}$$

and in the case $r = \infty$,

$$E(N; \infty, \infty; p, \infty) = \inf_{VB^*\varphi \leqslant N} \|A^*\varkappa - \varphi\|_{L_{p'}}. \tag{4.11}$$

We define the Fourier transform $\widetilde{x}$ of a function x by the relation

$$\widetilde{x}(t) = \int_{-\infty}^{\infty} e^{-i\xi t} x(\xi)\, d\xi.$$

Let $Z = \{y = \widetilde{x}; x \in K\}$. Then the Fourier transform $\widetilde{\theta}$ of a functional $\theta \in K'$ is the functional $\eta \in Z'$ acting on the function $y = \widetilde{x} \in Z$ according to the formula

$$(\eta, y) = 2\pi (\theta, x); \tag{4.12}$$

here, if η is an ordinary function, then

$$(\eta, y) = \int \overline{\eta(t)}\, y(t)\, dt, \quad y \in Z.$$

If $\theta \in L_s$, $1 < s \leqslant 2$, then

$$\widetilde{\theta}(t) = \int_{-\infty}^{\infty} e^{-i\xi t} \theta(\xi)\, d\xi,$$

where the integral is understood in the sense of L_s; then (see [19], Theorem 74) $\widetilde{\theta} \in L_{s'}$, and (4.12) is Parseval's equation

$$\int \overline{\widetilde{\theta}(t)}\, \widetilde{x}(t)\, dt = 2\pi \int \theta(t)\, x(t)\, dt.$$

We recall that for the convolution $z(t) = \int \theta(\zeta) x(t + \zeta) d\zeta$ we have the formula

$$\widetilde{z} = \overline{\eta}\widetilde{x}, \quad \eta = \widetilde{\theta}.$$

We denote the set of Fourier transforms $\widetilde{\theta}$ of functionals $\theta \in L^\beta_\alpha$ by M^β_α, and put $\|\tilde{\theta}\|_{M_\alpha^\beta} = \|\theta\|_{L_\alpha^\beta}$. Hörmander [13] proved that M_2^2 is the space L_∞ of bounded (complex-valued) functions (with even real part and odd imaginary part), with $\|\zeta\|_{M_2^2} = \|\zeta\|_{L_\infty}$. Hence it is not difficult to obtain the following proposition.

If $p = q = 2$, $r = 1$ and $1 < s \leqslant 2$, then

$$E(N) = \varepsilon(N) = \inf_{\|\theta\|_{L_s} \leqslant N} \left\| \frac{a(-it) - \tilde{\theta}(t)}{b(-it)} \right\|_{L_\infty}, \tag{4.13}$$

where a and b are the characteristic polynomials of the operators A and B.

In fact, we put

$$\varepsilon_0(N) = \inf_{\|\theta\|_{L_s} \leqslant N} \left\| \frac{a(-it) - \tilde{\theta}(t)}{b(-it)} \right\|_{L_\infty}$$

and first prove the inequality $\epsilon_0(N) \leqslant \epsilon(N)$. We may assume that $\epsilon(N) < \infty$. Let $\|A^*\kappa - \varphi\|_{L_2^2} < \infty$ and $B^*\varphi = \theta \in L_1^s$. Then $\theta \in L_s$, and the Fourier transform of the function $A^*\kappa - \varphi$ is a function $m \in L_\infty$.

If $x \in K$, then

$$(A^*\varkappa - \varphi, Bx) = \frac{1}{2\pi} \int \overline{m(t)}\, b(it)\, \tilde{x}(t)\, dt.$$

On the other hand,

$$(A^*\varkappa, Bx) = Ax(0) = \frac{1}{2\pi} \int a(it)\, \tilde{x}(t)\, dt,$$

$$(\varphi, Bx) = (B^*\varphi, x) = \int \theta x\, dt = \frac{1}{2\pi} \int \overline{\tilde{\theta}(t)}\, \tilde{x}(t)\, dt.$$

From the last three relations it follows that

$$\overline{m(t)}\; b(it) = a(it) - \overline{\tilde{\theta}(t)}$$

or

$$\overline{m(t)} = \frac{a(it) - \overline{\tilde{\theta}(t)}}{b(it)},$$

and thus

$$\| A^*\varkappa - \varphi \|_{L_2^2} = \left\| \frac{a(-it) - \tilde{\theta}(t)}{b(-it)} \right\|_{L_\infty}.$$

Hence $\epsilon_0(N) \leqslant \epsilon(N) = E(N)$.

We will show that $E(N) \leqslant \epsilon_0(N)$. By virtue of Lemma 3.2, it is enough to prove the inequality $E(K, N) \leqslant \epsilon_0(N)$. Again, we will assume that $\epsilon_0(N) < \infty$. Let $\theta \in L_s$ and

$$m(t) = \frac{a(-it) - \tilde{\theta}(t)}{b(-it)} \in L_\infty.$$

Then for $x \in K$ and $\zeta \in (-\infty, \infty)$ we have

$$\frac{1}{2\pi} \int e^{i\xi t}\, \overline{m(t)}\, b(it)\, \tilde{x}(t)\, dt = (Ax)(\xi) - (\theta * x)(\xi).$$

Hence, applying Plancherel's theorem, we obtain

$$\| Ax - \theta * x \|_2 = \frac{1}{\sqrt{2\pi}} \| m(t) b(it) \tilde{x}(t) \|_2 \leqslant \| m \|_\infty \| Bx \|_2.$$

Therefore, if we put $T_\theta x = \theta * x$, we will have

$$U(K, T_\theta) \leqslant \| m \|_{L_\infty} \quad \text{and} \quad \| T_\theta \|_{L_1}^{L_s} = \| \theta \|_{L_s'}.$$

Consequently $E(K, N) \leqslant \epsilon_0(N)$, and (4.13) is proved.

For $s = 2$, the last quantity in (4.13) can be computed. In fact, we set

$$g(\mu) = \inf_{\|\eta\|_{L_2} \leqslant \mu} \left\| \frac{a(-it) - \eta(t)}{b(-it)} \right\|_{L_\infty} \tag{4.14}$$

and show that if

$$\eta_\rho(t) = \operatorname{sgn} \; a(-it) \max\{0, |a(-it)| - \rho |b(-it)|\}, \tag{4.15}$$

then for $0 < \rho \leqslant \|a(-it)/b(-it)\|_{L_\infty}$

$$g(\| \eta_\rho \|_{L_2}) = \rho. \tag{4.16}$$

In fact, we have

$$g(\| \eta_\rho \|_{L_2}) \leqslant \left\| \frac{a(-it) - \eta_\rho(t)}{b(-it)} \right\|_{L_\infty} \leqslant \rho.$$

We now need to prove that

$$\| \eta \|_{L_2} \leqslant \| \eta_\rho \|_{L_2} \Rightarrow \left\| \frac{a(-it) - \eta(t)}{b(-it)} \right\|_{L_\infty} \geqslant \rho. \tag{4.17}$$

If $\rho = \|a(-it)/b(-it)\|_{L_\infty}$, then $\|\eta_\rho\|_{L_2} = 0$, and (4.17) is true. Therefore, let $\rho < \|a(-it)/b(-it)\|_{L_\infty}$, and let the function η satisfy the condition

$$\left\| \frac{a(-it) - \eta(t)}{b(-it)} \right\|_{L_\infty} < \rho.$$

Then $|a(-it) - \eta(t)| < \rho |b(-it)|$ for almost all t. Hence $|\eta(t)| > |a(-it)| - \rho |b(-it)|$, and since in addition $|\eta(t)| \geqslant 0$, this implies $\|\eta\|_{L_2} > \|\eta_\rho\|_{L_2}$. Consequently (4.17), and hence also (4.16), hold.

The quantities (4.13) and (4.14) are connected by the relation

$$\varepsilon(N) = g(\sqrt{2\pi}\, N). \tag{4.18}$$

Therefore we have the following assertion.

If $s = p = q = 2$ and $r = 1$ (or $r = p = q = 2$ and $s = \infty$), then for $0 < \rho \leqslant \|a(-it)/b(-it)\|_{L_\infty}$,

$$E\left(\frac{1}{\sqrt{2\pi}} \| \eta_\rho \|_{L_2} \right) = \rho, \tag{4.19}$$

where η_ρ is defined by (4.15).

For the given values of the parameters, we find in addition the values (1.33) and (1.34). We have

$$\lambda(\delta) = \sup_{\Phi(\delta)} \int yAx\,dt,$$
$$\Phi(\delta) = \{(y, x): \|y\|_2 \leqslant 1,\ x \in L_1,\ \|Bx\|_2 \leqslant 1,\ \nu(y, x) \leqslant \delta\}, \tag{4.20}$$

and, as is easily seen, $\nu(y, x) = \|y * x\|_2$. If $(y, x) \in \Phi(\delta)$, then by Parseval's equation

$$\int yAx\,dt = \frac{1}{2\pi}\int a(it)\,\tilde{x}(t)\,\overline{\tilde{y}(t)}\,dt,$$
$$\|b(it)\tilde{x}\|_2 = \|\widetilde{Bx}\|_2 \leqslant \sqrt{2\pi},\ \|\tilde{y}\|_2 \leqslant \sqrt{2\pi},$$
$$\|\tilde{x}\bar{\tilde{y}}\|_2 = \sqrt{2\pi}\,\|y * x\|_2 \leqslant \delta\sqrt{2\pi}.$$

The function $z = |\tilde{x}\bar{\tilde{y}}|/2\pi$ is even and nonnegative, and for it

$$\int_{-\infty}^{\infty} yAx\,dt \leqslant \int_{-\infty}^{\infty} |a(it)|\,z(t)\,dt = 2\int_0^{\infty} |a(it)|\,z(t)\,dt,$$
$$2\int_0^{\infty} |b(it)|\,z(t)\,dt = \|b(it)\,z\|_1 \leqslant \frac{1}{2\pi}\|b(it)\tilde{x}\|_2\,\|\tilde{y}\|_2 \leqslant 1,$$
$$\sqrt{2}\,\|z\|_{L_2(0,\infty)} = \|z\|_2 \leqslant \frac{\delta}{\sqrt{2\pi}}.$$

In $L_2(0, \infty)$, we distinguish the class of functions

$$\Psi(\mu) = \{\zeta \geqslant 0 : \|\zeta\|_{L_2(0,\infty)} \leqslant \mu,\ \|b(it)\zeta\|_{L_1(0,\infty)} \leqslant 1\} \tag{4.21}$$

and define

$$\bar{\lambda}(\mu) = \sup_{\zeta \in \Psi(\mu)} \int_0^{\infty} |a(it)|\,\zeta(t)\,dt. \tag{4.22}$$

The estimates for z imply the inequality $\lambda(\delta) \leqslant \bar{\lambda}(\delta/\sqrt{\pi})$. We will show that, in fact, these last two quantities coincide. For $\zeta \in \Psi(\delta\pi^{-1/2})$ and $t \in (-\infty, \infty)$ we put

$$u(t) = \sqrt{\pi\frac{\zeta(|t|)}{|b(it)|}},\qquad \tilde{y}(t) = \sqrt{\pi|b(it)|\,\zeta(|t|)}\ \operatorname{sgn}\ a(it).$$

Suppose that the function ζ and the polynomial b are such that the inverse Fourier transform $x = \hat{u}$ of the function u lies in $L_1(-\infty, \infty)$. Then the condition $\zeta \in \Psi(\delta\pi^{-1/2})$ implies the relations

$$\|Bx\|_{L_2} = \frac{1}{\sqrt{2\pi}}\|b(it)\,u\|_{L_2} = \|\zeta b(it)\|_{L_1(0,\infty)}^{1/2} \leqslant 1,$$
$$\|y\|_{L_2} = \frac{1}{\sqrt{2\pi}}\|\tilde{y}\|_{L_2} \leqslant 1,$$
$$\nu(y, x) = \|y * x\|_{L_2} = \frac{1}{\sqrt{2\pi}}\|\tilde{x}\bar{\tilde{y}}\|_{L_2} = \sqrt{\pi}\,\|\zeta\|_{L_2(0,\infty)} \leqslant \delta.$$

They signify that $(y, x) \in \Phi(\delta)$, and since

$$\int_{-\infty}^{\infty} yAx\,dt = \int_0^{\infty} |a(it)|\,\zeta(t)\,dt,$$

this implies the inequality

$$\lambda(\delta) \geqslant \int_0^\infty |a(it)|\,\zeta(t)\,dt. \tag{4.23}$$

If the condition $\hat{u} = x \in L_1(-\infty, \infty)$ is not satisfied, then by smoothing out the function u one can construct a sequence of functions x_ν which, in the limit as $\nu \longrightarrow \infty$, give (4.23). Hence it follows that $\lambda(\delta) \geqslant \bar{\lambda}(\delta\pi^{-1/2})$. Thus

$$\lambda(\delta) = \bar{\lambda}(\pi^{-1/2}\delta), \quad \delta > 0. \tag{4.24}$$

Relations (1.34) and (4.24) give the expression

$$\Lambda(N) = \sup_{\delta>0} \{\bar{\lambda}(\pi^{-1/2}\delta) - \delta N\}$$

or

$$\Lambda(\pi^{-1/2}N) = \sup_{\delta>0} \{\bar{\lambda}(\delta) - \delta N\}. \tag{4.25}$$

We introduce the notation

$$\begin{gathered} \zeta_\rho(t) = |\eta_\rho(t)|\,\|b(it)\,\eta_\rho\|^{-1}_{L_1(0,\infty)}, \\ \delta_\rho = \|\zeta_\rho\|_{L_2(0,\infty)} = \|\eta_\rho\|_{L_2(0,\infty)}\|\eta_\rho b(it)\|^{-1}_{L_1(0,\infty)}, \\ N_\rho = \frac{1}{\sqrt{2\pi}}\|\eta_\rho\|_2 = \frac{1}{\sqrt{\pi}}\|\eta_\rho\|_{L_2(0,\infty)}; \end{gathered} \tag{4.26}$$

it is easily verified that $\|b(it)\zeta_\rho\|_{L_1(0,\infty)} = 1$, and hence $\zeta_\rho \in \Psi(\delta_\rho)$. As a supplement to (4.19), we prove the following assertion.

If $s = p = q = 2$ *and* $r = 1$, *then for* $0 < \rho \leqslant \|a(-it)/b(-it)\|_{L_\infty}$

$$\Lambda(N_\rho) = E(N_\rho) = \rho, \tag{4.27}$$

$$\bar{\lambda}(\delta_\rho) = \lambda(\sqrt{\pi}\delta_\rho) = \int_0^\infty |a(it)|\,\zeta_\rho(t)\,dt. \tag{4.28}$$

In fact, by virtue of (4.19), (4.26), (1.35) and (4.25) we have

$$\rho = E(N_\rho) \geqslant \Lambda(N_\rho) \geqslant \bar{\lambda}(\delta_\rho) - \sqrt{\pi}\delta_\rho N_\rho.$$

But since $\zeta_\rho \in \Psi(\delta_\rho)$, we have

$$\bar{\lambda}(\delta_\rho) \geqslant \int_0^\infty |a(it)|\,\zeta_\rho(t)\,dt,$$

and therefore

$$\bar{\lambda}(\delta_\rho) - \sqrt{\pi}\,\delta_\rho N_\rho \geqslant M_\rho \|\eta_\rho b(it)\|^{-1}_{L_1(0,\infty)},$$

where

$$M_\rho = \int_0^\infty |a(it)|\,\eta_\rho(t)\,dt - \|\eta_\rho\|^2_{L_2(0,\infty)}.$$

Simple computations show that

$$M_\rho = \rho \int_0^\infty |b(it)|\, \eta_\rho(t)\, dt,$$

and thus

$$\int_0^\infty |a(it)|\, \zeta_\rho(t)\, dt - \sqrt{\pi}\, \delta_\rho N_\rho = \rho.$$

This implies (4.27) and (4.28).

In the particular case $Ax = x^{(k)}$, $Bx = x^{(n)}$ ($s = p = q = 2, r = 1$), by virtue of (4.19) and (4.27) we have

$$E(N) = \Lambda(N) = N^{-\frac{n-k}{k+1/2}} \left(\frac{2(n-k)}{\sqrt{2\pi(2n+1)(2k+1)(n+k+1)}} \right)^{\frac{n-k}{k+1/2}}. \tag{4.29}$$

(4.28) allows us to write out the value of (4.22), which in the given case has the form

$$\bar{\lambda}(\mu) = \sup_{\substack{\|\zeta\|_{L_2(0,\infty)} \leqslant \delta \\ \|t^n \zeta\|_{L_1(0,\infty)} \leqslant 1}} \int_0^\infty t^k \zeta(t)\, dt. \tag{4.30}$$

This quantity is related in an obvious way to sharp inequalities of Carlson type, and its value can be found in Levin [20].

We remark that (4.18), (4.19) and (4.29) imply

$$\| \zeta t^n \|_{L_2(0,\infty)} \| t^{k-n} - \zeta \|_{L_\infty(0,\infty)}^{\frac{n-k}{k+1/2}} \geqslant \frac{\sqrt{2}(n-k)}{\sqrt{(2n+1)(2k+1)(n+k+1)}} \tag{4.31}$$

for an arbitrary real function ζ for which both quantities on the left side are finite (and different from zero). It becomes an equality for the functions

$$\zeta_h(t) = t^{k-n} \max\{0,\ 1 - ht^{n-k}\},\ h > 0.$$

We note that (4.31) is equivalent to (4.7) for the given case.

Stečkin [1] observed that in the case $q = s$, $Ax = x^{(k)}$, $Bx = x^{(n)}$ the quantity $E(N)$ gives an upper bound for the best constant C in the inequality

$$\| x^{(k)} \|_q \leqslant C \| x \|_r^\alpha \| x^{(n)} \|_p^{1-\alpha}, \qquad \alpha = \frac{n-k+q^{-1}-p^{-1}}{n-p^{-1}+r^{-1}}. \tag{4.32}$$

Namely, one has the estimate

$$C \leqslant \left(\frac{N}{\alpha}\right)^\alpha \left(\frac{E(N)}{\beta}\right)^\beta, \quad \beta = 1 - \alpha. \tag{4.33}$$

In fact, from homogeneity considerations for the quantity

$$\omega(\delta) = \sup_{\substack{\|x\|_r \leqslant \delta \\ \|x^{(n)}\|_p \leqslant 1}} \| x^{(k)} \|_q$$

one easily obtains (see [21] and [22])

$$\omega(\delta) = \delta^\alpha C, \qquad C = \omega(1).$$

From this follows, for (1.6), the representation

$$\Omega(N) = \beta\alpha^{\alpha/\beta}C^{1/\beta}N^{-\alpha/\beta},$$

and this, by virtue of (1.5), implies (4.33).

This scheme for finding an upper bound for the best constant in inequality (4.32), following Stečkin [1], [2], was employed in the papers [3]–[7], [9]–[12], and others. Gabušin [6] proved that for $q = \infty$ this estimate is exact, i.e. (4.33) becomes equality. The equality $\Omega(n) = E(N)$ also holds in certain other cases, for example (see [3]) for $p = q = r = s = 2$. Stečkin, in an oral discussion, made the assertion that if in (3.2) one restricts oneself to operators T of convolution type, then (4.33), generally speaking, is strict. It therefore follows from Theorem 1.1 that in the general case, too, equality in (4.33) does not always hold.

Thus, we show that if $p = q = s = 2$, $r = 1$, $n = 1$ and $k = 0$, then $\Omega(N) < E(N)$. In fact, by a result of Sz.-Nagy [21]

$$\|x\|_2 \leqslant \left(\frac{3\sqrt{3}}{4\pi}\|x\|_1^2\|x'\|_2\right)^{1/3};$$

and hence

$$\Omega(N) = \frac{N^{-2}}{3\pi\sqrt{3}}.$$

On the other hand, by (4.29)

$$E(N) = \frac{N^{-2}}{3\pi}$$

and hence $\Omega(N) < E(N)$. We note that, by virtue of (4.29), in the given case (1.35) becomes an equality.

§5. Order of approximation and finiteness conditions

In this section we consider the questions of the finiteness and the orders of the best approximations of differential operators with constant coefficients on the real axis $J = (-\infty, \infty)$ and the half-line $J = [0, \infty)$.

Let

$$\begin{gathered} Ax = \sum_{\nu=0}^{k} a_\nu x^{(\nu)}, \quad Bx = \sum_{\nu=0}^{n} b_\nu x^{(\nu)}, \quad a_k \neq 0,\ b_n \neq 0, \\ W = W_{p,r} = W_{p,r}(B, J) = \{x \in L_r(J) : Bx \in L_p(J)\}, \\ Q = Q_{p,r}(B, J) = \{x \in W_{p,r}(B, J) : \|Bx\|_{L_p} \leqslant 1\}, \\ \mathcal{E}(N) = \mathcal{E}(N; J) = \inf_{\|T\|_{L_r(J)}^{L_s(J)} \leqslant N} \ \sup_{x \in Q_{p,r}(B, J)} \|Ax - Tx\|_{L_q(J)}. \end{gathered} \tag{5.1}$$

If $A = d^k/dt^k$ and $B = d^n/dt^n$, then in place of $\mathcal{E}(N)$ we will write $e(N)$ or $e(N, J)$. As has already been noted, the quantity $e(N)$ (for $q = s$) was studied in [1]–[12]. Following Stečkin ([1], Theorem 2), we can show that for any $\overline{N} > 0$ and $h > 0$

$$e(h^{-k+s^{-1}-r^{-1}}\overline{N}) = h^{n-k-p^{-1}+q^{-1}}e(\overline{N}) \tag{5.2}$$

and hence, if $k - s^{-1} + r^{-1} > 0$,

$$e(N)=N^{-\gamma}e(1),\qquad \gamma=\frac{n-k-p^{-1}+q^{-1}}{k-s^{-1}+r^{-1}}. \tag{5.3}$$

It is known that for $s = q$, the condition $q \geqslant \max(p, r)$ is necessary [7] and sufficient [11] for $e(N) < \infty$. The content of Gabušin [17], in essense, is the proper order as $N \to \infty$, and the finiteness, of $E(N)$ for the case $q = s = \infty$.

In the first assertion of this section we estimate the deviations

$$U(T)=\sup_{x\in Q_{p,r}(B,\,J)}\|(Ax-Tx)\|_{L_q(J)}$$

and the norms of the operators $T = T_h$, constructed as follows.

Let the numbers μ_l $(l = 0, 1, \dots, n-1)$ satisfy the system of equations

$$\sum_{l=0}^{n-j}(-1)^l b_{l+j}\mu_l=-a_{j-1},\qquad j=1,\dots,n, \tag{5.4}$$

where $a_\nu = 0$ for $\nu > k$. The matrix of this system is triangular, and the elements of the auxiliary diagonal are the numbers $(-1)^{n-j}b_n \neq 0$; therefore the system has a solution. We easily find that

$$\mu_0=\dots=\mu_{n-k-2}=0,\qquad \mu_{n-k-1}=(-1)^{n-k}\frac{a_k}{b_n}\neq 0.$$

We denote by η an arbitrary function defined and infinitely differentiable on the half-line $[0, \infty)$, equal to zero for $t \geqslant 1$ and to one for $0 \leqslant t \leqslant 1/2$. We put

$$\begin{gathered}\rho(t)=\rho_h(t)=\eta(ht)\sum_{l=n-k-1}^{n-1}\frac{\mu_l}{l!}t^l,\\ \lambda(t)=\lambda_h(t)=\sum_{\nu=0}^{n}(-1)^{\nu+1}b_\nu\rho^{(\nu)}(t).\end{gathered} \tag{5.5}$$

Clearly the functions λ and ρ have compact support. Finally we define operators T_h and F_h by the formulas

$$\begin{gathered}(T_hx)(\tau)=\int_0^\infty\lambda(t)\,x(t+\tau)\,dt,\\ (F_hx)(\tau)=\int_0^\infty\rho(t)\,x(t+\tau)\,dt.\end{gathered} \tag{5.6}$$

LEMMA 5.1. *If $s \geqslant r$, $q \geqslant p$, then for any $h_0 > 0$ there exist finite constants c_1 and c_2 such that for $h \geqslant h_0$*

$$\|T\|_{L_r}^{L_s}\leqslant c_1h^{k+\frac1r-\frac1s}, \tag{5.7}$$

$$U(T_h)\leqslant\|F_h\|_{L_p}^{L_q}\leqslant c_2h^{-n+k+\frac1p-\frac1q} \tag{5.8}$$

PROOF. We first show that on the set W_p we have

$$A-T_h=F_h\circ B, \tag{5.9}$$

i.e., for $x \in W_{p,r}(B, J)$ and $\tau \in J$,

$$(Ax)(\tau)-\int_0^\infty \lambda(t)\,x(t+\tau)\,dt=\int_0^\infty \rho(t)\,Bx(t+\tau)\,dt. \tag{5.10}$$

It is sufficient to prove this for $\tau = 0$. Integrating by parts, we obtain

$$\int_0^\infty \rho(t)\,x^{(\nu)}(t)\,dt=\sum_{i=0}^{\nu-1}(-1)^{i+1}\rho^{(i)}(0)x^{(\nu-i-1)}(0)+(-1)^\nu\int_0^\infty \rho^{(\nu)}(t)\,x(t)\,dt.$$

Hence

$$\int_0^\infty \rho Bx\,dt=\int_0^\infty x\sum_{\nu=0}^{n}(-1)^\nu b_\nu\rho^{(\nu)}\,dt+\sum_{\nu=1}^{n}b_\nu\sum_{i=0}^{\nu-1}(-1)^{i+1}\rho^{(i)}(0)x^{(\nu-i-1)}(0). \tag{5.11}$$

We denote the last sum in (5.11) by Σ. In it we replace the index of summation i by $j = \nu - i$, and change the order of summation; this gives

$$\Sigma=\sum_{\nu=1}^{n}\sum_{j=1}^{\nu}(-1)^{\nu+j-1}\rho^{(\nu-j)}(0)\,b_\nu x^{(j-1)}(0)=\sum_{j=1}^{n}(-1)^{j+1}x^{(j-1)}(0)\sum_{\nu=j}^{n}(-1)^\nu\rho^{(\nu-j)}(0)\,b_\nu.$$

By the choice (5.5) of the function ρ, we have $\rho^{(i)}(0) = \mu_i$. But since the numbers μ_i satisfy (5.4), we obtain, as a result,

$$\Sigma=\sum_{j=1}^{n}a_{j-1}x^{(j-1)}(0)=(Ax)(0).$$

Therefore (5.11) coincides with (5.10) for $\tau = 0$.

From (5.9), or, what is the same thing, (5.10), it follows that

$$U(T_h)\leqslant\|F_h\|_{L_p}^{L_q}.$$

For the proof of the lemma, we need to estimate the norms of the operators T_h and F_h. We first estimate the quantity $\|\rho^{(\nu)}\|_\epsilon = \|\rho^{(\nu)}\|_{L_\epsilon(0,\infty)}$ for $1 \leqslant \epsilon \leqslant \infty$. We have

$$\rho^{(\nu)}(t)=\sum_{l=n-k-1}^{n-1}\frac{\mu_l}{l!}(t^l\eta(ht))^{(\nu)}=\sum_{l=n-k-1}^{n-1}\frac{\mu_l}{l!}h^{\nu-l}\eta_l^{(\nu)}(ht),$$

where $\eta_l(\tau) = \tau^l\eta(\tau)$. Consequently

$$\|\rho^{(\nu)}\|_\varepsilon\leqslant\sum_{l=n-k-1}^{n-1}\frac{|\mu_l|}{l!}h^{\nu-l-\frac{1}{\varepsilon}}\|\eta_l^{(\nu)}\|_\varepsilon.$$

Hence, for $h \geqslant h_0 > 0$,

$$\|\rho^{(\nu)}\|_\varepsilon\leqslant d_\nu h^{\nu-n-k+1-\frac{1}{\varepsilon}}, \tag{5.12}$$

where $d_\nu = d_\nu(h_0)$.

For the operator F_h (of convolution type), for $q \geqslant p$ we have

$$\|F_h\|_{L_p}^{L_q}\leqslant\|\rho\|_\varepsilon,\qquad 1-\frac{1}{\varepsilon}=\frac{1}{p}-\frac{1}{q}.$$

But this, by virtue of (5.12), gives (5.8).

For $s \geqslant r$ we also have the inequality

$$\|T_h\|_{L_r}^{L_s}\leqslant\|\lambda\|_\varepsilon,\qquad 1-\frac{1}{\varepsilon}=\frac{1}{r}-\frac{1}{s};$$

and since

$$\|\lambda\|_\varepsilon \leqslant \sum_{\nu=0}^{n} |b_\nu| \|\rho^{(\nu)}\|_\varepsilon,$$

applying (5.12), we obtain (5.7). The lemma is proved.

We set

$$H_0 = \sup_{x,\, y\in K} \frac{1}{\|y\|_{q'} \|x^{(n)}\|_p} \left\{a_k \int y x^{(k)} dt - \|y\|_{s'} \|x\|_r\right\}, \tag{5.13}$$

where $K = K(J)$ is the set of infinitely differentiable functions with compact support on J, and $\|z\|_\epsilon = \|z\|_{L_\epsilon(J)}$. We will show that

$$H_0 > 0, \qquad \text{if} \qquad k + \frac{1}{r} - \frac{1}{s} > 0. \tag{5.14}$$

For $x, y \in K$ we put $x_h(t) = x(ht)$ and $y_h(t) = y(ht)$. We have

$$a_k \int y_h x_h^{(k)} dt - \|y_h\|_{s'} \|x_h\|_r = h^{\frac{1}{s} - \frac{1}{r} - 1} \left\{h^{k - \frac{1}{s} + \frac{1}{r}} a_k \int y x^{(k)} dt - \|y\|_{s'} \|x\|_r\right\}$$

and, thus, if $a_k \int yx^{(k)} dt > 0$, then, in the case $k - 1/s + 1/r > 0$, for sufficiently large h,

$$a_k \int y_h x_h^{(k)} dt - \|y_h\|_{s'} \|x_h\|_r > 0,$$

and hence (5.14) follows.

LEMMA 5.2.

$$\varliminf_{h\to\infty} h^{n-k+\frac{1}{q}-\frac{1}{p}} E\left(h^{k+\frac{1}{r}-\frac{1}{s}}\right) \geqslant \frac{1}{|b_n|} H_0. \tag{5.15}$$

PROOF. Let $h > 0$ and $N_h = h^{k+1/r-1/s}$. For $x, y \in K$ we put $x_h(t) = x(ht)$ and $y_h(t) = y(ht)$. Then we have

$$E(N_h) \geqslant \frac{1}{\|y_h\|_{q'} \|Bx_h\|_p} \left\{\int y_h A x_h\, dt - N_h \|y_h\|_{s'} \|x_h\|_r\right\}.$$

If $h \longrightarrow \infty$, then, as is easily verified,

$$h^{1-k} \int y_h A x_h\, dt \to a_k \int y x^{(k)} dt, \qquad h^{-n+\frac{1}{p}} \|Bx\|_p \to |b_n| \|x^{(n)}\|_p.$$

Hence

$$\varliminf_{h\to\infty} h^{n-k+\frac{1}{q}-\frac{1}{p}} E(N_h) \geqslant \frac{1}{\|y\|_{q'} \|x^{(n)}\|_p |b_n|} \left\{a_k \int y x^{(k)} dt - \|y\|_{s'} \|x\|_r\right\},$$

and this relation, by virtue of (5.13), implies (5.15).

From Lemmas 5.1 and 5.2 and relation (5.14) we have the following assertion.

THEOREM 5.1. *If*

$$s \geqslant r, \; q \geqslant p, \tag{5.16}$$

then there exists an $\bar{N} \geqslant 0$ *such that*

$$\mathrm{E}(N) < \infty \quad for \quad N > \overline{N}, \tag{5.17}$$

and if, moreover, $k + 1/r - 1/s > 0$, *then as* $N \to \infty$

$$\mathrm{E}(N) \asymp N^{-\gamma}, \qquad \gamma = \frac{n - k - p^{-1} + q^{-1}}{k - s^{-1} + r^{-1}}. \tag{5.18}$$

We will now indicate necessary conditions for the finiteness of the quantity $\mathrm{E}(N)$. We first verify the following simple proposition.

If $Bx = \Sigma_m^n b_\nu x^{(\nu)}$, *then in the case* $k = 0$, $s = r$, $m - 1/p + 1/q > 0$,

$$\mathrm{E}(N) = \begin{cases} 0 & for \quad N \geqslant |a_0|, \\ \infty & for \quad N < |a_0|. \end{cases} \tag{5.19}$$

In fact, we always have $\mathrm{E}(N) \geqslant 0$, and for the operator $T = a_0 I$, where I is the identity operator, we have $\|T\|_{L_r}^{L_r} = |a_0|$ and $U(T) = 0$. Therefore $\mathrm{E}(N) = 0$ for $N \geqslant |a_0|$. Let $N < |a_0|$. We choose functions $x, y \in K$ such that

$$v(x, y) = a_0 \int yx\, dt - N \|x\|_r \|y\|_{r'} > 0.$$

Then, putting $z_h(t) = z(ht)$ for any $h > 0$, we will have

$$\mathrm{E}(N) \geqslant \frac{v(x_h, y_h)}{\|Bx_h\|_p \|y_h\|_{p'}} = h^{-\left(m - \frac{1}{p} + \frac{1}{q}\right)} \frac{v(x, y)}{h^{-m+\frac{1}{p}} \|Bx_h\|_p \|y\|_{q'}}. \tag{5.20}$$

As $h \to 0$,

$$h^{-m+\frac{1}{p}} \|Bx_h\|_p = |b_m| \|x^{(m)}\|_p + O(h);$$

therefore, if $m - 1/p + 1/q > 0$, we obtain from (5.20) that $\mathrm{E}(N) = \infty$. Thus (5.19) is valid.

It is clear that if $q < p$ $(k = 0, s = r)$, then (5.19) holds.

In the case $J = (-\infty, \infty)$ we will denote $\mathrm{E}(N)$ by $E(N)$, and in the case $J = [0, \infty)$, by $E^+(N)$.

If $q < p$, *then, except in the case* $k = 0$, $s = r$,

$$E(N) = \infty \quad for\ any \quad N > 0. \tag{5.21}$$

In fact, we consider the quantity

$$\|A\|_{K_r}^{L_s} = \sup_{x \in K} \frac{\|Ax\|_s}{\|x\|_r}.$$

Putting $x_h(t) = x(ht)$, we have, as $h \to \infty$,

$$\|A\|_{K_r}^{L_s} \geqslant \frac{\|Ax_h\|_s}{\|x_h\|_r} = h^{k - \frac{1}{s} + \frac{1}{r}} |a_k| \frac{\|x^{(k)}\|_s + O(h^{-1})}{\|x\|_r}.$$

Hence, if $k - 1/s + 1/r > 0$, then $\|A\|_{K_r}^{L_s} = \infty$. By virtue of Corollary 2.1, (5.21) follows in the case $k - 1/s + 1/r > 0$.

The condition $k - 1/s + 1/r \leqslant 0$ can occur only in the cases when either $k = 1$, $s = 1$ and $r = \infty$, or $k = 0$ and $r \geqslant s$. Let $x \in K$, $\|x\|_r = 1$, and $x(t) = 0$ outside the interval $[0, 1]$. We put

$$x_\mu(t) = \sum_{\nu=0}^{\mu-1} x(t+\nu).$$

Then

$$\| Ax_\mu \|_s = \mu^{1/s} \| Ax \|_s, \qquad \| x_\mu \|_r = \mu^{1/r}, \qquad \| A \|_{K_r}^{L_s} \geqslant \mu^{\frac{1}{s}-\frac{1}{r}} \| Ax \|_s,$$

and consequently, if $r > s$, then $\|A\|_{K_r}^{L_s} = \infty$, and thus (5.21) is fulfilled. In particular, it holds for $k = 1, s = 1, r = \infty$ and $k = 0, s < r$. Assertion (5.21) is proved.

Now let $s < r$. By virtue of Corollary 1.1,

$$E(N) = E(0) = \sup_{x \in W} \frac{\| Ax \|_q}{\| Bx \|_p}. \tag{5.22}$$

The last quantity may be finite. So, let $q \geqslant p$, and

$$A^*\varkappa \in L_\varepsilon, \quad \text{where} \quad 1 - \frac{1}{\varepsilon} = \frac{1}{p} - \frac{1}{q}, \tag{5.23}$$

and where κ is a fundamental solution for the operator B^*.

Then we have

$$Ax = A^*\varkappa * Bx, \ x \in W_{p,r};$$

whence

$$E(N) = E(0) \leqslant \| A^*\varkappa \|_\varepsilon, \ N \geqslant 0. \tag{5.24}$$

If

$$Bx = \sum_{\nu=m}^{n} b_\nu x^{(\nu)}, \quad Ax = \sum_{\nu=l}^{k} a_\nu x^{(\nu)}, \ a_l \neq 0, \ b_m \neq 0 \tag{5.25}$$

and if $m - l + 1/q - 1/p > 0$, then

$$E(0) = \sup_{x \in W_{p,r}} \frac{\| Ax \|_q}{\| Bx \|_p} = \infty. \tag{5.26}$$

In fact, let $x \in K$ and $x_h(t) = x(ht)$. As $h \to \infty$, we have

$$E(0) \geqslant \frac{\| Ax_h \|_q}{\| Bx_h \|_p} = h^{-m+l-\frac{1}{q}+\frac{1}{p}} \left\{ \frac{| a_l | \| x^{(l)} \|_q}{| b_m | \| x^{(m)} \|_p} + O(1) \right\},$$

which implies (5.26).

By virtue of Corollary 1.1, this implies the following assertion for operators A and B of the form (5.25).

If

$$s < r, \quad m - l - \frac{1}{p} + \frac{1}{q} > 0, \tag{5.27}$$

then

$$E(N) = \infty \qquad \textit{for any} \qquad N > 0. \tag{5.28}$$

The following rather obvious proposition allows us to carry over assertions (5.21) and (5.28) to the case $J = [0, \infty)$.

If

$$B = (-1)^n B^*, \ A = (-1)^k A^*, \tag{5.29}$$

then, for any $N > 0$,

$$E(2^{1/s}N) \leqslant 2^{1/q}E^+(N). \tag{5.30}$$

In fact, let T^+ be a bounded operator from $L_r[0, \infty)$ to $L_s[0, \infty)$. We define the operator T on $L_r(-\infty, \infty)$ by the formula

$$(Tx)(\tau) = \begin{cases} (T^+x_+)(\tau), & \text{if} \quad \tau > 0, \\ (-1)^k (T^+x_-)(-\tau), & \text{if} \quad \tau < 0, \end{cases}$$

where $x_+(t) = x(t)$ and $x\_(t) = x(-t)$ for $t \geqslant 0$.

When $q < \infty$, for $x \in W_{p,r}(B)$, using the equality $A = (-1)^k A^*$, we obtain

$$\| Ax - Tx \|_{L_q(-\infty, \infty)} = \left\{ \int_0^\infty | Ax_+ - T^+x_+ |^q \, dt + \int_0^\infty | Ax_- - T^+x_- |^q \, dt \right\}$$
$$\leqslant U(T^+) (\| Bx_+ \|_p^q + \| Bx_- \|_p^q)^{1/q} \leqslant U(T^+) 2^{1/q} \| Bx \|_{L_p(-\infty, \infty)}.$$

Hence

$$U(T) \leqslant 2^{1/q} U(T^+).$$

It is easily seen that this also holds for $q = \infty$.

In an analogous way, we obtain

$$\| T \| \leqslant 2^{1/s} \| T^+ \|.$$

The last two inequalities imply (5.30).

BIBLIOGRAPHY

1. S. B. Stečkin, *Best approximation of linear operators,* Mat. Zametki **1** (1967), 137–148 = Math. Notes **1** (1967), 91–99. MR **35** #2051.

2. ———, *Inequalities between norms of derivatives of arbitrary functions,* Acta Sci. Math. (Szeged) **26** (1965), 225–230. (Russian) MR **32** #2534.

3. Ju. N. Subbotin and L. D. Taĭkov, *Best approximation of the differentiation operator in the space* L_2, Mat. Zametki **3** (1968), 157–164 = Math. Notes **3** (1968), 100–105. MR **37** #4471.

4. L. D. Taĭkov, *Inequalities of Kolmogorov type and best formulas for numerical differentiation,* Mat. Zametki **4** (1968), 233–238 = Math. Notes **4** (1968), 631–634. MR **38** #1459.

5. V. N. Gabušin, *Best approximation of a differentiation operator on the half-line,* Mat. Zametki **6** (1969), 573–582 = Math. Notes **6** (1969), 804–810. MR **41** #4080.

6. ———, *Best approximation of functionals on certain sets,* Mat. Zametki **8** (1970), 551–562 = Math. Notes **8** (1970), 780–785. MR **43** #2407.

7. ———, *Best approximation of the differentiation operator in the* L_p*-metric,* Mat. Zametki **12** (1972), 531–538 = Math. Notes **12** (1972), 756–760. MR **48** #2629.

8. ———, *Best approximation of operators,* Proc. Central Regional Union of Math. Dept's., no. 1: Functional Anal. and Function Theory, Kalinin Gos. Ped. Inst., Kalinin, 1970, pp. 76–79. (Russian) MR **43** #3714.

9. V. I. Berdyšev, *Best approximation in* $L_{[0,\infty)}$ *of the differentiation operator,* Mat. Zametki **9** (1971), 477–481 = Math. Notes **9** (1971), 275–277. MR **44** #3064.

10. V. V. Arestov, *Best uniform approximation of differentiation operators*, Mat. Zametki **5** (1969), 273–284 = Math. Notes **5** (1969), 167–173. MR **40** #4651.

11. ———, *On best approximation of differentiation operators in the uniform metric*, Candidate's Dissertation, Moscow, 1969. (Russian)

12. ———, *Sharp inequalities between the norms of functions and their derivatives*, Acta Sci. Math. (Szeged) **33** (1972), 243–267. (Russian) MR **47** #9263.

13. Lars Hörmander, *Estimates for translation invariant operators in L^p-spaces*, Acta Math. **104** (1960), 93–140. MR **22** #12389.

14. G. E. Šilov, *Mathematical analysis: Second special course*, "Nauka", Moscow, 1965; English transl., *Generalized functions and partial differential equations*, Gordon and Breach, New York, 1968. MR **36** #2943; **37** #5694.

15. V. V. Arestov, *Approximation of linear operators and related extremal problems*, Trudy Mat. Inst. Steklov. **138** (1975), 29–42 = Proc. Steklov Inst. Math. **138** (1975), 31–44.

16. Arthur Sard, *Integral representations of remainders*, Duke Math. J. **15** (1948), 333–345. MR **10**, 197.

17. V. N. Gabušin, *Estimates for the extrema of a function and its derivatives*, Search for Extrema (Proc. Third All-Union Sympos. Extremal Problems), Tomsk. Gos. Univ., Tomsk, 1969, pp. 243–244. (Russian) RŽMat. **1970** #6585.

18. I. P. Natanson, *Theory of functions of a real variable*, 3rd ed., "Nauka", Moscow, 1974; English transl. of Chapters I–IX of 1st ed., Ungar, New York, 1955. MR **16**, 804.

19. E. C. Titchmarsh, *Introduction to the theory of Fourier integrals*, Clarendon Press, Oxford, 1937.

20. V. I. Levin, *Exact constants in inequalities of the Carlson type*, Dokl. Akad. Nauk SSSR **59** (1948), 635–638. (Russian) MR **9**, 415.

21. B. Sz.-Nagy, *Über Integralungleichungen zwischen einer Funktion und ihrer Ableitung*, Acta. Sci. Math. (Szeged) **10** (1941/43), 64–74. MR **2**, 351.

22. V. N. Gabušin, *Inequalities for norms of a function and its derivatives in L_p-metrics*, Mat. Zametki 1 (1967), 291–298 = Math. Notes 1 (1967), 194–198. MR **34** #6518.

Translated by M. M. ARCHER

Trudy Mat. Inst. Steklov.
138 (1975)

Proc. Steklov Inst. Math.
138 (1975)

CONVERGENCE OF AN INTERPOLATION PROCESS FOR PARABOLIC AND CUBIC SPLINES

UDC 517:518.12

N. L. ZMATRAKOV

ABSTRACT. The convergence of Lagrange interpolation processes is studied for cubic and parabolic splines on an arbitrary sequence of meshes on the class of continuous functions. Necessary and sufficient conditions on the ratios of adjacent balls of the mesh are found in order that the process converge for any continuous function.

Bibliography: 8 titles.

§1. Introduction. Statement of the problem

Suppose that we are given a continuous function $f(x)$ and a mesh

$$\Delta_n \cdot a = x_0^{(n)} < x_1^{(n)} < \ldots < x_{N_n}^{(n)} = b.$$

on a segment $[a, b]$. We set $h_i^{(n)} = x_i^{(n)} - x_{i-1}^{(n)}$ and $\delta_n = \max_i h_i^{(n)}$.

DEFINITION 1. A function $S_n(x)$ is called a *cubic spline* if

a) $S_n(x)$ is a cubic polynomial on every segment $[x_{i-1}^{(n)}, x_i^{(n)}]$;

b) $S_n(x) \in C^2[a, b]$.

DEFINITION 2. A function $S_n(x)$ is called a *parabolic spline* if

a) $S_n(x)$ is a polynomial of degree two on every segment

$$\left[\frac{x_{i-1}^{(n)} + x_i^{(n)}}{2}, \frac{x_i^{(n)} + x_{i+1}^{(n)}}{2}\right], \left[a, \frac{x_0^{(n)} + x_1^{(n)}}{2}\right], \left[\frac{x_{N_n-1}^{(n)} + x_{N_n}^{(n)}}{2}, b\right];$$

b) $S_n(x) \in C^1[a, b]$.

We consider the following interpolation problem. It is required to construct a spline $S_n(x)$ (cubic or parabolic) which interpolates $f(x)$ at the knots of the mesh Δ_n; that is,

$$S_n(x_i^{(n)}) = f(x_i^{(n)}) \quad (i = 0, 1, \ldots, N_n),$$

and satisfies the boundary conditions

$$\left.\begin{aligned} tm_0 + \mu_n m_1 &= c_0^{(n)}, \qquad m_i = S_n'(x_i^{(n)}), \\ \lambda_n m_{N_n-1} + tm_{N_n} &= c_{N_n}^{(n)} \end{aligned}\right\}, \tag{1.1}$$

at the end points of $[a, b]$ or, if $f(x)$ is a $(b - a)$-periodic function, the conditions

$$S_n'(a) = S_n'(b), \quad S_n''(a) = S_n''(b), \tag{1.2}$$

where the parameter t equals 2 in the cubic case and equals 3 in the parabolic case. In the periodic case $S_n(x)$ may also be regarded as a $(b - a)$-periodic function, and in the

AMS (MOS) subject classifications (1970). Primary 41A15.

case of cubic splines $S_n(x) \in C^2(-\infty, \infty)$, while in the parabolic case $S_n(x) \in C^1(-\infty, \infty)$. For this it is necessary to extend the mesh Δ_n and $S_n(x)$ to the whole line by periodicity.

DEFINITION 3. The interpolation process is said to be *convergent in* $C[a, b]$ if for all $f(x) \in C[a, b]$ we have $\lim_{n\to\infty} S_n(x) = f(x)$ for every $x \in [a, b]$, and *uniformly convergent* if the convergence here is uniform.

The convergence of an interpolation process with boundary conditions (1.1) or (1.2) for cubic splines was studied by Ahlberg, Nilson and Walsh [1], §2.4. They imposed the restriction $\max_i \delta_n/h_i^{(n)} < K < \infty$ on the sequence of meshes Δ_n, where K is an absolute constant. In 1964 Schoenberg [2] asked about the existence of a divergent process. Nord [3] gave such an example. In that example $\max_{|i-j|=1} h_i^{(n)}/h_j^{(n)} \longrightarrow \infty$ as $n \longrightarrow \infty$. Meir and Sharma [4] considered the restriction

$$\lim_{n\to\infty} \delta_n = 0, \qquad \max_{|i-j|=1} \frac{h_i^{(n)}}{h_j^{(n)}} \leqslant \rho, \qquad 1 \leqslant \rho < \infty. \tag{1.3}$$

on the sequence of meshes Δ_n. Under this restriction they obtained convergence of the process for cubic splines when $\rho < \sqrt{2}$. Later, almost simultaneously, Cheney and Schurer [5] and Zav'jalov [6] obtained convergence for $\rho < 2$ and $\rho < 1 + \sqrt{2} \approx 2.414$, respectively. Later Hall [7] repeated Zav'jalov's result. The problem in this form was not considered for parabolic splines.

The question arises of the existence of ρ^* such that for $\rho < \rho^*$ a process would be convergent and for $\rho > \rho^*$ there would exist a continuous function and a sequence of meshes satisfying (1.3) such that the process diverges. In our article this question is resolved affirmatively. In §5 it is proved that for cubic splines $\rho^* = (3 + \sqrt{5})/2 \approx 2.618$ and for parabolic splines $\rho^* = 2 + \sqrt{3}$. Moreover, for $\rho = \rho^*$ the process diverges.

We briefly set forth the contents of the article. In §2 the basic equations defining cubic and parabolic splines are given. Theorems of Jackson type are proved in §3 for the deviation $\|f - S_n\|$ under the restrictions

$$\lim_{n\to\infty} \delta_n = 0, \qquad \max_{|i-j|\leqslant s} \frac{h_i^{(n)}}{h_i^{(n)}} \leqslant \rho, \tag{1.4}$$

$$\sup_n \max (|\mu_n|, |\lambda_n|) = \gamma < t^2 + \frac{t(t-1)}{\rho}, \tag{1.5}$$

where $1 \leqslant \rho < \infty$, s is a natural number and the parameter t comes from condition (1.1). In this case a lower bound for ρ^* is also obtained, and from this lower bound it follows that $\rho^* > 2^s$ for cubic splines and $\rho^* > 3^s$ for parabolic splines. The convergence of a process is considered under the restriction

$$\delta_n (|c_0^{(n)}| + |c_{N_n}^{(n)}|) \to 0 \qquad \text{as} \qquad n \to \infty. \tag{1.6}$$

Some auxiliary results are established in §4. The possibility of divergence for a process with sufficiently large ρ and functions in Lip α, $0 < \alpha < 1$, is considered in §6, and the possibility of divergence in L_p, $p \geqslant 1$, is considered in §7.

The author thanks Ju. N. Subbotin for the statement of the problem and help in the work.

§2. Basic equations and formulas for cubic and parabolic splines

In the case of interpolating cubic splines for $m_i = S'_n(x_i^{(n)})$ we have (see, for example, [1], formula 2.1.14)

$$h_{i+1}m_{i-1} + 2(h_i + h_{i+1})m_i + h_i m_{i+1} = 3\frac{h_i}{h_{i+1}}[f(x_{i+1}) - f(x_i)] + 3\frac{h_{i+1}}{h_i}[f(x_i) - f(x_{i-1})]. \tag{2.1}$$

We establish an analogous equality for parabolic splines. On the segment

$$\left[\frac{x_{i-1}+x_i}{2}, \frac{x_i + x_{i+1}}{2}\right]$$

we have

$$S_n(x) = S_n(x_i) + S'_n(x_i)(x - x_i) + S''_n(x_i)\frac{(x-x_i)^2}{2}.$$

From Definition 2 we obtain the equalities

$$S_n(x_{i-1}) + S'_n(x_{i-1})\frac{h_i}{2} + S''_n(x_{i-1})\frac{h_i^2}{8} = S_n(x_i) - S'_n(x_i)\frac{h_i}{2} + S''_n(x_i)\frac{h_i^2}{8},$$

$$S_n(x_i) + S'_n(x_i)\frac{h_{i+1}}{2} + S''_n(x_i)\frac{h_{i+1}^2}{8} = S_n(x_{i+1}) - S'_n(x_{i+1})\frac{h_{i+1}}{2} + S''_n(x_{i+1})\frac{h_{i+1}^2}{8}$$

$$S'_n(x_{i-1}) + S''_n(x_{i-1})\frac{h_i}{2} = S'_n(x_i) - S''_n(x_i)\frac{h_i}{2},$$

$$S'_n(x_i) + S''_n(x_i)\frac{h_{i+1}}{2} = S'_n(x_{i+1}) - S''_n(x_{i+1})\frac{h_{i+1}}{2}.$$

We multiply the equations by $4h_{i+1}/h_i$, $4h_i/h_{i+1}$, $-h_{i+1}$ and h_i, respectively, and add the resulting equations to obtain

$$h_{i+1}m_{i-1} + 3(h_i + h_{i+1})m_i + h_i m_{i+1} = 4\frac{h_i}{h_{i+1}}[S_n(x_{i+1}) - S_n(x_i)] + 4\frac{h_{i+1}}{h_i}[S_n(x_i) - S_n(x_{i-1})] = 4\frac{h_i}{h_{i+1}}[f(x_{i+1}) - f(x_i)] + 4\frac{h_{i+1}}{h_i}[f(x_i) - f(x_{i-1})], \tag{2.2}$$

for interpolating parabolic splines.

Taking (2.1) and (2.2) into account, we will write a system of equations for the definition of the parameters m_i of cubic ($t = 2$) and parabolic ($t = 3$) interpolating splines. In the periodic case we have

$$\frac{h_{i+1}}{h_i + h_{i+1}}m_{i-1} + tm_i + \frac{h_i}{h_i + h_{i+1}}m_{i+1} = \frac{t+1}{h_i + h_{i+1}}\left\{\frac{h_i}{h_{i+1}}[f(x_{i+1}) - f(x_i)] + \frac{h_{i+1}}{h_i}[f(x_i) - f(x_{i-1})]\right\} = c_i \quad (i = 1, 2, \ldots, N), \quad m_0 = m_N, \quad m_{N+1} = m_1, \tag{2.3}$$

and in the case of the boundary conditions (1.1)

$$tm_0 + \mu_n m_1 = c_0,$$

$$\frac{h_{i+1}}{h_i + h_{i+1}} m_{i-1} + tm_i + \frac{h_i}{h_i + h_{i+1}} m_{i+1} = \frac{t+1}{h_i + h_{i+1}} \left\{ \frac{h_i}{h_{i+1}} [f(x_{i+1}) - f(x_i)] \right. \tag{2.4}$$

$$\left. + \frac{h_{i+1}}{h_i} [f(x_i) - f(x_{i-1})] \right\} = c_i \quad (i = 1, 2, \ldots, N-1), \quad \lambda_n m_{N-1} + tm_N = c_N.$$

Later on we will need the following representation of interpolating cubic splines ([16], §4) on the segment $[x_{i-1}, x_i]$:

$$S_n(x) = u_i(x) m_{i-1} - v_i(x) m_i + w_i(x) \frac{f(x_i) - f(x_{i-1})}{h_i} + \frac{f(x_i) + f(x_{i-1})}{2}, \tag{2.5}$$

where

$$u_i(x) = \frac{(x - x_{i-1})(x_i - x)^2}{h_i^2}, \qquad v_i(x) = \frac{(x - x_{i-1})^2 (x_i - x)}{h_i^2},$$

$$w_i(x) = \frac{1}{h_i^2} \left(x - \frac{x_{i-1} + x_i}{2} \right) [(x_i - x)^2 + 4(x_i - x)(x - x_{i-1}) + (x - x_{i-1})^2].$$

On the segment $[x_{i-1}, x_i]$ the estimates

$$u_i(x) \geqslant 0, \quad v_i(x) \geqslant 0, \quad -\frac{1}{2} h_i \leqslant w_i(x) \leqslant \frac{1}{2} h_i,$$

$$\max_{x_{i-1} \leqslant x \leqslant x_i} (u_i(x) + v_i(x)) = \frac{1}{4} h_i. \tag{2.6}$$

hold.

For the second derivative of a cubic spline on $[x_{i-1}, x_i]$ we have (see, for example, [1], formula 2.1.12)

$$S_n''(x) = -2m_{i-1} \frac{2x_i + x_{i-1} - 3x}{h_i^2} - 2m_i \frac{2x_{i-1} + x_i - 3x}{h_i^2}$$

$$+ 6 \frac{f(x_i) - f(x_{i-1})}{h_i^3} (x_i + x_{i-1} - 2x). \tag{2.7}$$

For parabolic interpolating splines it is not difficult to obtain on $[x_{i-1}, x_i]$ the representation

$$S_n(x) = u_i(x) m_{i-1} - v_i(x) m_i + w_i(x) \frac{f(x_i) - f(x_{i-1})}{h_i} + \frac{f(x_{i-1}) + f(x_i)}{2}, \tag{2.8}$$

where

$$u_i(x) = (x - x_{i-1}) - \frac{3}{2} \frac{(x - x_{i-1})^2}{h_i} + 2 \frac{(x - y_i)_+^2}{h_i},$$

$$v_i(x) = \frac{(x - x_{i-1})^2}{2h_i} - 2 \frac{(x - y_i)_+^2}{h_i},$$

$$w_i(x) = 2 \frac{(x - x_{i-1})^2}{h_i} - 4 \frac{(x - y_i)_+^2}{h_i} - \frac{h_i}{2},$$

$$y_i = \frac{x_{i-1} + x_i}{2}, \qquad \varphi(x - a)_+ = \begin{cases} 0, & \text{if} \quad x < a, \\ \varphi(x - a), & \text{if} \quad x \geqslant a. \end{cases}$$

The estimates (2.6), obtained for the cubic case, remain valid in the parabolic case, and for the second derivative we have

$$S_n''(x)=\begin{cases} 4\dfrac{f(x_i)-f(x_{i-1})}{h_i^2}-\dfrac{3m_{i-1}+m_i}{h_i}, & \text{if} \quad x<y_i, \\ -4\dfrac{f(x_i)-f(x_{i-1})}{h_i^2}+\dfrac{m_{i-1}+3m_i}{h_i}, & \text{if} \quad x>y_i. \end{cases} \tag{2.9}$$

We mention examples of unifofm meshes and boundary conditions (1.1) for which an interpolating spline does not exist.

For cubic splines ($t = 2$) Zav'jalov [6] noted that for $N = 1$ and $\lambda = \mu = 2$ the matrix of the system (2.4) is singular. It is easy to check that for $N = 2$, $\lambda = \mu = 4$ and $N = 3$, $\lambda = \mu = 6$ we also obtain a singular matrix.

For parabolic splines ($t = 3$), setting $\lambda = \mu = 3$, $\lambda = \mu = 9$ and $\lambda = \mu = 15$ respectively, we obtain a singular matrix for the system of equations (2.4).

Later on we will consider meshes with $N \geqslant 3$.

§3. A theorem of Jackson type

For a matrix Q of order n with elements a_{ij}, where a_{ij} is the element in the ith row and jth column, we set

$$\|Q\|=\max_{1\leqslant i\leqslant 1}\Big(\sum_{j=1}^{n}|a_{ij}|\Big).$$

From the general theory it is known (see, for example, [1], §2.3) that if a matrix Q has a dominating main diagonal–that is, if

$$\Theta_i=|a_{ii}|-\sum_{\substack{j=1\\ j\neq i}}^{n}|a_{ij}|>0 \qquad (i=1,2,\ldots,n)$$

–then for the norm of the inverse matrix Q^{-1} we have

$$\|Q^{-1}\|\leqslant(\min_i\Theta_i)^{-1}. \tag{3.1}$$

Here we will consider matrices of a special form, where $a_{ij} = 0$ if $|i - j| \geqslant 2$ with the possible exception of the elements a_{1n} and a_{n1}. Suppose we are given numbers $t > 1$, α_i and β_i ($i = 1, 2, n - 1, n$) satisfying the inequalities

$$\Theta_i=t-|\alpha_i|-|\beta_i|>0 \quad (i=1,2,n-1,n).$$

Further, suppose the positive numbers h_m ($m = 0, \pm 1, \pm 2, \ldots$) satisfy the conditions

$$\max_{|l-k|\leqslant s}\frac{h_l}{h_k}\leqslant\rho,$$

where $1 \leqslant \rho < \infty$ and s is a natural number; for $i = 3, 4, \ldots, n - 2$ we set

$$\alpha_i=\frac{h_{i+1}}{h_i+h_{i+1}}\sqrt[2j]{\frac{h_{i-j+1}\cdots h_{i+j}}{h_{i-j}\cdots h_{i+j-1}}}, \qquad \beta_i=\frac{h_i}{h_i+h_{i+1}}\sqrt[2j]{\frac{h_{i-j+1}\cdots h_{i+j}}{h_{i-j+2}\cdots h_{i+j+1}}},$$

where j is a natural number.

We define the matrix

$$G_n = \begin{pmatrix} t & \beta_1 & 0 & \dots & 0 & \alpha_1 & \\ \alpha_2 & t & \beta_2 & & 0 & 0 & \\ 0 & \alpha_3 & t & & \cdot & \cdot & \\ \cdot & & & \dots & \cdot & \cdot & \\ \cdot & & & & \cdot & \cdot & \\ \cdot & & & & \alpha_{n-1} & t & \beta_{n-1} \\ \beta_n & 0 & & \dots & & 0 & \alpha_n \; t \end{pmatrix}$$

and in the following lemma we estimate $\|G^{-1}\|$. Here we designate by $\rho_s(t) > 1$ a root of the equation

$$t\sqrt[s]{\rho}(\rho+1)-(1+\sqrt[s]{\rho^2}\,\rho)=0. \tag{3.2}$$

It is obvious that for $t > 1$ the equation has a positive root, and, since by the substitution $\sqrt[s]{\rho} = z$ the equation (3.2) reduces to the recursive equation

$$z^{s+2} - tz^{s+1} - tz + 1 = 0$$

with two changes of sign, it follows that the root $\rho_s(t) > 1$ exists and is unique.

LEMMA 1. *If $t > 1$ and the number ρ satisfies the inequality*

$$1 \leqslant \rho < \rho_s(t),$$

then there exists J, not depending on n, such that for all $j \geqslant J$

$$t\rho^{\nu}(\rho+1)-(1+\sqrt[j]{\rho}\,\rho^{2\nu+1})>0, \tag{3.3}$$

and

$$\|G_n^{-1}\| \leqslant \max\left\{\frac{\rho^{\nu}(\rho+1)}{t\rho^{\nu}(\rho+1)-(1+\sqrt[s]{\rho}\,\rho^{2\nu+1})}, \frac{1}{\theta_1}, \frac{1}{\theta_2}, \frac{1}{\theta_{n-1}}, \frac{1}{\theta_n}\right\},$$

where

$$\nu = \frac{1}{2j}\left(\left[\frac{2j}{s}\right] + \operatorname{sgn}\left(\frac{2j}{s} - \left[\frac{2j}{s}\right]\right)\right).$$

PROOF. Setting

$$u = \frac{h_i}{h_{i+1}}, \qquad y = \sqrt[2j]{\frac{h_{i-j+1}\cdots h_{i+j}}{h_{i-j}\cdots h_{i+j-1}}},$$

we obtain from (1.4)

$$\rho^{-1} \leqslant u \leqslant \rho, \qquad \rho^{-\nu} \leqslant y \leqslant \rho^{\nu},$$

$$\sqrt[2j]{\frac{h_{i-j+1}\cdots h_{i+j}}{h_{i-j+2}\cdots h_{i+j+1}}} = \sqrt[2j]{\frac{h_{i-j+1}}{h_{i+j+1}}} = \sqrt[2j]{\frac{h_{i-j+1}}{h_{i-j}}\cdot\frac{h_{i+j}}{h_{i+j+1}}\cdot\frac{h_{i-j}}{h_{i+j}}} \leqslant \frac{\sqrt[j]{\rho}}{y} \tag{3.4}$$

and consequently, for $i = 3, 4, \ldots, n-2$,

$$\begin{aligned}\Theta_i = t - \alpha_i - \beta_i &\geqslant t - \frac{h_{i+1}}{h_i + h_{i+1}}y - \frac{h_i}{h_i + h_{i+1}}\frac{\sqrt[j]{\rho}}{y} \\ &= t - \frac{1}{1+u}\left(y + \frac{u\sqrt[j]{\rho}}{y}\right) = t - \Psi(u, y);\end{aligned}$$

that is,

$$\Theta_i \geqslant t - \Psi(u, y) \quad (i = 3, 4, \ldots, n-2),$$
$$\Theta_i = t - |\alpha_i| - |\beta_i| > 0 \quad (i = 1, 2, n-1, n). \tag{3.5}$$

It is obvious that for an estimate of Θ_i it is sufficient to estimate from above the function

$$\Psi(u, y) = \frac{1}{1+u}\left(y + \frac{u\sqrt[j]{\rho}}{y}\right)$$

on the domain (3.4).

We consider the two possible cases:

1) If $u\sqrt[j]{\rho} \geqslant 1$, then

$$\Psi(u, y) \leqslant \Psi(u, \rho^{-\nu}) = \rho^{-\nu}\,\frac{1 + \sqrt[j]{\rho}\,\rho^{2\nu}u}{1+u},$$

and since $\sqrt[j]{\rho}\,\rho^{2\nu} \geqslant 1$ we have

$$\Psi(u, \rho^{-\nu}) \leqslant \Psi(\rho, \rho^{-\nu}) = \frac{1 + \sqrt[j]{\rho}\,\rho^{2\nu+1}}{\rho^{\nu}(1+\rho)}.$$

2) If $u\sqrt[j]{\rho} \leqslant 1$, then

$$\Psi(u, y) \leqslant \Psi(u, \rho^{\nu}) = \rho^{\nu}\,\frac{1 + \sqrt[j]{\rho}\,\rho^{-2\nu}u}{1+u},$$

and since $\sqrt[j]{\rho}\,\rho^{-2\nu} \leqslant 1$ we have

$$\Psi(u, \rho^{\nu}) \leqslant \Psi(\rho^{-1}, \rho^{\nu}) = \frac{\sqrt[j]{\rho} + \rho^{2\nu+1}}{\rho^{\nu}(1+\rho)}.$$

For $\rho \geqslant 1$ we have $\psi(\rho, \rho^{-\nu}) \geqslant \psi(\rho^{-1}, \rho^{\nu})$; hence on the domain (3.4)

$$\Psi(u, y) \leqslant \Psi(\rho, \rho^{-\nu}) = \frac{1 + \sqrt[j]{\rho}\,\rho^{2\nu+1}}{\rho^{\nu}(1+\rho)}. \tag{3.6}$$

Taking (3.5) into account, we obtain

$$\Theta_i \geqslant t - \Psi(\rho, \rho^{-\nu}) = \frac{t\rho^{\nu}(\rho+1) - (1 + \sqrt[j]{\rho}\,\rho^{2\nu+1})}{\rho^{\nu}(1+\rho)} \quad (i = 3, 4, \ldots, n-2). \tag{3.7}$$

It remains for us to prove that for a suitable choice of j the right-hand side of (3.7) is positive. Indeed, since $\nu \longrightarrow 1/s$ as $j \longrightarrow \infty$, we obtain from (3.2) and the fact that $\rho < \rho_s(t)$ the inequality

$$t\rho^{\nu}(\rho+1) - (1 + \sqrt[j]{\rho}\,\rho^{2\nu+1}) > 0$$

for all j greater than some J. Taking into account (3.1), (3.5) and (3.7), we obtain the assertion of the lemma.

Let $\omega(f, \delta)$ be the modulus of continuity of the function $f(x)$:

$$\omega(f, \delta) = \max_{|x'-x''| \leqslant \delta} |f(x') - f(x'')|, \; x', x'' \in [a, b].$$

For simplicity of notation the upper index n of the elements of the mesh Δ_n in conditions (1.1)–(1.6) will be dropped.

THEOREM 1. *Suppose given a continuous function* $f(x)$ *on the segment* $[a, b]$, *and a cubic* ($t = 2$) *or parabolic* ($t = 3$) *spline* $S_{n,t}(x)$ *interpolating* $f(x)$ *on the mesh* Δ_n *and satisfying the boundary conditions* (1.1) *with restriction* (1.5). *Let the sequence of meshes* Δ_n *satisfy the conditions in* (1.4), *where* ρ *satisfies the condition of Lemma* 1. *Furthermore, let the fixed number* γ_1 *and natural number* j *be chosen according to the conditions*

1) $\gamma < t\gamma_1 < t^2 + \dfrac{t(t-1)}{\rho}$,

2) $t\rho^\nu(\rho+1) - (1 + \sqrt[j]{\rho}\,\rho^{2\nu+1}) > 0$,

3) *if* $\sqrt[j]{\rho} \geqslant \gamma_1$, *then* $t(\rho+1) - (\gamma_1 + \sqrt[j]{\rho}\,\rho) > 0$,

4) *if* $\sqrt[j]{\rho} \leqslant \gamma_1$, *then* $t(\rho+1) - (\gamma_1\rho + \sqrt[j]{\rho}) > 0$,

where

$$\nu = \frac{1}{2j}\left(\left[\frac{2j}{s}\right] + \operatorname{sign}\left(\frac{2j}{s} - \left[\frac{2j}{s}\right]\right)\right),$$

Then

$$\| f - S_{n,t} \| \leqslant \omega(f, \delta_n) + A_0(t) A_n(t), \tag{3.8}$$

where

$$A_0(t) = \max\left\{\frac{\rho^\nu(\rho+1)}{t\rho^\nu(\rho+1) - (1 + \sqrt[j]{\rho}\,\rho^{2\nu+1})}, \frac{\gamma_1}{t\gamma_1 - \gamma}, \frac{\rho+1}{t(\rho+1) - (\gamma_1 + \sqrt[j]{\rho}\,\rho)}, \right.$$
$$\left. \frac{\rho+1}{t(\rho+1) - (\gamma_1\rho + \sqrt[j]{\rho})}\right\} \max\{1, \gamma_1\} \frac{\rho^{2\mu}}{4},$$

$$A_n(t) = \max\left\{(t+1)\,\omega(f, \delta_n), |c_0| \frac{\delta_n}{\gamma_1\rho^\mu}, |c_N| \frac{\delta_n}{\gamma_1\rho^\mu}\right\},$$

$$\mu = \frac{(k+1)(j+l)}{2j}, \quad j - 1 = ks + l, \quad 0 \leqslant l < s.$$

In the periodic case j *is chosen according to condition* 2), *and in* (3.8)

$$A_0(t) = \frac{(\rho+1)\rho^{2\mu-\nu}}{t\rho^\nu(\rho+1) - (1 + \sqrt[j]{\rho}\,\rho^{2\nu+1})} \frac{1}{4}, \quad A_n(t) = (t+1)\,\omega(f, \delta_n).$$

PROOF. Taking (2.5), (2.8) and (2.6) into account, on the interval $[x_{i-1}, x_i]$ we obtain

$$\begin{aligned} |f(x) - S_{n,t}(x)| \leqslant &\left|\left(\frac{1}{2} - \frac{w_i(x)}{h_i}\right)[f(x) - f(x_{i-1})]\right. \\ &\left. + \left(\frac{1}{2} + \frac{w_i(x)}{h_i}\right)[f(x) - f(x_i)]\right| + |v_i(x) m_i - u_i(x) m_{i-1}| \\ &\leqslant \omega(f, \delta_n) + \frac{1}{4} h_i \max(|m_{i-1}|, |m_i|). \end{aligned} \tag{3.9}$$

Thus it is clear that for the proof of the theorem it is sufficient to estimate

$$h_i \max(|m_{i-1}|, |m_i|).$$

We consider the periodic case first. We multiply both sides of the ith equation of the system (2.3) by $\sqrt[2j]{h_{i-j+1}\cdots h_{i+j}}$ and set

$$a_i = \sqrt[2j]{h_{i-j+1}\ldots h_{i+j}}\, m_i. \tag{3.10}$$

Since $h_0 = h_N$, $h_1 = h_{N+1}$ and so on, we have $a_0 = a_N$ and $a_{N+1} = a_1$.

In (2.3) we substitute for the variables according to (3.10) and obtain

$$\begin{pmatrix} t & \beta_1 & 0 & \ldots & 0 & \alpha_1 \\ \alpha_2 & t & \beta_2 & & 0 & 0 \\ 0 & \alpha_3 & t & & \vdots & \vdots \\ \vdots & & \ldots & & \vdots & \vdots \\ 0 & 0 & \ldots & \alpha_{N-1} & t & \beta_{N-1} \\ \beta_N & 0 & \ldots & 0 & \alpha_N & t \end{pmatrix} \begin{pmatrix} a_1 \\ a_2 \\ a_3 \\ \vdots \\ a_{N-1} \\ a_N \end{pmatrix} = \begin{pmatrix} q_1 \\ q_2 \\ q_3 \\ \vdots \\ q_{N-1} \\ q_N \end{pmatrix},$$

where

$$\alpha_i = \frac{h_{i+1}}{h_i + h_{i+1}} \sqrt[2j]{\frac{h_{i-j+1}\cdots h_{i+j}}{h_{i-j}\cdots h_{i+j-1}}}, \qquad \beta_i = \frac{h_i}{h_i + h_{i+1}} \sqrt[2j]{\frac{h_{i-j+1}\cdots h_{i+j}}{h_{i-j+2}\cdots h_{i+j+1}}}, \tag{3.11}$$

$$q_i = \sqrt[2i]{h_{i-j+1}\cdots h_{i+j}}\, c_i \quad (i = 1, 2, \ldots, N).$$

We write this system more briefly in the matrix form

$$Qa = q. \tag{3.12}$$

Since the h_i satisfy condition (1.4), it is easy to see that the matrix Q satisfies the condition of Lemma 1; hence

$$\|Q^{-1}\| \leqslant \frac{\rho^{\nu}(\rho+1)}{t\rho^{\nu}(\rho+1) - (1 + \sqrt[j]{\rho}\,\rho^{2\nu+1})}.$$

We estimate the norm $\|q\|$. Since

$$\frac{h_{i-j+1}\cdots h_{i+j}}{h_{i+1}\cdots h_{i+1}} \leqslant \rho^{2j\mu}, \qquad \frac{h_{i-j+1}\cdots h_{i+j}}{h_i\cdots h_i} \leqslant \rho^{2j\mu},$$

we obtain

$$|q_i| = |c_i| \sqrt[2i]{h_{i-j+1}\cdots h_{i+j}} \leqslant (t+1)\frac{h_i}{h_{i+1}} \frac{\sqrt[2j]{h_{i-j+1}\cdots h_{i+j}}}{h_i + h_{i+1}}$$

$$\times |f(x_{i+1}) - f(x_i)| + (t+1)\frac{h_{i+1}}{h_i}\frac{\sqrt[2j]{h_{i-j+1}\cdots h_{i+j}}}{h_i + h_{i+1}} |f(x_i) - f(x_{i-1})|$$

$$\leqslant (t+1)\,\omega(f, \delta_n)\left[\frac{h_i}{h_i + h_{i+1}}\sqrt[2j]{\frac{h_{i-j+1}\cdots h_{i+j}}{h_{i+1}\cdots h_{i+1}}}\right.$$

$$\left. + \frac{h_{i+1}}{h_i + h_{i+1}}\sqrt[2j]{\frac{h_{i-j+1}\cdots h_{i+j}}{h_i\cdots h_i}}\right] \leqslant (t+1)\,\omega(f, \delta_n)\,\rho^{\mu}$$

and consequently

$$\|q\| \leqslant (t+1)\omega(f, \delta_n)\rho^{\mu}.$$

For the solution of the system (3.12) we have

$$\| a \| \leqslant \| Q^{-1} \| \| q \|. \tag{3.16}$$

Taking (3.10) into account, we obtain

$$h_i | m_i | \leqslant \frac{\| a \| h_i}{\sqrt[2j]{h_{i-j+1} \cdots h_{i+j}}} \leqslant \| a \| \rho^{\mu},$$

$$h_i | m_{i-1} | \leqslant \frac{\| a \| h_i}{\sqrt[2j]{h_{i-j} \cdots h_{i+j-1}}} \leqslant \| a \| \rho^{\mu}.$$

From this we have

$$h_i \max (| m_i | , | m_{i-1} |) \leqslant \| a \| \rho^{\mu}, \tag{3.17}$$

and by virtue of (3.9), (3.13), (3.15) and (3.16) for $x \in [x_{i-1}, x_i]$ we have

$$| f(x) - S_{n,t}(x) | \leqslant \omega (f, \delta_n) + \frac{1}{4} \frac{(t+1)(\rho+1) \rho^{2\mu+\nu} \omega (f, \delta_n)}{t\rho^2 (\rho+1) - (1 + \sqrt[j]{\rho}\, \rho^{2\nu+1})}. \tag{3.18}$$

The right-hand side of (3.18) does not depend on i; therefore the estimate is valid on the whole segment $[a, b]$. The theorem is proved in the periodic case.

In order to use the above method of proof in the case of the boundary conditions (1.1), it is necessary to define h_i for $i \leqslant 0$ and $i \geqslant N+1$. Setting successively for $i = 1, 2, \ldots, j-1$

$$h_{-i+1} = h_{i+1}, \qquad h_{N+j-i} = h_{N-j+i}$$

and

$$\gamma_1^{2j} h_{-j+1} = h_{j+1}, \qquad \gamma_1^{2j} h_{N+j} = h_{N-j},$$

for a given j we obtain a completely well-defined mesh. It is obvious that the h_i $(-j+1 < i < N+j)$ satisfv (1.4).

Again we multiply both sides of the ith equation of (2.4) by $\sqrt[2j]{h_{i-j+1} \cdots h_{i+j}}$ and substitute according to (3.10) to obtain

$$\begin{pmatrix} t & \beta_0 & 0 & 0 & 0 & 0 \\ \alpha_1 & t & \beta_1 & \cdot & \cdot & \cdot \\ 0 & \alpha_2 & t & \cdot & \cdot & \cdot \\ \vdots & \cdots & \cdots & \cdots & \cdots & \cdots \\ \vdots & & & \alpha_{N-1} & t & \beta_{N-1} \\ 0 & 0 & 0 \ldots & 0 & \alpha_N & t \end{pmatrix} \begin{pmatrix} a_0 \\ a_1 \\ a_2 \\ \vdots \\ a_{N-1} \\ a_N \end{pmatrix} = \begin{pmatrix} q_0 \\ q_1 \\ q_2 \\ \vdots \\ q_{N-1} \\ q_N \end{pmatrix},$$

where α_i, β_i and q_i $(i = 1, \ldots, N-1)$ are as in (3.11), and

$$\beta_0 = \mu_n \sqrt[2j]{\frac{h_{-j+1} \cdots h_j}{h_{-j+2} \cdots h_{j+1}}} = \frac{\mu_n}{\gamma_1}, \quad \alpha_N = \lambda_n \sqrt[2j]{\frac{h_{N-j+1} \cdots h_{N+j}}{h_{N-j} \cdots h_{N+j-1}}} = \frac{\lambda_n}{\gamma_1}, \tag{3.19}$$

$$q_0 = c_0 \sqrt[2j]{h_{-j+1} \cdots h_j}, \quad q_N = c_N \sqrt[2j]{h_{N-j+1} \cdots h_{N+j}}.$$

We write this system more briefly in the matrix form

$$Qa = q. \tag{3.20}$$

We show that the matrix Q satisfies the conditions of Lemma 1. Indeed,

$$\Theta_0 = t - |\beta_0| = t - \frac{|\mu_n|}{\gamma_1} \geqslant \frac{t\gamma_1 - \gamma}{\gamma_1} > 0,$$

$$\Theta_N = t - |\alpha_N| = t - \frac{|\lambda_n|}{\gamma_1} \geqslant \frac{t\gamma_1 - \gamma}{\gamma_1} > 0.$$

Furthermore, taking into account that $N \geqslant 3$ and setting $u = h_1/h_2$, we obtain

$$\Theta_1 = t - |\alpha_1| - |\beta_1| \geqslant t - \frac{\gamma_1}{1+u} - \frac{\sqrt[j]{\rho}\, u}{1+u}.$$

We find the maximum of the function

$$\Psi(u) = \frac{\gamma_1 + \sqrt[j]{\rho}\, u}{1+u}$$

in the domain $\rho^{-1} \leqslant u \leqslant \rho$.

We consider two possible cases:

1) If $\sqrt[j]{\rho} \geqslant \gamma_1$, then $\Psi(u) \leqslant \Psi(\rho) = (\gamma_1 + \sqrt[j]{\rho}\rho)/(1+\rho)$.

2) If $\sqrt[j]{\rho} \leqslant \gamma_1$, then $\Psi(u) \leqslant \Psi(\rho^{-1}) = (\gamma_1\rho + \sqrt[j]{\rho})/(1+\rho)$.

From this and the hypotheses of the theorem we have

$$\Theta_1 \geqslant \begin{cases} t - \dfrac{\gamma_1 + \sqrt[j]{\rho}\,\rho}{1+\rho} > 0, & \text{if} \quad \sqrt[j]{\rho} \geqslant \gamma_1, \\[2ex] t - \dfrac{\gamma_1\rho + \sqrt[j]{\rho}}{1+\rho} > 0, & \text{if} \quad \sqrt[j]{\rho} \leqslant \gamma_1. \end{cases}$$

Analogously we find exactly the same estimate for Θ_{N-1}. It is not difficult to see that for a fixed γ_1 satisfying condition 1) it is possible to find a number J such that for all $j \geqslant J$ one of the conditions 3) or 4) of the theorem is satisfied. This says that the hypotheses of the theorem are compatible. Consequently, in view of Lemma 1 we have

$$\|Q^{-1}\| \leqslant \max \left\{ \frac{\rho^{\nu}(\rho+1)}{t\rho^{\nu}(\rho+1) - (1 + \sqrt[j]{\rho}\,\rho^{2\nu+1})}, \frac{\gamma_1}{t\gamma_1 - \gamma}, \right. \tag{3.21}$$
$$\left. \frac{\rho+1}{t(\rho+1) - (\gamma_1 + \sqrt[j]{\rho}\,\rho)}, \frac{\rho+1}{t(\rho+1) - (\gamma_1\rho + \sqrt[j]{\rho})} \right\} = B_0.$$

In order to estimate $\|q\|$ it suffices to estimate $|q_0|$ and $|q_N|$. For $|q_i|$ $(i = 1, \ldots, N-1)$ the estimates were obtained in (3.14). In view of (3.19),

$$|q_0| \leqslant |c_0| \frac{\delta_n}{\gamma_1}, \quad |q_N| \leqslant |c_N| \frac{\delta_n}{\gamma_1},$$

and consequently

$$\|q\| \leqslant \max \left\{ (t+1)\,\omega(f, \delta_n)\,\rho^{\mu}, \quad |c_0| \frac{\delta_n}{\gamma_1}, \quad |c_N| \frac{\delta_n}{\gamma_1} \right\} = B_n. \tag{3.22}$$

Then for the solution a of the system (3.20) we obtain

$$\|a\| \leqslant \|Q^{-1}\| \|q\| \leqslant B_0 B_n. \tag{3.23}$$

Since

$$h_1 |m_0| \leqslant \frac{\|a\| h_1}{\sqrt[2j]{h_{-j+1} \cdots h_j}} \leqslant \|a\| \gamma_1 \rho^\mu, \qquad h_N |m_N| \leqslant \|a\| \gamma_1 \rho^\mu,$$

on taking (3.17) into account we find that

$$h_i \max (|m_{i-1}|, |m_i|) \leqslant \max \{1, \gamma_1\} \|a\| \rho^\mu.$$

From this, (3.9) and (3.23) we obtain

$$|f(x) - S_{n,t}(x)| \leqslant \omega (f, \delta_n) + \frac{1}{4} \max \{1, \gamma_1\} B_0 B_n \rho^\mu,$$

for $x \in [x_{i-1}, x_i]$; and, since the right-hand side of the inequality does not depend on i, the estimate is valid on the whole segment $[a, b]$. Theorem 1 is proved.

§4. Convergence of interpolating splines for functions of a special form

For the demonstration of a divergent process we will need some properties of interpolating splines for specially constructed functions and sequence of meshes.

a) Suppose we are given a real number $\rho \geqslant 2$, an integer $m \geqslant 2$ and a sequence of positive numbers H_n satisfying the conditions

$$\begin{aligned} &H_n \to 0 && (n \to \infty), \\ &H_{n-2} < (2m - 1) H_n && (n = 6, 8, \ldots). \end{aligned} \tag{4.1}$$

We define a function $f_{\rho,m}(x)$ on the segment $[-1, 1]$. We will construct it successively on the interval $[1/2^r, 1/2^{r-2}]$ $(r = 4, 6, 8, \ldots)$. We divide the interval $[1/2^r, 1/2^{r-2}]$ at the points

$$\frac{1}{2^r} = \xi_0 < \xi_1 < \cdots < \xi_{M_r} \leqslant \frac{1}{2^{r-2}},$$

choosing M_r and ξ_i according to the conditions

$$\frac{1}{2^{r-2}} - \xi_{M_r} < \xi_{M_r} - \xi_{M_r - 1}, \quad \xi_i - \xi_{i-1} = \begin{cases} \dfrac{1}{2^{r-1}\rho^i} & (i = 1, 2, \ldots, r-1), \\ \dfrac{1}{2^{r-1}\rho^r} & (i = r, r+1, \ldots, M_r). \end{cases}$$

Such a partitioning is possible since

$$\xi_r < \frac{1}{2^r} + \frac{1}{2^{r-1}} \sum_{i=1}^{\infty} \frac{1}{\rho^i} = \frac{1}{2^r} \left(1 + \frac{2}{\rho - 1}\right) < \frac{1}{2^{r-2}}.$$

From the construction of the ξ_i it is evident that the number of equal segments grows as $r \to \infty$; therefore there exists an even number R such that for all $r \geqslant R$ the condition

$$4 \leqslant 2m \leqslant M_r - r. \tag{4.2}$$

is fulfilled. We will consider $r = R, R + 2, \ldots$. Let

$$M = \begin{cases} M_r - 2m, & \text{if } M_r \text{ is even,} \\ M_r - 2m - 1, & \text{if } M_r \text{ is odd.} \end{cases}$$

We set

$$f_{\rho,m}(\xi_0) = f_{\rho,m}(\xi_2) = \ldots = f_{\rho,m}(\xi_M) = H_r,$$
$$f_{\rho,m}(\xi_1) = f_{\rho,m}(\xi_3) = \ldots = f_{\rho,m}(\xi_{M+1}) = 0, \qquad f_{\rho,m}\left(\frac{1}{2^{r-2}}\right) = H_{r-2}.$$

We extend $f(x)$ linearly on the intervals $[\xi_i, \xi_{i+1}]$ $(i = 0, 1, \ldots, M)$ and $[\xi_{M+1}, 1/2^{r-2}]$. Furthermore, we set $f_{\rho,m}(½) = f_{\rho,m}(1) = 0$ and extend $f_{\rho,m}(x)$ linearly on the intervals $[1/2^{R-2}, ½]$ and $[½, 1]$. For continuity we set $f_{\rho,m}(0) = 0$. For $x \in [-1, 0]$ we set $f_{\rho,m}(x) = f_{\rho,m}(-x)$. We obtain a continuous even function on $[-1, 1]$.

From (4.1) it follows that for all $x', x'' \in [-1, 1]$ such that $|x' - x''| = \xi_{M_n} - \xi_{M_n - 1}$ the inequality

$$|f_{\rho,m}(x') - f_{\rho,m}(x'')| \leqslant H_n \tag{4.3}$$

is satisfied.

b) We define the mesh $\overline{\Delta}_n$, n being an even number, $n \geqslant R$, where R is defined for the constructed function $f_{\rho,m}(x)$. Let

$$\overline{\Delta}_n: -1 = x_{-(n+k+1)} < \ldots < x_{-1} < x_0 < \ldots < x_{n+k} = 1, \tag{4.4}$$

where $x_0 = 1/2^n$, $x_{-i} = -x_{i-1}$, the points $x_1, \ldots, x_{n+k-1}$ are defined by the assignment of the steps $h_i = x_i - x_{i-1}$ according to the formulas

$$\begin{aligned} h_i &= \frac{1}{2^{n-1}\rho^i} \quad (i = 0, 1, \ldots, n-1), \\ h_i &= \frac{1}{2^{n-1}\rho^n} \quad (i = n, n+1, \ldots, n+k-1), \\ h_{n+k} &= \frac{d_n}{2^{n-1}\rho^n}, \quad 1 \leqslant d_n < 2, \\ h_{-i} &= h_i \ (i = 1, 2, \ldots, n+k), \end{aligned} \tag{4.5}$$

and the numbers $k = k(n)$ and d_n are found from the conditions

$$1 \leqslant d_n < 2, \quad \frac{1}{2} h_0 + h_1 + \ldots + h_{n+k} = 1.$$

It is clear that for $\rho \geqslant 2$ and the above d_n the mesh $\overline{\Delta}_n$ satisfies (1.3). We note further that for $k = k(n)$

$$k > 2^{n-2}\rho^n. \tag{4.6}$$

We estimate the number σ_n of segments $[x_{i-1}, x_i]$ with constant length lying on the segment $[1/2^n, 1/2^{n-2}]$. It is clear that $\sigma_n = [l_1/h_n]$, where $[\alpha]$ is the integral part of the number α and

$$l_1 = \frac{1}{2^{n-2}} - \frac{1}{2^{n-1}}\left(\frac{1}{2} + \frac{1}{\rho} + \frac{1}{\rho^2} + \dots + \frac{1}{\rho^{n-1}}\right) = \frac{1}{2^n}\left(3 - \frac{2\left(1 - \frac{1}{\rho^n}\right)}{\rho - 1}\right).$$

$$\frac{l_1}{h_n} = \frac{\rho^n}{2}\left(3 - \frac{2}{\rho-1} + \frac{2}{\rho^{n-1}(\rho-1)}\right) = \rho^n\left(\frac{3}{2} - \frac{1}{\rho-1}\right) + \frac{\rho}{\rho-1}.$$

Hence

$$\sigma_n > \frac{1}{2}\rho^n. \tag{4.7}$$

We estimate the number ν_n of all intervals $[x_{i-1}, x_i]$ lying on the interval $[1/2^{n-2}, 1/2]$. We have $\nu_n = [l_2/h_n]$, where

$$l_2 = \frac{1}{2} - \frac{1}{2^{n-2}} = \frac{1}{2^{n-2}}(2^{n-3} - 1), \quad \frac{l_2}{h_n} = 2\rho^n(2^{n-3} - 1),$$

from which we obtain

$$\nu_n < \frac{1}{4}(2\rho)^n. \tag{4.8}$$

c) For the mesh $\overline{\Delta}_n$ constructed in (4.4) we rewrite the boundary conditions (1.1) with restrictions

$$\begin{gathered} t m_{-(n+k+1)} + \mu_n m_{-(n+k)} = c_{-(n+k+1)}, \\ \lambda_n m_{n+k-1} + t m_{n+k} = c_{n+k}, \\ \sup_n \max(|\mu_n|, |\lambda_n|) = \gamma < t^2 + \frac{t(t-1)}{\rho}, \\ \delta_n(|c_{-(n+k-1)}| + |c_{n+k}|) \to 0 \quad \text{for} \quad n \to \infty. \end{gathered} \tag{4.9}$$

We continue the function $f_{\rho,m}(x)$ to the whole line with period 2 and construct the periodic spline (cubic or parabolic) interpolating it on the mesh $\overline{\Delta}_n$. We prove that on the segment $[-1, 1]$ it coincides with the interpolating spline satisfying the condition (4.9) for $\mu_n = \lambda_n = 0$ and $c_{n+k} = c_{-(n+k+1)} = 0$.

Indeed, since $f_{\rho,m}(x)$ is an even function and the nodes of interpolation are symmetric with respect to the origin, $S_n(x)$ will also be an even function. For cubic splines this fact is proved in [3], Lemma 3 and for parabolic splines it is proved analogously. It is not difficult to see that $f_{\rho,m}(x)$ and $S_n(x)$ are even, relative to $x = 1$ and $x = -1$. It follows that $S_n'(-1) = S_n'(1) = 0$, and from the uniqueness of interpolating splines which satisfy (4.9) we have what we need.

d) LEMMA 2. *Suppose given a mesh $\overline{\Delta}_n$ (4.4) on the segment $[-1, 1]$ and a function $f_{\rho,m}(x)$, where m and H_n satisfy the requirements (4.1) and $S_{n,t}(x)$ is the cubic ($t = 2$) or parabolic ($t = 3$) spline satisfying the conditions (4.9) and interpolating $f_{\rho,m}$ at the knots of the mesh $\overline{\Delta}_n$, where $\rho \geqslant \rho_t^*$ ($\rho_2^* = (3 + \sqrt{5})/2$; $\rho_3^* = 2 + \sqrt{3}$). Then there exists a number T depending only on ρ such that for all $n > T$*

$$S_{n,t}(0) \geqslant \begin{cases} \beta n H_n, & \text{if} \quad \rho = \rho_t^*, \\ \beta\left(\frac{\rho^2}{\varepsilon_t}\right)^n H_n, & \text{if} \quad \rho > \rho_t^* \end{cases}$$

and

$$m_{-1} = S'_{n,t}(x_{-1}) > 0, \qquad m_0 = S'_{n,t}(x_0) < 0,$$

where β *is a positive constant depending only on* ρ *and*

$$\varepsilon_t = \frac{t(\rho+1) + \sqrt{t^2(\rho+1)^2 - 4\rho}}{2}.$$

PROOF. From (2.5) and (2.8) it is clear that

$$S_{n,t}(0) = H_n + \frac{1}{4}\frac{1}{2^n}(m_{-1} - m_0) \geqslant \frac{1}{2^{n+2}}(m_{-1} - m_0);$$

hence to estimate $S_{n,t}(0)$ it suffices to estimate m_{-1} and m_0.

We rewrite the system (2.4) for the mesh $\bar{\Delta}_n$ and function $f_{\rho,m}(x)$:

$$tm_{-(n+k+1)} + \mu_n m_{-(n+k)} = c_{-(n+k+1)},$$

$$h_{i+1}m_{i-1} + t(h_i + h_{i+1})m_i + h_i m_{i+1} = (t+1)\left\{\frac{h_i}{h_{i+1}}[f_{\rho,m}(x_{i+1}) - f_{\rho,m}(x_i)] + \frac{h_{i+1}}{h_i}[f_{\rho,m}(x_i) - f_{\rho,m}(x_{i-1})]\right\} = c_i$$

$$(i = 0, \pm 1, \pm 2, \ldots, \pm(n+k-1), -(n+k))$$

$$\lambda_n m_{n+k-1} + tm_{n+k} = c_{n+k}. \tag{4.10}$$

Substituting in (4.10) the values (4.5) of h_i, we obtain

$$\begin{aligned}
&\frac{1}{2^{n-1}}[tm_{-(n+k+1)} + \mu_n m_{-(n+k)}] = \frac{1}{2^{n-1}}c_{-(n+k+1)},\\
&\frac{1}{2^{n-1}}\left[\frac{1}{d_n}m_{-(n+k+1)} + t\left(1+\frac{1}{d_n}\right)m_{-(n+k)} + m_{-(n+k-1)}\right] = c_{-(n+k)}\rho^n,\\
&\frac{1}{2^{n-1}}[m_{i-1} + 2tm_i + m_{i+1}] = c_i\rho^n\ (i = -(n+k-1), \ldots, -(n+1)),\\
&\frac{1}{2^{n-1}}[\rho m_{i-1} + t(\rho+1)m_i + m_{i+1}] = c_i\rho^{-i}\ (i = -n, \ldots, -1),\\
&\frac{1}{2^{n-1}}[m_{i-1} + t(\rho+1)m_i + \rho m_{i+1}] = c_i\rho^{i+1}\ (i = 0, 1, \ldots, n-1),\\
&\frac{1}{2^{n-1}}[m_{i-1} + 2tm_i + m_{i+1}] = c_i\rho^n\ (i = n, n+1, \ldots, n+k-2),\\
&\frac{1}{2^{n-1}}\left[m_{n+k-2} + t\left(1+\frac{1}{d_n}\right)m_{n+k-1} + \frac{1}{d_n}m_{n+k}\right] = c_{n+k-1}\rho^n,\\
&\frac{1}{2^{n-1}}[\lambda_n m_{n+k-1} + tm_{n+k}] = \frac{1}{2^{n-1}}c_{n+k}.
\end{aligned} \tag{4.11}$$

We compute the right-hand sides of (4.11) and denote them by b_i:

$$
\begin{aligned}
&b_0 = -(t+1)\rho^2 H_n, \qquad b_{-1} = (t+1)\rho^2 H_n,\\
&b_i = (-1)^{i+1}(t+1)\rho^{i-1}\left(\rho - \frac{1}{\rho}\right) H_n \, (i = 1, 2, \ldots, n-1),\\
&b_{-i} = (-1)^{i+1}(t+1)\rho^{i}\left(\rho - \frac{1}{\rho}\right) H_n \, (i = 2, 3, \ldots, n),\\
&b_i = 0 \, (i = n, \ \pm(n+1), \ldots, \pm(n+\sigma_n - 2m - 1), \ -(n+\sigma_n - 2m)),\\
&b_i = 0 \, (i = n+\sigma_n+\nu_n+3, \pm(n+\sigma_n+\nu_n+4), \ldots,\\
&\qquad\qquad \pm(n+k-1), -(n+k)),
\end{aligned}
\tag{4.12}
$$

$$
\begin{aligned}
&b_i = (t+1)\rho^n [f_{\rho,m}(x_{i+1}) - f_{\rho,m}(x_{i-1})] (i = (n+\sigma_n - 2m),\\
&\qquad \pm(n+\sigma_n - 2m+1), \ldots, \pm(n+\sigma_n+\nu_n+2), \ -(n+\sigma_n+\nu_n+3)),\\
&\qquad b_{n+k} = \frac{1}{2^{n-1}} c_{n+k}, \qquad b_{-(n+k+1)} = \frac{1}{2^{n-1}} c_{-(n+k+1)},
\end{aligned}
$$

where σ_n and ν_n are defined in the construction of the mesh $\bar{\Delta}_n$.

We denote the basic determinant of the system of equations (4.11) by A, and the determinants corresponding to the unknowns m_{-1} and m_0 by B and C respectively. These determinants have order $2(n+k+1)$. By Cramer's formula we have $m_{-1} = B/A$ and $m_0 = C/A$. Before estimating the determinants A, B and C we estimate some auxiliary determinants.

We need a relation for determinants of the form

$$
P_{n+k} = \begin{vmatrix}
b & c & 0 & \cdots & & & 0\\
a & b & c & & & & \vdots\\
0 & a & b & & & & \vdots\\
& & & \ddots & & & \\
& & & a & b & c & \\
\vdots & & & & r & d & u\\
\vdots & & & & & \ddots & \\
\vdots & & & & r & d & u\\
0 & \cdots & & & 0 & r & d
\end{vmatrix}
\quad \left.\right\} n
$$

Expanding the determinant P_{n+k} by elements of the $(n+1)$th row, we obtain

$$
\begin{aligned}
P_{n+k} &= dQ_nR_{k-1} - rcQ_{n-1}R_{k-1} - urQ_nR_{k-2}\\
&= Q_n(dR_{k-1} - urR_{k-2}) - rcQ_{n-1}R_{k-1} = Q_nR_k - rcQ_{n-1}R_{k-1},
\end{aligned}
\tag{4.13}
$$

where

$$
Q_n = \begin{vmatrix}
b & c & 0 & \cdots & 0\\
a & b & c & & \vdots\\
0 & a & b & & \vdots\\
& \cdots & & & \\
\vdots & & a & b & c\\
0 & \cdots & 0 & a & b
\end{vmatrix}, \qquad
R_k = \begin{vmatrix}
d & u & 0 & \cdots & 0\\
r & d & u & & \vdots\\
0 & r & d & & \vdots\\
& \cdots & & & \\
\vdots & & r & d & u\\
0 & \cdots & 0 & r & d
\end{vmatrix}.
$$

For the determinant

$$F_j = F_j(\rho) = \begin{vmatrix} t(\rho+1) & \rho & 0 & \dots & 0 \\ 1 & t(\rho+1) & \rho & & \vdots \\ 0 & 1 & t(\rho+1) & & \vdots \\ \vdots & & \dots & & \\ \vdots & & 1 & t(\rho+1) & \rho \\ 0 & \dots & 0 & 1 & t(\rho+1) \end{vmatrix}$$

the recursion relation

$$F_j = t(\rho+1)F_{j-1} - \rho F_{j-2}, \quad j \geqslant 1, \quad F_{-1} = 0, \quad F_0 = 1,$$

holds. Solving this difference equation, we find that

$$F_j = \frac{1}{\sqrt{t^2(\rho+1)^2 - 4\rho}}\left[\left(\frac{t(\rho+1) + \sqrt{t^2(\rho+1)^2 - 4\rho}}{2}\right)^{j+1} - \left(\frac{t(\rho+1) - \sqrt{t^2(\rho+1)^2 - 4\rho}}{2}\right)^{j+1}\right]. \tag{4.14}$$

We set

$$E_j = F_j(1), \quad \varepsilon = \varepsilon_t(\rho) = \frac{t(\rho+1) + \sqrt{t^2(\rho+1)^2 - 4\rho}}{2},$$

$$\varepsilon_1 = \varepsilon_t(1) = t + \sqrt{t^2 - 1}.$$

Since for $\rho \geqslant 1$ and $t \geqslant 2$

$$\varepsilon \geqslant 2 + \sqrt{3}, \quad \frac{t(\rho+1) - \sqrt{t^2(\rho+1)^2 - 4\rho}}{2} = \frac{\rho}{\varepsilon} < \frac{1}{2},$$

on taking (4.14) into account we obtain

$$\frac{3}{4}\frac{\varepsilon^{j+1}}{\sqrt{t^2(\rho+1)^2 - 4\rho}} \leqslant F_j \leqslant \frac{\varepsilon^{j+1}}{\sqrt{t^2(\rho+1)^2 - 4\rho}}, \tag{4.15}$$

$$\frac{3}{8}\frac{\varepsilon_1^{j+1}}{\sqrt{t^2 - 1}} \leqslant E_j \leqslant \frac{1}{2}\frac{\varepsilon_1^{j+1}}{\sqrt{t^2 - 1}}. \tag{4.16}$$

For the determinants

$$M_j = \begin{vmatrix} t & g & 0 & \dots & 0 \\ d & t(1+d) & 1 & & \vdots \\ 0 & 1 & 2t & 1 & \vdots \\ \vdots & & \dots & & \\ \vdots & & 1 & 2t & 1 \\ 0 & \dots & 0 & 1 & 2t \end{vmatrix}, \quad K_j = \begin{vmatrix} 2t & 1 & 0 & & \dots & & 0 \\ 1 & 2t & 1 & & & & \vdots \\ 0 & 1 & 2t & 1 & & & \vdots \\ \vdots & & \dots & & & & \\ \vdots & & 1 & 2t & 1 & & \\ & & & 1 & t(1+d) & d & \\ 0 & \dots & & 0 & g & t & \end{vmatrix},$$

taking (4.13) into account we have

$$K_j = M_j = [t^2 + (t^2 - g)d]E_{j-2} - E_{j-3}. \tag{4.17}$$

Since for $\rho > 2$, $\frac{1}{2} \leqslant d \leqslant 1$ and $|g| < t^2 + t(t-1)/\rho$ there hold the conditions $M_1 <$

M_2 and $M_j = 2tM_{j-1} - M_{j-2} = (2t-1)M_{j-1} + (M_{j-1} - M_{j-2})$, we have the inequality

$$M_{j-1} \leqslant M_j \tag{4.18}$$

and obviously the estimates

$$M_j \leqslant 4t^2 E_{j-2} \leqslant \frac{2t^2}{\sqrt{t^2-1}} \cdot \varepsilon_1^{j-1}. \tag{4.19}$$

We estimate the determinants

$$\Delta_{k+1+i} = \begin{vmatrix} t & g & 0 & & \cdots & & & 0 \\ d & t(1+d) & 1 & & & & & \\ 0 & 1 & 2t & 1 & & & & \cdot \\ & & \cdots & & & & & \cdot \\ & & 1 & 2t & 1 & & & \cdot \\ \cdot & & & \rho & t(\rho+1) & 1 & & \\ \cdot & & & & \cdots & & & \\ \cdot & & & & \rho & t(\rho+1) & 1 & \\ 0 & \cdots & & & 0 & \rho & t(\rho+1) & \end{vmatrix} \Bigg\} k+1,$$

$$D_{k+1+i} = \begin{vmatrix} t(\rho+1) & \rho & 0 & 0 & & \cdots & 0 \\ 1 & t(\rho+1) & \rho & 0 & & & \cdot \\ 0 & 1 & t(\rho+1) & \rho & & & \cdot \\ & & \cdots & & & & \cdot \\ & & 1 & t(\rho+1) & \rho & & \\ & & & 1 & 2t & 1 & \\ \cdot & & & & 1 & 2t & 1 & 0 \\ \cdot & & & & & 1 & t(1+d) & d \\ 0 & \cdots & & & & 0 & g & t \end{vmatrix} \Bigg\} i,$$

where $i, k \geqslant 0$, and the restrictions on d, g and ρ are the same as for the determinant M_j. According to (4.13) we obtain

$$\Delta_{k+1+i} = F'_i M_{k+1} - \rho F'_{i-1} M_k,$$
$$D_{k+1+i} = F_i K_{k+1} - \rho F_{i-1} K_k,$$

where F'_i is the transpose of F_i; therefore $F'_i = F_i$. In view of (4.17) we have

$$\Delta_{k+1+i} = D_{k+1+i}, \tag{4.20}$$

and since $1/\epsilon_1 < 1/3$ and $\rho/\epsilon < 1/2$, from (4.15) and (4.17) we deduce

$$\Delta_{k+1+i} \geqslant \frac{\varepsilon^{i+1}}{\sqrt{t^2(\rho+1)^2 - 4\rho}} \left[\frac{3}{4} M_{k+1} - \frac{\rho}{\varepsilon} M_k\right] = \frac{\varepsilon^{i+1}}{\sqrt{t^2(\rho+1)^2 - 4\rho}}$$
$$\times \left[\frac{3}{4}(t^2 + t^2 d - gd) E_{k-1} - \frac{3}{4} E_{k-2} - \frac{\rho}{\varepsilon}(t^2 + t^2 d - gd) E_{k-2} + \frac{\rho}{\varepsilon} E_{k-3}\right]$$
$$\geqslant \frac{\varepsilon^{i+1}\varepsilon_1^k}{\sqrt{t^2(\rho+1)^2 - 4\rho}\,\sqrt{t^2-1}} \left[\frac{3}{8}\,\frac{3}{4}(t^2 + t^2 d - gd) - \frac{3}{8\varepsilon_1} - \frac{\rho}{2\varepsilon\varepsilon_1}(t^2 + t^2 d - gd)\right]$$
$$\geqslant \frac{\varepsilon^{i+1}\varepsilon_1^k}{\sqrt{t^2(\rho+1)^2 - 4\rho}\,\sqrt{t^2-1}} \left[\frac{19}{96}(t^2 + t^2 d - gd) - \frac{1}{8}\right] \geqslant \beta_1 \varepsilon_1^k \varepsilon^i, \tag{4.21}$$

where β_1 is a positive constant depending only on ρ. The estimate

$$\Delta_{k+1+i} \leqslant F_i M_{k+1} \leqslant \alpha_1 \varepsilon_1^k \varepsilon^i \tag{4.22}$$

is also obvious, where α_1 is a positive constant depending only on ρ.

We return now to the estimate of the determinants A, B and C of the system (4.11). Having set $g = \mu_n$ for the determinants M and Δ and $g = \lambda_n$, $d = 1/d_n$ for the determinants K and D, we estimate the determinant $A = 1/(2^{2(n+k)(n-1)})A_1$, where

$$A_1 = \left|\begin{array}{cccccccccccccc}
t & \mu_n & 0 & & & \cdots & & & & & & & 0 \\
\frac{1}{d_n} & t\left(1+\frac{1}{d_n}\right) & 1 & & & & & & & & & & \vdots \\
0 & 1 & 2t & 1 & & & & & & & & & \vdots \\
 & & \cdots & & & & & & & & & & \\
 & & 1 & 2t & 1 & & & & & & & & \\
\vdots & & & \rho & t(\rho+1) & 1 & & & & & & & \\
 & & & & \cdots & & & & & & & & \\
 & & & & \rho & t(\rho+1) & 1 & & & & & & \\
 & & & & & 1 & t(\rho+1) & \rho & & & & & \\
 & & & & & & \cdots & & & & & & \\
 & & & & & & 1 & t(\rho+1) & \rho & & & & \\
 & & & & & & & 1 & 2t & 1 & & & \\
 & & & & & & & & \cdots & & & & \\
 & & & & & & & & 1 & 2t & 1 & 0 & \\
 & & & & & & & & & & 1 & t\left(1+\frac{1}{d_n}\right) & \frac{1}{d_n} \\
0 & \cdots & & & & & & & & & 0 & \lambda_n & t
\end{array}\right| \begin{array}{l} \left.\right\} k+1 \\ \left.\right\} n \\ \left.\right\} n \end{array}.$$

Expanding this determinant by the elements of the $(k+n+1)$th row analogous to (4.13) we obtain

$$0 < A < \Delta_{k+1} D_{k+1+n} \cdot \frac{1}{2^{2(n+k)(n-1)}} \leqslant \alpha \frac{\varepsilon^{2n}\varepsilon_1^{2k}}{2^{2(n+k)(n-1)}}, \tag{4.23}$$

where α is a positive constant depending only on ρ.

We estimate the determinant $C = 1/2^{(n-1)(2n+2k-1)}C_1$, where

$$C_1 = \left|\begin{array}{ccccccccccccc}
t & \mu_n & 0 & \cdots & & b_{-(n+k+1)} & & \cdots & & & & & 0 \\
\frac{1}{d_n} & t\left(1+\frac{1}{d_n}\right) & 1 & & & b_{-(n+k)} & & & & & & & \\
0 & 1 & 2t & 1 & & \vdots & & & & & & & \vdots \\
 & & \cdots & & & & & & & & & & \\
 & & 1 & 2t & 1 & & & & & & & & \\
 & & & \rho & t(\rho+1) & 1 & & & & & & & \\
\vdots & & & \rho & t(\rho+1) & b_{-1} & & & & & & & \\
 & & & & 1 & b_0 & \rho & & & & & & \\
 & & & & & b_1 & t(\rho+1) & \rho & & & & & \\
 & & & & & & \cdots & & & & & & \\
 & & & & & & 1 & t(\rho+1) & \rho & & & & \\
 & & & & & & & 1 & 2t & 1 & & & 0 \\
 & & & & & b_{n+k-1} & & & & 1 & t\left(1+\frac{1}{d_n}\right) & & \frac{1}{d_n} \\
0 & \cdots & & & & b_{n+k} & & & & 0 & & \lambda_n & t
\end{array}\right|$$

and differs from A_1 in the $(n+k+2)$th column.

Expanding it by the elements b_i, we obtain

$$C = -2^{-(n-1)(2n+2k-1)}(L_1 + L_2 + L_3 + L_4), \tag{4.24}$$

where

$$L_1 = -\Delta_{n+k-1} \sum_{i=0}^{n-1} (-1)^i b_i D_{n+k-i} \rho^i,$$

$$L_2 = D_{n+k} \sum_{i=0}^{n-1} (-1)^i b_{-i-1} \Delta_{n+k-i} \rho^i,$$

$$L_3 = (-1)^{n+\sigma_n-1} \rho^n \Delta_{n+k+1} \sum_{i=0}^{\nu_n+4} (-1)^i b_{n+\sigma_n-2m+i} K_{k-\sigma_n+2m-i} + (-1)^{n+k+1} b_{n+k} \frac{1}{d_n} \rho^n \Delta_{n+k+1},$$

$$L_4 = (-1)^{n+\sigma_n} \rho^n D_{n+k} \sum_{i=0}^{\nu_n+4} (-1)^i b_{-(n+\sigma_n-2m+i+1)} M_{k-\sigma_n+2m+1-i} + (-1)^{n+k+2} b_{-(n+k+1)} \frac{1}{d_n} \rho^n D_{n+k}.$$

Since $b_{n+k} = c_{n+k}/2^{n-1} = \delta_n c_{n+k} \to 0$ as $n \to \infty$ (4.9), in view of (4.3), (4.7), (4.8), (4.19) and (4.22) we have for large n

$$|L_3| \leqslant (t+1)\rho^n \max_i |f_{\rho,m}(x_{i+1}) - f_{\rho,m}(x_{i-1})| (\nu_n + 2m + 2) K_{k-\sigma_n+2m} \rho^n \Delta_{n+k+1}$$
$$+ |b_{n+k}| \rho^n \Delta_{n+k+1} \leqslant 4(t+1) H_n \nu_n \rho^{2n} K_{k-\sigma_n+2m} \Delta_{n+k+1} \leqslant \alpha_2 \varepsilon^n \varepsilon_1^{2k} H_n \frac{(2\rho)^n \rho^{2n}}{\varepsilon_1^{\frac{1}{2}\rho^n}},$$

where α_2 is a positive constant depending only on ρ. Estimating L_4 analogously, we finally obtain

$$|L_i| \leqslant \alpha_2 \varepsilon^n \varepsilon_1^{2k} H_n \frac{(2\rho)^n \rho^{2n}}{\varepsilon_1^{\frac{1}{2}\rho^n}} \qquad (i = 3, 4), \tag{4.25}$$

where α_2 is a positive constant depending only on ρ. Substituting in L_1 the values for b_i in (4.12) and taking (4.21) into account, we obtain

$$L_1 = (t+1)(\rho^2 - 1) H_n \left[\frac{\rho^2}{\rho^2 - 1} D_{n+k} + \rho^2 D_{n+k-1} + \ldots + \rho^{2(n-1)} D_{k+1}\right] \Delta_{n+k+1}$$
$$\geqslant 3(\rho^2 - 1) H_n \beta_1 \varepsilon_1^k \varepsilon [\varepsilon^{n-1} + \rho^2 \varepsilon^{n-2} + \ldots + \rho^{2(n-1)}] \beta_1 \varepsilon^{n+1} \varepsilon_1^k.$$

For L_2 an analogous estimate holds. Finally, we have

$$L_i \geqslant \beta_2 \varepsilon^n \varepsilon_1^{2k} H_n [\varepsilon^{n-1} + \rho^2 \varepsilon^{n-2} + \ldots + \rho^{2(n-1)}] \qquad (i = 1, 2), \tag{4.26}$$

where β_2 is a positive constant depending only on ρ.

Since in (4.25)

$$\frac{(2\rho)^n \rho^{2n}}{\varepsilon_1^{\frac{1}{2}\rho^n}} \to 0 \qquad \text{for} \qquad n \to \infty,$$

on considering (4.24) we find the estimate

$$C \leqslant -\frac{1}{2^{(n-1)(2n+2k-1)}}(L_1 + L_2 - |L_3| - |L_4|)$$

$$\leqslant -\frac{2\beta_2 H_n \varepsilon^n \varepsilon_1^{2k}}{2^{(n-1)(2n+2k-1)}}[\varepsilon^{n-1} + \rho^2\varepsilon^{n-2} + \ldots + \rho^{2(n-1)}] + \frac{2\alpha_2 H_n \varepsilon^n \varepsilon_1^{2k}}{2^{(n-1)(2n+2k-1)}} \frac{(2\rho)^n \rho^{2n}}{\frac{1}{\varepsilon_1^2}\rho^n} \tag{4.27}$$

$$\leqslant -\frac{\beta_2 H_n \varepsilon^n \varepsilon_1^{2k}}{2^{(n-1)(2n+2k-1)}}[\varepsilon^{n-1} + \rho^2\varepsilon^{n-2} + \ldots + \rho^{2(n-1)}]$$

for sufficiently large n.

Thus the estimate

$$m_0 = \frac{C}{A} \leqslant -\frac{2^{n-1}\beta_2 H_n}{\alpha\varepsilon^n}[\varepsilon^{n-1} + \rho^2\varepsilon^{n-2} + \ldots + \rho^{2(n-1)}].$$

holds for m_0. Using the representation $m_{-1} = B/A$ and estimating B from below and A from above, we finally obtain

$$\begin{aligned} m_0 &\leqslant -\beta_3 2^{n-1} H_n \frac{\varepsilon^{n-1} + \rho^2\varepsilon^{n-2} + \ldots + \rho^{2(n-1)}}{\varepsilon^n}, \\ m_{-1} &\geqslant \beta_3 2^{n-1} H_n \frac{\varepsilon^{n-1} + \rho^2\varepsilon^{n-2} + \ldots + \rho^{2(n-1)}}{\varepsilon^n}, \end{aligned} \tag{4.28}$$

where β_3 is a positive constant depending only on ρ.

We note that the number ρ_t^* satisfies the condition

$$2\rho^2 = t(\rho + 1) + \sqrt{t^2(\rho + 1)^2 - 4\rho} = 2\varepsilon,$$

and $\rho^2 > \epsilon$ for $\rho > \rho_t^*$, while $\rho^2 < \epsilon$ for $1 \leqslant \rho < \rho_t^*$.

We note the inequality

$$\varepsilon^{n-1} + \rho^2\varepsilon^{n-2} + \ldots + \rho^{2(n-1)} = \begin{cases} n\varepsilon^{n-1}, & \text{if} \quad \rho^2 = \varepsilon, \\ \dfrac{\rho^{2n} - \varepsilon^n}{\rho^2 - \varepsilon}, & \text{if} \quad \rho^2 \neq \varepsilon. \end{cases} \tag{4.29}$$

If $\rho^2 > \epsilon$, then from (4.28) we have

$$\begin{aligned} m_0 &\leqslant -\beta_3 2^{n-1} H_n \frac{1}{\rho^2 - \varepsilon}\left[\left(\frac{\rho^2}{\varepsilon}\right)^n - 1\right] \leqslant -\beta_4 2^{n-1} H_n \left(\frac{\rho^2}{\varepsilon}\right)^n, \\ m_{-1} &\geqslant \beta_4 2^{n-1} H_n \left(\frac{\rho^2}{\varepsilon}\right)^n \end{aligned} \tag{4.30}$$

and consequently, in view of the estimate for $S_{n,t}(0)$ adduced at the beginning of the proof of the lemma,

$$S_{n,t}(0) \geqslant \frac{1}{2^{n+2}}(|m_{-1}| + |m_0|) \geqslant \beta H_n \left(\frac{\rho^2}{\varepsilon}\right)^n. \tag{4.31}$$

If $\rho^2 = \epsilon$, then

$$m_0 \leqslant -\beta_4 2^{n-1} H_n n, \qquad m_{-1} \geqslant \beta_4 2^{n-1} H_n n \tag{4.32}$$

and

$$S_{n,t}(0) \geqslant \beta n H_n, \tag{4.33}$$

where β is a positive constant depending only on ρ. Lemma 2 is proved.

§5. A convergence criterion for the interpolation process with cubic and parabolic splines

THEOREM 2. *Suppose given a sequence of meshes on the interval* $[a, b]$ *and suppose* $S_n(x)$ *is the cubic (parabolic) spline interpolating a continuous function* $f(x)$ *at the knots of the mesh* Δ_n; $S_n(x)$ *satisfies the boundary conditions* (1.1) *with restrictions* (1.5) *and* (1.6), *or it is* $(b-a)$*-periodic together with* $f(x)$. *In order that the interpolation process converge in* $C[a, b]$ *for any sequence of meshes* Δ_n *satisfying condition* (1.3), *it is necessary and sufficient that* $\rho < (3+\sqrt{5})/2 = \rho^*$ $(\rho < 2+\sqrt{3} = \rho^*)$. *If convergence does obtain, it is uniform.*

Necessity. Without loss of generality it is possible to take the segment $[-1, 1]$ in place of $[a, b]$. We construct the sequence of meshes $\overline{\Delta}_n$ (4.4) on $[-1, 1]$ and take $f_{\rho,m}(x)$ in the capacity of $f(x)$.

We suppose we are given a number $\rho \geqslant \rho^*$, and for $f_{\rho,m}$ we assume $m = 2$ and $H_n = 1/\sqrt{n}$. It is easy to see that the conditions (4.1) and (4.2), for this choice of m and H_n, are fulfilled for $n \geqslant 4$. Then from Lemma 2 for cubic and parabolic splines we have

$$S_n(0) - f_{\rho,2}(0) = S_n(0) \geqslant \begin{cases} \beta\sqrt{n}, & \text{if} \quad \rho = \rho^*, \\ \beta\left(\dfrac{\rho^2}{\varepsilon}\right)^n \dfrac{1}{\sqrt{n}}, & \text{if} \quad \rho > \rho^*. \end{cases}$$

Since $\rho^2 > \epsilon$ for $p > \rho^*$, we obviously have $S_n(0) \longrightarrow \infty$ as $n \longrightarrow \infty$.

Sufficiency. In the hypotheses of Theorem 1 put $s = 1$ and

$$\gamma_1 = \frac{1}{2}\left(\gamma + t^2 + \frac{t(t-1)}{\rho}\right)$$

It is not difficult to verify that all the hypotheses of Theorem 1 are fulfilled, and consequently the interpolation process converges uniformly.

REMARK 1. It is possible to consider the more general problem where the restrictions (1.3) are replaced by the conditions

$$\lim_{n\to\infty} \delta_n = 0, \qquad \max_{|i-j|=1} \frac{h_i^{(n)}}{h_j^{(n)}} = \rho_n < \rho, \qquad 1 < \rho < \infty. \tag{5.1}$$

In particular, if $\rho = \rho_t^*$ and $\rho_n^2/\epsilon_n = 1 - 1/n$, where

$$\varepsilon_n = \frac{t(\rho_n+1) + \sqrt{t^2(\rho_n+1)^2 - 4\rho_n}}{2},$$

then Lemma 2 does not give an estimate for $S_n(0)$. In order to obtain an analogous estimate it is necessary to make a change in the construction of $f_{\rho,m}$ and $\overline{\Delta}_n$.

In the construction of $f_{\rho,m}(x)$ on the segment $[1/2^r, 1/2^{r-2}]$ we take ρ_r $(r = 4, 6, 8, \ldots)$ in place of ρ, and in the construction of $\overline{\Delta}_n$ we take ρ_n $(n = 4, 6, 8, \ldots)$. Setting $m = 2$ and $H_n = 1/\sqrt{n}$, it is obvious that conditions (4.1) and (4.2) are fulfilled.

Since in this case we have

$$\varepsilon_n^{n-1} + \rho_n^2\varepsilon_n^{n-1} + \ldots + \rho_n^{2(n-1)} = \frac{\varepsilon_n^n - \rho_n^{2n}}{\varepsilon_n - \rho_n^2} = \frac{\varepsilon_n^n\left(1 - \left(\frac{\rho_n^2}{\varepsilon_n}\right)^n\right)}{\varepsilon_n\left(1 - \frac{\rho_n^2}{\varepsilon_n}\right)}$$

$$= n\varepsilon_n^{n-1}\left(1 - \left(\frac{\rho_n^2}{\varepsilon_n}\right)^n\right),$$

in place of (4.29) and taking into account that $(\rho_n^2/\epsilon_n)^n = (1 - 1/n)^n \longrightarrow e^{-1} - 0$ as $n \longrightarrow \infty$, we obtain, analogously to (4.30), that

$$m_0 \leqslant -\beta_2 2^{n-1} H_n \frac{1}{\rho_n^2 - \varepsilon_n}\left[\left(\frac{\rho_n^2}{\varepsilon_n}\right)^n - 1\right] \leqslant -\beta_4 H_n 2^{n-1} n, \qquad m_{-1} \geqslant \beta_4 H_n 2^{n-1} n,$$

where β_4 is a positive constant depending only on ρ. Then we have

$$S_n(0) \geqslant \beta n H_n, \tag{5.2}$$

and since

$$S_n(0) - f_{\rho,m}(0) = S_n(0) \geqslant \beta n \frac{1}{\sqrt{n}} \to \infty \quad \text{for} \quad n \to \infty,$$

we obtain the divergence of the process.

§6. Divergence of the interpolation process for functions from Lip α

THEOREM 3. *For any* α $(0 < \alpha < 1)$ *there exists a number* ρ_α *such that for any* $\rho \geqslant \rho_\alpha$ *there is on* $[-1, 1]$ *a sequence of meshes* Δ_n *satisfying condition* (1.3) *and a function* $f(x) \in$ Lip α *such that* $S_n(0) - f(0) \not\to 0$ *as* $n \longrightarrow \infty$, *where* $S_n(x)$ *is the cubic or parabolic spline interpolating* $f(x)$ *at the knots of the mesh* Δ_n *and satisfying the boundary conditions* (1.1) *with restrictions* (1.5) *and* (1.6) *or is 2-periodic together with* $f(x)$.

PROOF. In Lemma 2 we choose the numbers m, H_n and ρ so that $f_{\rho,m}$ will lie in the class Lip α $(0 < \alpha < 1)$.

Set

$$H_n = \left(\frac{\varepsilon}{\rho^2}\right)^n (n = 6, 8, \ldots), \qquad m > \frac{1}{2}\left(\frac{\rho^2}{\varepsilon}\right)^2, \tag{6.1}$$

where $\epsilon = (t(\rho + 1) + \sqrt{t^2(\rho + 1)^2 - 4\rho})/2$ and t equals 2 or 3.

We show that it is possible to choose ρ_α such that for all $\rho \geqslant \rho_\alpha$ the function $f_{\rho,m}(x) \in$ Lip α, that is,

$$|f_{\rho,m}(x') - f_{\rho,m}(x'')| \leqslant |x' - x''|^\alpha \qquad \text{for all} \qquad x', x'' \in [-1, 1]. \tag{6.2}$$

Since $f_{\rho,m}(x)$ is piecewise linear, we see in view of (4.1) and (6.1) that condition (6.2) is equivalent to the following:

$$\left(\frac{\varepsilon}{\rho^2}\right)^n = H_n \leqslant h_n^\alpha = \left(\frac{1}{2^{n-1}\rho^n}\right)^\alpha (n = R, R+2, \ldots), \tag{6.3}$$

where R is defined in the construction of $f_{\rho,m}(x)$. We write this condition in the form

$$\left(\frac{\varepsilon}{\rho^{2-\alpha}}\right)^{n/\alpha} \leqslant \frac{1}{2^{n-1}}.$$

From the explicit form of ϵ it is obvious that $\epsilon/\rho^{2-\alpha} \longrightarrow 0$ as $\rho \longrightarrow \infty$. Hence it is possible to find ρ_α such that for any $\rho \geqslant \rho_\alpha$

$$\left(\frac{\varepsilon}{\rho^{2-\alpha}}\right)^{1/\alpha} < \frac{1}{3},$$

and consequently condition (6.3) will be fulfilled for such ρ. Then from Lemma 2 we obtain the divergence of the process for cubic and parabolic splines. The theorem is proved.

We note that if $\alpha = 1$ and $\gamma < t$ in (1.5) and the sequence of meshes Δ_n satisfies the condition $\delta_n \longrightarrow 0$ as $n \longrightarrow \infty$, then the interpolation process converges. Indeed, writing the system (2.4) in brief matrix form

$$Qm = q,$$

we obviously have

$$\|Q^{-1}\| \leqslant \max\left\{(t-1), \frac{1}{t-\gamma}\right\} = B_0, \qquad \|q\| \leqslant \max\{|c_0|, |c_N|, (t+1)\} = B_n$$

and consequently

$$\|m\| \leqslant \|Q^{-1}\| \cdot \|q\| \leqslant B_0 B_n.$$

Since $\delta_n B_n \longrightarrow 0$ as $n \longrightarrow \infty$ (1.6), in view of (3.9) we obtain uniform convergence. For $\gamma = 0$ this result was obtained by Hall [7].

§7. Divergence of the process in L_p

Let $f(x)$ be a function on the segment $[-1, 1]$ with $\|f\|_p < \infty$, where

$$\|f\|_p = \left(\int_{-1}^{1} |f(x)|^p\, dx\right)^{1/p}, \quad p \geqslant 1.$$

THEOREM 4. *For any $p \geqslant 1$ there exists a number ρ_p such that for any $\rho \geqslant \rho_p$ there is a sequence of meshes satisfying condition* (1.3) *and a continuous function $f(x)$ such that*

$$\|f - S_n\|_p \not\to 0, \qquad n \to \infty,$$

where $S_n(x)$ is the cubic or parabolic spline interpolating $f(x)$ at the knots of mesh Δ_n and satisfying the boundary conditions (1.1) *with restrictions* (1.5) *and* (1.6) *or $S_n(x)$ is 2-periodic together with $f(x)$.*

PROOF. As in the previous theorem, we take the mesh $\overline{\Delta}_n$ (4.4) and the function $f_{\rho,m}(x)$. We set $m = 2$ and $H_n = 1/\sqrt{n}$. Then obviously conditions (4.1) and (4.2) are satisfied.

We first consider cubic splines. From (2.5) it is obvious that $S_n(x) \geqslant H_n$ on the segment $[-1/2^n, 1/2^n]$. If $|m_{-1}| \leqslant |m_0|$, then on the segment $[0, 1/2^n]$

$$S_n''(x) = -2m_{-1}\frac{x_0 - 3x}{h_0^2} - 2m_0\frac{-x_0 - 3x}{h_0^2}$$

$$\leqslant -2|m_{-1}|\frac{x_0 - 3x}{h_0^2} - 2|m_{-1}|\frac{x_0 + 3x}{h_0^2} \leqslant -\frac{4|m_{-1}|x_0}{h_0^2} < 0.$$

It follows that on $[0, 1/2^{n+1}]$ we have

$$S_n(x) - f_{\rho,2}(x) \geqslant S_n(x) - H_n \geqslant \frac{1}{2}(S_n(0) - H_n)$$

and consequently

$$\int_{-1}^{1} |S_n(x) - f_{\rho,2}(x)|^p\, dx \geqslant \int_0^{1/2^{n+1}} \left[\frac{1}{2}(S_n(0) - H_n)^p\right] dx \geqslant \left(\frac{1}{3} S_n(0)\right)^p \frac{1}{2^{n+1}}$$

or, in view of Lemma 2,

$$\|f_{\rho,2} - S_n\|_p \geqslant \frac{1}{3} S_n(0)\left(\frac{1}{2^{n+1}}\right)^{1/p} \geqslant \frac{1}{3\cdot 2^{1/p}}\beta\frac{1}{\sqrt{n}}\left(\frac{\rho^2}{\varepsilon 2^{1/p}}\right)^n,$$

and, since $\rho^2\epsilon^{-1} \longrightarrow \infty$ as $\rho \longrightarrow \infty$, there obviously exists a number ρ_p such that for all $\rho \geqslant \rho_p$ the inequality $\rho^2 > \epsilon 2^{1/\rho}$ will be fulfilled. The divergence of the process in L_p follows from this. In the case $|m_{-1}| \geqslant |m_0|$ we obtain an analogous estimate on the segment $[-1/2^n, 0]$.

In the parabolic case the proof is analogous, since we obtain from (2.8) that $S_n(x) \geqslant H_n$ on $[-1/2^n, 1/2^n]$ and since $S_n''(x) < 0$ either on $[0, 1/2^n]$ or on $[-1/2^n, 0]$ (2.9), since $S_n''(+0) + S_n''(-0) < 0$.

Added in proof. After this paper was sent to the printer, the article of Marsden [8] appeared in which similar questions are investigated for cubic splines.

BIBLIOGRAPHY

1. J. H. Ahlberg, E. N. Nilson and J. L. Walsh, *The theory of splines and their application*, Academic Press, New York, 1967. MR 39 #684.

2. I. J. Schoenberg, *On interpolation by spline functions and its minimal properties*, On Approximation Theory (Proc. Conf. Oberwolfach, 1963), Birkhäuser, Basel, 1964, pp. 109–129. MR 31 #5015.

3. Stig Nord, *Approximation properties of the spline fit*, Nordisk Tidskr. Informations–Behandling (BIT) 7 (1967), 132–144. MR 36 #1887.

4. Amram Meir and Ambikeskwar Sharma, *On uniform approximation by cubic splines*, J. Approximation Theory 2 (1969), 270–274. MR 40 #3137.

5. E. W. Cheney and F. Schurer, *Convergence of cubic spline interpolants*, J. Approximation Theory 3 (1970), 114–116. MR 40 #7680.

6. Ju. S. Zav'jalov, *Interpolation by cubic splines*, Vyčisl. Sistemy Vyp. 38 (1970), 23–73. (Russian) MR 44 #5662.

7. C. A. Hall, *Uniform convergence of cubic spline interpolants*, J. Approximation Theory 7 (1973), 71–75. MR 50 #837.

8. Martin Marsden, *Cubic spline interpolation of continuous functions*, J. Approximation Theory 10 (1974), 103–111. MR 50 #2760.

Translated by LE BARON O. FERGUSON

Trudy Mat. Inst. Steklov.
138 (1975)

Proc. Steklov. Inst. Math.
138 (1975)

KOLMOGOROV ESTIMATES FOR DERIVATIVES IN $L_2[0, \infty)$

UDC 517.5

N. P. KUPCOV

ABSTRACT. Let k and n be natural numbers such that $1 \leqslant k < n$, and let $f(x)$ be a function defined on the half-line $[0, \infty)$ having locally absolutely continuous $(n-1)$th derivative $f^{(n-1)}$. Set

$$\|f\| = \left\{\int_0^\infty |f(x)|^2 \, dx\right\}^{1/2}.$$

In this paper, the following inequality is established:

$$\|f^{(k)}\| \leqslant M_{n,k} \|f\|^{(n-k)/n} \|f^{(n)}\|^{k/n}$$

with the best possible constant $M_{n,k}$. The asymptotic behavior of $M_{n,k}$ as $n \to \infty$ is analyzed.

Bibliography: 9 titles.

Introduction

This paper studies inequalities between the norms of a function and its derivatives, a problem first considered by Landau and Hadamard ([1], Russian p. 388).

Let I denote the real line or half-line. Then for each twice differentiable function $f(x)$, $x \in I$, the Landau-Hadamard inequality holds:

$$\|f'\|_I^2 \leqslant 4\|f\|_I \|f''\|_I, \quad \|f\|_I = \sup_{x \in I} |f(x)|.$$

In the case when I is the half-line, there is a function for which equality holds (the inequality is sharp).

Subsequently many authors analyzed similar inequalities. The first really complete result was obtained by Kolmogorov [2]; he established that the inequality

$$\|f^{(k)}\|_I \leqslant C_{n,k} \|f\|_I^{\frac{n-k}{n}} \|f^{(n)}\|_I^{\frac{k}{n}}$$

is sharp for arbitrary natural numbers k and n, where $k < n$, on the class of n-times differentiable functions defined on the real line. Other exact inequalities for derivatives of arbitrary order of functions defined on the real line are given in [3]–[5]. Interest in similar inequalities increased because of their close connection with problems of best approximation of the differentiation operator by bounded operators [6]–[8], as well as the best approximation of one class of functions by another [8], [5].

We shall consider functions which for a given natural number n have a locally absolutely continuous derivative $f^{(n-1)}(x)$ and a finite norm

AMS (MOS) subject classifications (1970). Primary 46E30, 26A84; Secondary 47E05, 34B05, 15A42.

$$\|f\|=\left\{\int_0^\infty |f(x)|^2\,dx\right\}^{1/2}.$$

The sharp inequality of Hardy-Littlewood ([1], §7.8) is well known:

$$\|f'\|\leqslant\sqrt{2}\,\|f\|^{1/2}\|f''\|^{1/2},$$

In the proof of this inequality using the calculus of variations, a difficulty arises in the computation of the influence of the end of the half-line. This circumstance was confirmed in a paper of Schoenberg [9] which analyzed a sharp inequality of Kolmogorov type on the half-line and was able to establish only a process for constructing an extremal function. Finally, let us mention the sharp inequality of Gabušin [9] on the half-line:

$$\|f^{(k)}\|_I\leqslant Q_{n,k}\|f\|^{1-\frac{2k+1}{2n}}\|f^{(n)}\|^{\frac{2k+1}{2n}}\quad(0\leqslant k<n),$$

where the constant $Q_{n,k}$ is expressed by means of the solution of a certain linear system of n equations; however, its behavior as k and n vary was not obtained.

Here we shall analyze the sharp inequality

$$\|f^{(k)}\|\leqslant M_{n,k}\|f\|^{\frac{n-k}{n}}\|f^{(n)}\|^{\frac{k}{n}},$$

which is equivalent to

$$\|f^{(k)}\|^2\leqslant\frac{1}{\gamma_{n,k}}\{\|f\|^2+\|f^{(n)}\|^2\},$$

where

$$M_{n,k}^2=\frac{1}{\gamma_{n,k}}\left\{\left(\frac{n-k}{n}\right)^{\frac{k}{n}}+\left(\frac{k}{n-k}\right)^{\frac{n-k}{n}}\right\}.$$

The order of presentation is as follows. After some auxiliary propositions (§1), the basic theorem is proved in §2; it establishes a connection between the constants $\gamma_{n,k}$ and the least eigenvalues of certain auxiliary matrices by taking into account the influence of the end of the half-line. This will permit us to calculate the quantity $\gamma_{n,k}$ directly for small values of the parameter n. Then the asymptotic behavior of the quantities $\gamma_{n,k}$ with k fixed and $n\longrightarrow\infty$ is analyzed.

§1. Auxiliary propositions

A two-sided infinite sequence of real numbers $\{\omega_j\}_{-\infty}^{\infty}$ is called a *Hurwitz sequence of order n* if $\omega_0=\omega_n=1$, $\omega_j=0$ for $j\neq 0,1,\ldots,n$, and all roots of the polynomial

$$p(\lambda)=\omega_0+\omega_1\lambda+\cdots+\omega_n\lambda^n=\sum_{j=-\infty}^{\infty}\lambda^j\omega_j \tag{1}$$

have negative real part.

LEMMA 1. *Suppose that n and k are given natural numbers* ($n=2,\ldots,\infty$; $k=1,\ldots,n-1$). *For each real number*

$$\gamma \in G_{nk} = \left(0, \frac{n}{k^{\frac{k}{n}} (n-k)^{\frac{n-k}{n}}}\right)$$

there exists one and only one Hurwitz sequence $\{\omega_j\}_{-\infty}^{\infty}$ *of order n having the property*

$$\sum_{\nu=-\infty}^{\infty} (-1)^{\nu} \omega_{j-\nu}\omega_{j+\nu} = \begin{cases} 0 & \text{for} \quad j \neq 0, k, n, \\ 1 & \text{for} \quad j = 0, n, \\ -\gamma & \text{for} \quad j = k. \end{cases}$$

PROOF. Consider the polynomial $q(\lambda) = (-1)^n \lambda^{2n} - \gamma(-1)^k \lambda^{2k} + 1$. Calculating the minimum of $q(ix)$ for $x \in (-\infty, \infty)$ we obtain

$$q(ix) = x^{2n} - \gamma x^{2k} + 1 \geqslant 1 - \gamma^{\frac{n}{n-k}} \frac{k^{\frac{k}{n-k}} (n-k)}{n^{\frac{n}{n-k}}}.$$

Consequently for $\gamma \in G_{nk}$ the polynomial $q(\lambda)$ does not have pure imaginary roots. Denote the roots of $q(\lambda)$ by $-z_1, \ldots, -z_n, z_1, \ldots, z_n$ (Real $z_j > 0$ for $j = 1, \ldots, n$), and set

$$p(\lambda) = (\lambda + z_1)(\lambda + z_2) \ldots (\lambda + z_n) = \lambda^n + \omega_{n-1}\lambda^{n-1} + \ldots + \omega_1 \lambda + \omega_0. \tag{2}$$

Then

$$q(\lambda) = p(\lambda) p(-\lambda) = (-1)^n \lambda^{2n} - \gamma(-1)^k \lambda^{2k} + 1. \tag{3}$$

It follows from this that $\omega_0^2 = 1$. Since $\omega_0 = z_1 \cdots z_n$ is positive, we obtain that $\omega_0 = 1$.

To the set of numbers $1, \omega_1, \omega_2, \ldots, \omega_{n-1}, 1$, add zeros out to infinity on both sides to obtain a sequence $\{\omega_j\}_{-\infty}^{\infty}$. From the construction it follows that $\{\omega_j\}_{-\infty}^{\infty}$ is a Hurwitz sequence of order n.

Equation (3) yields

$$(-1)^n \lambda^{2n} - \gamma(-1)^k \lambda^{2k} + 1 = \sum_{\nu=-\infty}^{\infty} \left(\sum_{j+s=\nu} \omega_j \omega_s (-1)^s \right) \lambda^{\nu}.$$

Equating coefficients, we obtain

$$\sum_{j+s=\nu} (-1)^s \omega_j \omega_s = \begin{cases} 0 & \text{for} \quad \nu \neq 0, 2k, 2n, \\ 1 & \text{for} \quad \nu = 0, \\ -\gamma(-1)^k & \text{for} \quad \nu = 2k, \\ (-1)^n & \text{for} \quad \nu = 2n. \end{cases}$$

Equation (1) follows from this. For $\gamma \in G_{nk}$, the existence of a Hurwitz sequence $\{\omega_j\}_{-\infty}^{\infty}$ of order n having property (1) is established.

Now let $\{\widetilde{\omega}_j\}_{-\infty}^{\infty}$ be a Hurwitz sequence of order n having property (1). Set $\widetilde{p}(\lambda) = \Sigma_{-\infty}^{\infty} \widetilde{\omega}_j \lambda^j$. Then on the basis of (1) we easily conclude that

$$\widetilde{p}(\lambda) \widetilde{p}(-\lambda) = (-1)^n \lambda^{2n} - \gamma(-1)^k \lambda^{2k} + 1 = q(\lambda). \tag{4}$$

It follows from this that $\widetilde{p}(\lambda)$ is constructed from the roots of the polynomial $q(\lambda)$ in the same manner that $p(\lambda)$ was, and so $\omega_j = \widetilde{\omega}_j$ for each integer j. Lemma 1 is established.

LEMMA 2. *If $f(x) \in L_2[0, \infty)$ and $f(x)$ is absolutely continuous together with $f'(x), f''(x), \ldots, f^{(n-1)}(x)$, and if $f^{(n)}(x) \in L_2[0, \infty)$, then*

$$\lim_{x\to\infty} f(x) = \lim_{x\to\infty} f'(x) = \cdots = \lim_{x\to\infty} f^{(n-1)}(x) = 0.$$

Lemma 2 does not contain new results, and is given here for completeness of presentation. For example, it follows immediately from an inequality of Gabušin.

In what follows, it will be necessary to deal with finite sets of numbers $\omega_0, \omega_1, \ldots, \omega_n$. Let us agree each time to assume that zeros are added on both sides out to infinity to give a sequence $\{\omega_j\}_{-\infty}^{\infty}$.

For a given set of numbers $\omega_0, \omega_1, \ldots, \omega_n$ one can construct the numbers

$$\Gamma_j = \sum_{\nu=-\infty}^{\infty} (-1)^{\nu} \omega_{j-\nu}\omega_{j+\nu}; \tag{5}$$

$$\begin{aligned} q_{js} &= \sum_{\nu=0}^{\infty} (-1)^{\nu} \omega_{j+\nu}\omega_{s-\nu-1} && \text{for} \quad 1 \leqslant s \leqslant j \leqslant n, \\ q_{js} &= q_{sj} && \text{for} \quad 1 \leqslant j < s \leqslant n. \end{aligned} \tag{6}$$

From the numbers q_{js} we form the symmetric matrix

$$Q = \begin{Vmatrix} q_{11} q_{12} & \cdots & q_{1n} \\ q_{21} q_{22} & \cdots & q_{2n} \\ \cdots & \cdots & \cdots \\ q_{n1} q_{n2} & \cdots & q_{nn} \end{Vmatrix} \tag{7}$$

If $a = \{a_1, \ldots, a_n\}$ is a given vector, then Q_a will denote the vector $b = \{b_1, \ldots, b_n\}$ whose coordinates are given by the formula $b_\nu = \Sigma_1^n q_{\nu j} a_j$ ($\nu = 1, \ldots, n$). We denote the scalar product of vectors a and b from n-dimensional space by (a, b).

LEMMA 3. *If $f(t)$ from $L_2[0, \infty)$ is absolutely continuous along with its derivatives up to order $(n-1)$ on each finite interval of the half-line $[0, \infty)$ and $f^{(n)}(t) \in L_2[0, \infty)$, then $f'(t), f''(t), \ldots, f^{(n-1)}(t)$ belong to $L_2[0, \infty)$, and for each set of real numbers $\omega_0, \omega_1, \ldots, \omega_n$*

$$\int_0^\infty \Big(\sum_{j=0}^n \omega_j f^{(j)}(t) \Big)^2 dt = \sum_{j=0}^n \Gamma_j \int_0^\infty [f^{(j)}(t)]^2 dt - (Qf_0, f_0), \tag{8}$$

where

$$f_0 = \{f(0), f'(0), \ldots, f^{(n-1)}(0)\}.$$

PROOF. Let $T > 0$. Then

$$\int_0^T \Big(\sum_{j=0}^n \omega_j f^{(j)}(t) \Big)^2 dt = \sum_{j=0}^n \omega_j^2 \int_0^T [f^{(j)}(t)]^2 dt + 2 \sum_{j=1}^n \sum_{s=0}^{j-1} \omega_j\omega_s \int_0^T f^j(t) f^{(s)}(t)\, dt. \tag{9}$$

After an integration by parts, we obtain for $j > s$ that

$$
\int_0^T f^{(j)} f^{(s)}\,dt = \begin{cases} [f^{(2\nu+s-1)} f^{(s)} - f^{(2\nu+s-2)} f^{(s+1)} + \cdots + (-1)^{\nu-1} f^{(s+\nu)} f^{(s+\nu-1)}]_0^T \\ \qquad + (-1)^{\nu} \int_0^T [f^{(s+\nu)}(t)]^2\,dt, \quad \text{if} \quad j - s = 2\nu, \\ [f^{(2\nu+s)} f^{(s)} - f^{(2\nu+s-1)} f^{(s+1)} + \cdots + (-1)^{\nu-1} f^{(s+\nu+1)} f^{(s+\nu-1)} \\ \qquad + \frac{1}{2}(-1)^{\nu} f^{(s+\nu)} f^{(s+\nu)}]_0^T, \quad \text{if} \quad j - s = 2\nu + 1. \end{cases} \tag{10}
$$

Denote the vectors $\{f(0), f'(0), \ldots, f^{(n-1)}(0)\}$ and $\{f(T), f'(T), \ldots, f^{(n-1)}(T)\}$ by f_0 and f_T, respectively. Then (10) can be rewritten (for $j > s$) as

$$
\int_0^T f^{(j)} f^{(s)}\,dt = \begin{cases} \frac{1}{2}(P_{js} f_T f_T) - \frac{1}{2}(P_{js} f_0, f_0) + (-1)^{\nu} \int_0^T [f^{\nu+s}(t)]^2\,dt, & \text{if } j = s + 2\nu, \\ \frac{1}{2}(P_{js} f_T, f_T) - \frac{1}{2}(P_{js} f_0, f_0), & \text{if } j = s + 2\nu + 1, \end{cases}
$$

where the matrix P_{js} has the form

$$
\left\| \begin{array}{cccccccc}
 & & & & & & & 0 \\
 & & & & & & \cdot^{\cdot^{\cdot}} & \\
 & & & & & 0 & & \\
 & & 0 & & 1 & & & \\
 & & & -1 & & & & \\
 & & -1 & & & & & \\
 & & \cdot^{\cdot^{\cdot}} & & & & & \\
 & -1 & & & & & & \\
 & -1 & & & & & & \\
 1 & & & & 0 & & & \\
 0 & & & & & & & \\
 \cdot^{\cdot^{\cdot}} & & & & & & & \\
 0 & & & & & & &
\end{array} \right\| \begin{array}{l} \\ \\ \\ s+1 \\ \\ \\ \\ \\ \\ j \\ \\ \\ \\ \end{array}
$$

$$
\qquad s+1 \qquad\qquad\qquad j
$$

Then (9) assumes the form

$$
\int_0^T \left(\sum_{j=0}^{n} \omega_j f^{(j)}(t)\right)^2 dt = \sum_{j=0}^{n} \omega_j^2 \int_0^T [f^{(j)}(t)]^2\,dt
$$
$$
+ 2 \sum_{\substack{j=1 \\ s-j}}^{n} \sum_{\substack{s=0 \\ \text{even}}}^{j-1} \omega_j \omega_s (-1)^{\frac{j-s}{2}} \int_0^T [f^{\left(\frac{j+s}{2}\right)}(t)]^2\,dt + (P f_T, f_T) - (P f_0, f_0), \tag{11}
$$

where

$$
P = \sum_{j=1}^{n} \sum_{s=0}^{j-1} \omega_j \omega_s P_{js}. \tag{11a}
$$

We have

$$2\sum_{\substack{j=1\\ j-s \text{ even}}}^{n}\sum_{s=0}^{j-1}\omega_j\omega_s(-1)^{\frac{s-j}{2}}\int_0^T[f^{\left(\frac{j+s}{2}\right)}(t)]^2\,dt = 2\sum_{\nu=1}^{n-1}\int_0^T[f^{(\nu)}(t)]^2\sum_{j=\nu+1}^{2\nu}\omega_j\omega_{2\nu-j}(-1)^{j-\nu}$$

$$=\sum_{\nu=1}^{n-1}\int_0^T[f^{(\nu)}(t)]^2\,dt\left\{\sum_{j=1}^{\nu}(-1)^j\omega_{\nu+j}\omega_{\nu-j}+\sum_{j=-1}^{-\nu}(-1)^j\omega_{\nu+j}\omega_{\nu-j}\right\}$$

$$=\sum_{\nu=1}^{n-1}(\Gamma_\nu-\omega_\nu^2)\int_0^T[f^{(\nu)}(t)]^2\,dt.$$

Formula (11) assumes the form

$$\int_0^T\left(\sum_{j=0}\omega_jf^{(j)}(t)\right)dt=\sum_{j=0}^{n}\Gamma_j\int_0^T[f^{(j)}(t)]^2\,dt+(Pf_T,f_T)-(Pf_0,f_0). \tag{12}$$

In order to prove that the functions $f'(t), f''(t), \ldots, f^{(n-1)}(t)$ belong to $L_2[0,\infty)$, we use the fact that the choice of numbers $\omega_0, \omega_1, \ldots, \omega_n$ is arbitrary. Let k be one of the numbers $1, \ldots, n-1$ and let γ be a arbitrary number from the interval

$$\left(0,\ \frac{n}{k^{\frac{k}{n}}(n-k)^{\frac{n-k}{n}}}\right).$$

On the basis of Lemma 1, construct a Hurwitz sequence $\{\hat{\omega}_j\}_{-\infty}^{\infty}$ of order n having property (1). Then from (12) we obtain

$$\int_0^T f^2(t)\,dt-\gamma\int_0^T[f^{(k)}(t)]^2\,dt+\int_0^T f^{(n)}(t)]^2+(\hat{P}f_T,f_T)-(\hat{P}f_0,f_0)\geqslant 0. \tag{13}$$

By Lemma 2, we have that $(\hat{P}f_T, f_T)\longrightarrow 0$ as $T\longrightarrow\infty$. Passing to the limit as $T\longrightarrow +\infty$ in (13), we obtain that $f^{(k)}(t)\in L_2[0,\infty)$ for $k=1,\ldots,n-1$.

Once again let us return to (12) with an arbitrary choice of numbers $\omega_0, \omega_1, \ldots, \omega_n$. As T tends to $+\infty$, we obtain

$$\int_0^\infty\left(\sum_{j=0}^{n}\omega_jf^{(j)}(t)\right)^2dt=\sum_{j=0}^{n}\Gamma_j\int_0^\infty[f^{(j)}(t)]^2\,dt-(Pf_0,f_0). \tag{14}$$

In order to complete the proof of the lemma, it is sufficient to establish the equation $P=Q$.

The matrix P is symmetric since all of the matrices P_{js} are symmetric. Let us calculate the elements of the matrix P. For this, note that of all of the matrices P_{rt}, only $P_{j,s-1}, P_{j+1,s-2}, \ldots$ $(j\geqslant s)$ have nonzero elements in the jth row and sth column. Indeed, $P_{j,s-1}$ has the number $+1$ in the indicated place, $P_{j+1,s-2}$ has -1, and so forth. This means that for $j\geqslant s$ the matrix P has the following number in its jth row and sth column:

$$\sum_{\nu=0}^{\infty}(-1)^{\nu}\omega_{j+\nu}\omega_{s-\nu-1},$$

i.e. (cf. (7)) $P=Q$. The lemma is proved.

The class of functions satisfying the requirements of Lemma 3 will be denoted by $L_2^{(n)}[0,\infty)$.

For $n = 2, \ldots, \infty$ and $k = 1, \ldots, n-1$, set

$$\Upsilon_{nk} = \inf_{f \in L_2^{(n)}[0,\infty)} \frac{\|f\|^2 + \|f^{(n)}\|^2}{\|f^{(k)}\|^2}, \quad \text{where } \|\varphi\| = \sqrt{\int_0^\infty \varphi^2(t)\,dt}. \tag{15}$$

LEMMA 4. *The following inequality holds*:

$$0 < \Upsilon_{nk} < \frac{n}{k^{\frac{k}{n}} (n-k)^{\frac{n-k}{n}}}. \tag{16}$$

PROOF. Let λ be an arbitrary positive number. From Lemma 3 for $\omega_0 = \lambda$, $\omega_1 = 1$ and $\omega_2 = 1/\lambda$ it follows for each function from $L_2^{(n)}[0, \infty)$ that

$$\left\| \lambda f + f' + \frac{1}{\lambda} f'' \right\|^2 = \lambda^2 \|f\|^2 - \|f'\|^2 + \frac{1}{\lambda^2}\|f''\|^2 - \left[\sqrt{\lambda} f(0) + \frac{1}{\sqrt{\lambda}} f'(0)\right]^2.$$

Hence

$$\|f'\|^2 \leqslant \lambda^2 \|f\|^2 + \frac{1}{\lambda^2}\|f''\|^2.$$

For $\lambda = \sqrt{\|f''\|/\|f\|}$, the previous inequality implies that

$$\|f'\|^2 \leqslant 2\|f\| \cdot \|f''\|.$$

Consequently for each function $f(t)$ from $L_2^{(n)}[0, \infty)$

$$\|f'\|^2 \leqslant 2\|f\| \cdot \|f''\|,$$
$$\|f''\|^2 \leqslant 2\|f'\| \cdot \|f'''\|,$$
$$\cdots\cdots\cdots\cdots$$
$$\|f^{(n-1)}\|^2 \leqslant 2\|f^{(n-2)}\| \cdot \|f^{(n)}\|.$$

From this we obtain, by setting $p_1 = (n-k)/n$, $p_2 = 2(n-k)/n, \ldots, p_k = k(n-k)/n$, $p_{k+1} = k(n-k-1)/n$, $p_{k+2} = k(n-k-2)/n, \cdots, p_{n-1} = k/n$, that

$$\|f'\|^{2p_1}\|f''\|^{2p_2}\ldots\|f^{(n-1)}\|^{2p_{n-1}} \leqslant 2^{p_1+p_2+\cdots+p_{n-1}}\|f\|^{p_1}\|f'\|^{p_2}\|f''\|^{p_1+p_3}\ldots$$
$$\ldots\|f^{(k)}\|^{p_{k-1}+p_{k+1}}\ldots\|f^{(n-1)}\|^{p_{n-2}}\|f^{(n)}\|^{p_{n-1}}$$

or

$$\|f^{(k)}\|^{2p_k - p_{k-1} - p_{k+1}} \leqslant 2^{p_1+\cdots+p_{n-1}}\|f\|^{p_1}\|f^{(n)}\|^{p_{n-1}}.$$

Finally,

$$\|f^{(k)}\|^2 \leqslant 2^{k(n-k)}\|f\|^{2\cdot\frac{n-k}{n}}\|f^{(n)}\|^{2\frac{k}{n}}.$$

By the use of the inequality $1 \leqslant C_\beta(y^\beta + 1/y^{2-\beta})$ (where $y > 0$, $0 < \beta < 2$, and C_β does not depend on y), this can be rewritten

$$\|f^{(k)}\|^2 \leqslant 2^{k(n-k)} C_\beta \|f\|^{2\frac{n-k}{n}}\|f^{(n)}\|^{2\frac{k}{n}}\left[\left(\frac{\|f\|}{\|f^{(n)}\|}\right)^{2\frac{k}{n}} + \left(\frac{\|f^{(n)}\|}{\|f\|}\right)^{2\frac{n-k}{n}}\right]$$
$$\leqslant 2^{k(n-k)} C_\beta [\|f\|^2 + \|f^{(n)}\|^2].$$

Consequently $\gamma_{nk} \geqslant 2^{-k(n-k)}/C_\beta > 0$. Now let us bound γ_{nk} above. With this aim, consider the function

$$f(t) = e^{i\Phi - zt} + e^{-i\Phi - \bar{z}t},$$

where $z = e^{i\theta}$, $\Phi = \pi/2 + \pi/(n+1)$ and $\theta = \pi/(n+1)$. We have

$$\int_0^\infty [f^{(s)}(t)]^2\, dt = \frac{1}{2} e^{2i\Phi} z^{2s-1} + \frac{1}{2} e^{-2i\Phi} \bar{z}^{2s-1} + \frac{2}{z+\bar{z}}$$

$$= \cos(2\Phi + (2s-1)\theta) + \frac{1}{\cos\theta} = \frac{\cos(2\Phi + 2s\theta) + \cos(2\Phi + 2(s-1)\theta) + 2}{2\cos\theta}$$

$$= \frac{\cos^2(\Phi + s\theta) + \cos^2(\Phi + (s-1)\theta)}{\cos\theta}.$$

From this we obtain

$$\frac{\|f\|^2 + \|f^{(n)}\|^2}{\|f^{(k)}\|^2} = \frac{\cos^2\Phi + \cos^2(\theta - \Phi) + \cos^2(\Phi + n\theta) + \cos^2(\Phi + (n-1)\theta)}{\cos^2(\Phi + k\theta) + \cos^2(\Phi + (k-1)\theta)}$$

$$= \frac{\cos^2\left(\frac{\pi}{2} + \frac{\pi}{n+1}\right) + \cos^2\left(\frac{\pi}{2} + \frac{\pi n}{n+1}\right)}{\cos^2\left(\frac{\pi}{2} + \frac{\pi(k+1)}{n+1}\right) + \cos^2\left(\frac{\pi}{2} + \frac{\pi k}{n+1}\right)} = \frac{2\sin^2\frac{\pi}{n+1}}{\sin^2\frac{\pi(k+1)}{n+1} + \sin^2\frac{\pi k}{n+1}} \leqslant 1$$

since $\sin(\pi k/(n+1)) \geqslant \sin(\pi/(n+1))$ for $k = 1, \ldots, n$. Thus we have shown that $\gamma_{nk} \leqslant 1$. It remains to be shown that

$$\frac{n}{k^{\frac{k}{n}}(n-k)^{\frac{n-k}{n}}} > 1.$$

Indeed,

$$\frac{n}{k^{\frac{k}{n}}(n-k)^{\frac{n-k}{n}}} \geqslant \inf_{\frac{1}{n} \leqslant \Theta \leqslant 1 - \frac{1}{n}} \frac{1}{\Theta^\Theta (1-\Theta)^{1-\Theta}} = \inf_{\frac{1}{n} \leqslant \Theta \leqslant 1 - \frac{1}{n}} e^{-[\Theta \ln\Theta + (1-\Theta)\ln(1-\Theta)]} > 1.$$

The lemma is proved.

§2. The main theorem

From the definition of the quantities γ_{nk}, we have for each function $f(t)$ in $L_2^{(n)}[0, \infty)$ that

$$\|f^{(k)}\|^2 \leqslant \frac{1}{\gamma_{nk}} [\|f\|^2 + \|f^{(n)}\|^2];$$

moreover, the constant $1/\gamma_{nk}$ in this inequality cannot be improved.

Now a method will be established for estimating the quantity γ_{nk}. Let γ be an arbitrary number from the interval

$$G_{nk} = \left(0, \frac{n}{k^{\frac{k}{n}}(n-k)^{\frac{n-k}{n}}}\right).$$

Let $\omega_j(\gamma)$ denote the elements of the Hurwitz sequence $\{\omega_j\}_{-\infty}^{\infty}$ of order n having property (1). The existence of this sequence is guaranteed by Lemma 1. Further, let

$$D(\gamma) = \begin{vmatrix} q_{11}(\gamma) & q_{12}(\gamma) & \dots & q_{1n}(\gamma) \\ q_{21}(\gamma) & q_{22}(\gamma) & \dots & q_{2n}(\gamma) \\ \dots & \dots & \dots & \dots \\ q_{n1}(\gamma) & q_{n2}(\gamma) & \dots & q_{nn}(\gamma) \end{vmatrix}, \tag{17}$$

where

$$q_{js}(\gamma) = q_{sj}(\gamma) = \sum_{\nu=1}^{\infty} (-1)^{\nu} \omega_{j+\nu}(\gamma) \omega_{s-\nu-1} \quad \text{for} \quad j \geqslant s.$$

THEOREM 1. *γ_{nk} is the least positive root of the equation $D(\gamma) = 0$.*

PROOF. Substituting the numbers $\omega_j(\gamma)$ in place of ω_j in (8), we obtain for $f \in L_2^{(n)}[0, \infty)$

$$\left\| \sum_{j=0}^{n} \omega_j(\gamma) f^{(j)}(t) \right\|^2 = \|f\|^2 + \|f^{(n)}\|^2 - \gamma \|f^{(k)}\|^2 - (Qf_0, f_0). \tag{18}$$

Here the matrix Q depends on γ, and therefore we will write $Q(\gamma)$ instead of Q.

Let $0 < \gamma < \gamma_{nk}$ (the existence of such a γ follows from Lemma 4), and let x be a nonzero vector with coordinates $x_0, x_1, \dots, x_{n-1}$. Let us solve the Cauchy problem

$$\sum_{j=0}^{n} \omega_j(\gamma) y^{(j)}(t) = 0,$$
$$y^{(j)}(0) = x_j \qquad (j = 0, 1, \dots n-1).$$

The solution of this problem together with all of its derivatives belongs to $L_2[0, \infty)$, since $\{\omega_j(\gamma)\}_{-\infty}^{\infty}$ is a Hurwitz sequence. Applying (18) to the function $y(t)$, we obtain

$$\|y\|^2 + \|y^{(n)}\|^2 - \gamma \|y^{(k)}\|^2 = (Q(\gamma)x, x) > 0.$$

This means that the minimal eigenvalue of the matrix $Q(\gamma)$ is positive for $\gamma < \gamma_{nk}$. From this it follows that $D(\gamma) = \det Q(\gamma) \neq 0$ for $\gamma \in (0, \gamma_{nk})$.

Now let

$$\gamma_{nk} < \gamma < \frac{n}{k^{\frac{k}{n}} (n-k)^{\frac{n-k}{n}}}$$

(the existence of such a γ follows from Lemma 4). Then for each function $f \in L_2^{(n)}[0, \infty)$ we have (cf. (18))

$$\|f\|^2 + \|f^{(n)}\|^2 - \gamma \|f^k\|^2 \geqslant (Q(\gamma)f_0, f_0). \tag{19}$$

By the definition of γ_{nk}, there is a function $y(t) \in L_2^{(n)}[0, \infty)$ such that $\|y\|^2 + \|y^{(n)}\|^2 - \gamma \|y^{(k)}\|^2 < 0$. Inequality (19) applied to $y(t)$ gives $(Q(\gamma)y_0, y_0) < 0$, where $y_0 = \{y(0), y'(0), \dots, y^{n-1}(0)\}$. This means that the minimal eigenvalue of the matrix $Q(\gamma)$ is negative for all

$$\gamma \in \left(\gamma_{nk}, \frac{n}{k^{\frac{n}{k}} (n-k)^{\frac{n-k}{n}}} \right).$$

Consequently for $\gamma = \gamma_{nk}$ the minimal eigenvalue of the matrix $Q(\gamma)$ is zero, and $D(\gamma) = \det Q(\gamma) = 0$. The theorem is proved.

COROLLARY 1. *For all* $n = 2, \ldots, \infty$ *and all* $k = 1, \ldots, n-1$, *the equation* $\gamma_{nk} = \gamma_{n,n-k}$ *holds.*

PROOF. Let $\{\omega_j(\gamma)\}_{-\infty}^{\infty}$ be the sequence constructed in Lemma 1 for given natural numbers n and k ($n = 2, \ldots, \infty$; $k = 1, \ldots, n-1$). Consider the sequence $\{\tilde{\omega}_j(\gamma)\}_{-\infty}^{\infty}$, where $\tilde{\omega}_j(\gamma) = \omega_{n-j}(\gamma)$. Clearly it is a Hurwitz sequence of order n, and

$$\sum_{\nu=-\infty}^{\infty} (-1)^{\nu} \tilde{\omega}_{j-\nu}(\gamma)\, \tilde{\omega}_{j+\nu}(\gamma) = \sum_{\nu=-\infty}^{\infty} (-1)^{\nu} \omega_{n-j+\nu}(\gamma)\, \omega_{n-j-\nu}(\gamma)$$

$$= \begin{cases} 0 & \text{for} \quad n-j \neq 0, k, n, \\ 1 & \text{for} \quad n-j = 0, n, \\ -\gamma & \text{for} \quad n-j = k. \end{cases}$$

This means that $\{\tilde{\omega}_j(\gamma)\}_{-\infty}^{\infty}$ is the sequence constructed in accordance with Lemma 1 for the natural numbers n and $n-k$.

For $j \geqslant s$, set

$$\tilde{q}_{js}(\gamma) = \sum_{\nu=0}^{\infty} (-1)^{\nu} \tilde{\omega}_{j+\nu}(\gamma)\, \tilde{\omega}_{s-\nu-1}(\gamma).$$

Then

$$\tilde{q}_{js}(\gamma) = \sum_{\nu=0}^{\infty} (-1)^{\nu} \omega_{n-j-\nu}(\gamma)\, \omega_{n-s+\nu+1}(\gamma) = q_{n-s+1,\, n-j+1}(\gamma).$$

Thus

$$\tilde{Q}(\gamma) = \begin{Vmatrix} \tilde{q}_{11}(\gamma) & \tilde{q}_{12}(\gamma) & \cdots & \tilde{q}_{1n}(\gamma) \\ \tilde{q}_{21}(\gamma) & \tilde{q}_{22}(\gamma) & \cdots & \tilde{q}_{2n}(\gamma) \\ \cdot & \cdot & \cdot & \cdot \\ \tilde{q}_{n1}(\gamma) & \tilde{q}_{n2}(\gamma) & \cdots & \tilde{q}_{nn}(\gamma) \end{Vmatrix} = \begin{Vmatrix} q_{nn}(\gamma) & q_{n-1,\,n}(\gamma) & \cdots & q_{1n}(\gamma) \\ q_{n,\,n-1}(\gamma) & q_{n-1,\,n-1}(\gamma) & \cdots & q_{1,\,n-1}(\gamma) \\ \cdot & \cdot & \cdot & \cdot \\ q_{n1}(\gamma) & q_{n-1,\,1}(\gamma) & \cdots & q_{11}(\gamma) \end{Vmatrix}.$$

This means that $\det \tilde{Q}(\gamma) = \det Q(\gamma)$. On the basis of Theorem 1, we conclude from this that $\gamma_{n,n-k} = \gamma_{nk}$. Corollary 1 is proved.

COROLLARY 2. *The following assertions hold*:

a) $\gamma_{21} = 1$;

b) $\gamma_{31} = \gamma_{32} = \sqrt[3]{3 - 2\sqrt{2}}$;

c) γ_{41} (*and consequently* γ_{43}) *is the least positive root of the equation*

$$\gamma^8 - 6\gamma^4 - 8\gamma^2 + 1 = 0;$$

d) γ_{42} *is the least positive root of the equation*

$$\gamma^4 - 2\gamma^2 - 4\gamma + 1 = 0.$$

The proof is accomplished by simple but tedious algebraic calculations based on Theorem 1. So in case $n = 2$ and $k = 1$, we have from (1) $\omega_1^2 - 2 = -\gamma$ and, from (7), $D(\gamma) = \omega_1^2 - \gamma$; consequently $\gamma_{21} = 1$. In case $n = 3$ and $k = 1$, the system (1) has the form

$$\omega_1^2 - 2\omega_2 = -\gamma, \qquad \omega_2^2 - 2\omega_1 = 0,$$

and $D(\gamma) = 1 - \gamma\omega_2$.

From this we obtain assertion b) without difficulty. To determine γ_{41}, we have the system

$$\omega_1^2 - 2\omega_2 = -\gamma, \ \omega_2^2 - 2\omega_1\omega_3 + 2 = 0, \quad \omega_3^2 - 2\omega_2 = 0, \ \gamma(\omega_1^2 - \omega_2) = 1,$$

from which assertion c) follows. Assertion d) can be obtained from the system

$$\omega_1^2 - 2\omega_2 = 0, \quad \omega_2^2 - 2\omega_1\omega_3 + 2 = -\gamma, \quad \omega_3^2 - 2\omega_2 = 0,$$
$$\gamma^2 + (2 - \omega_1\omega_3)\gamma + 1 = 0.$$

We conclude this section with the following assertion: *all of the γ_{nk} are algebraic numbers*.

§3. A transformation of $D(\gamma)$

This section will be devoted to a simplification of the determinant $D(\gamma)$. For this the following lemma will be useful.

LEMMA 5. *Let*

$$\Delta_0 = 1, \ \Delta_1 = \omega_1, \ \Delta_s = \begin{vmatrix} \omega_1\omega_0 & 0 & \dots 0 \\ \omega_2\omega_1 & \omega_0 & \dots 0 \\ \dots & \dots & \dots \\ \omega_s\omega_{s-1}\omega_{s-2} & \dots & 0 \end{vmatrix}$$

for $s = 2, \dots, \infty$ and $\Delta_s = 0$ for $s = -\infty, \dots, -1$. Then

$$\Delta_j = \omega_j + \gamma(-1)^k \Delta_{j-2k} \qquad for \ \ j = 0, 1, \dots, 2n-1, \tag{20}$$

$$\Delta_j = \omega_j + \gamma(-1)^k \omega_{j-2k} + \gamma^2\omega_{j-4k} + \dots \qquad for \ \ j = 0, \dots, 2n-1. \tag{21}$$

PROOF. First note that (21) is a direct corollary of (20).

Expand the function $1/p(\zeta)$ in a power series, where $p(\zeta)$ is defined by (2):

$$\frac{1}{p_*(\zeta)} = \sigma_0 + \sigma_1\zeta + \sigma_2\zeta^2 + \dots$$

From this we get

$$\sum_{j+s=\nu} \omega_j\sigma_s = \begin{cases} 1 & \text{for} \quad \nu = 0, \\ 0 & \text{for} \quad \nu > 0. \end{cases}$$

From this, in turn, we obtain without difficulty that

$$\sigma_0 = \omega_0 = 1, \ \sigma_1 = -\omega_1, \ \sigma_s = (-1)^s\Delta \ \ (\text{for} \quad s = 2, 3, \dots).$$

Thus

$$\frac{1}{p(\zeta)} = \Delta_0 - \Delta_1\zeta + \Delta_2\zeta^2 - \dots .$$

Consequently (cf. (3))

$$p(-\zeta) = [(-1)^n\zeta^{2n} - \gamma(-1)^k\zeta^{2k} + 1][\Delta_0 - \Delta_1\zeta + \Delta_2\zeta^2 - \dots].$$

Multiplying out the polynomials on the right side and equating coefficients, we obtain

$$\omega_0=\Delta_0,\quad \omega_1=\Delta_1,\ \omega_2=\Delta_2\ldots\omega_{2k-1}=\Delta_{2k-1};$$

$$(-1)^j\omega_j=(-1)^j\Delta_j-\gamma(-1)^{k+j+2k}\Delta_{j-2k}\qquad\text{for}\qquad j=2k,\ 2k+1,\ldots,2n-1.$$

Equation (20) follows from this if we consider that $\Delta_s = 0$ for $s < 0$.

Now let us proceed to the transformation of the determinant $D(\gamma)$. For $j \geqslant s$ the elements of $Q(\gamma)$ are defined by (7). For $j < s$ we will have

$$q_{js}=q_{sj}=\sum_{\nu=0}^{\infty}(-1)^\nu\omega_{s+\nu}\omega_{j-1-\nu}=\sum_{\nu=-\infty}^{\infty}(-1)^\nu\omega_{s+\nu}\omega_{j-1-\nu}$$

$$-\sum_{\nu=1}^{\infty}(-1)^\nu\omega_{s-\nu}\omega_{j-1+\nu}=A_{js}+\sum_{\nu=0}^{\infty}(-1)^\nu\omega_{j+\nu}\omega_{s-\nu-1},$$

where because of (1) it is easy to obtain that

$$A_{js}=\sum_{\nu=-\infty}^{\infty}(-1)^\nu\,\omega_{s+\nu}\omega_{j-1-\nu}=\begin{cases}0, & s+j=2,3\ldots,2k,2k+2,\ldots,2n,\\ (-1)^s, & s+j=1,\\ (-1)^{n+s}, & s+j=2n+1,\\ -\gamma(-1)^{k+s}, & s+j=2k+1.\end{cases}$$

In particular, for $1 \leqslant j < s \leqslant n$ we have

$$A_{js}=\gamma(-1)^{k+s+1}\delta_{j+s,\,2k+1}.$$

Thus for all $j = 1, \ldots, n$ and $s = 1, \ldots, n$ one can write

$$q_{js}=\sum_{\nu=0}^{\infty}(-1)^\nu\omega_{j+\nu}\omega_{s-\nu-1}+A_{js},$$

where

$$A_{js}=\begin{cases}0 & \text{for}\quad j\geqslant s\quad\text{or}\quad j+s\neq 2k+1,\\ (-1)^{k+j}\gamma & \text{for}\quad j<s\ \text{and}\ j+s=2k+1.\end{cases}\tag{22}$$

In what follows, the following formula will be more suitable:

$$A_{js}=\begin{cases}0 & \text{for}\quad j=k+1,\ k+2,\ldots n,\\ (-1)^{k+j}\gamma\Delta_{j+s-2k-1}\delta_{j+s,\,2k+1} & \text{for}\quad j=1,2,\ldots k.\end{cases}\tag{22a}$$

The equivalence of (22) and (22a) can be established without difficulty (it follows by taking into account that $\Delta_0 = 1$ and $\Delta_s = 0$ for $s < 0$).

Using the quantities A_{js}, one can rewrite $D(\gamma)$ in the form

$$D(\gamma)=\begin{vmatrix}\omega_1\omega_0+A_{11} & \omega_1\omega_1-\omega_2\omega_0+A_{12} & \ldots & \omega_1\omega_{n-1}-\omega_2\omega_{n-2}+\ldots+(-1)^{n-1}\omega_n\omega_0+A_{1n}\\ \omega_2\omega_0+A_{21} & \omega_2\omega_1-\omega_3\omega_0+A_{22} & \ldots & \omega_2\omega_{n-1}-\omega_3\omega_{n-2}+\ldots+(-1)^{n-1}\omega_{n+1}\omega_0+A_{2n}\\ \ldots & \ldots & \ldots & \ldots\\ \omega_n\omega_0+A_{n1} & \omega_n\omega_1-\omega_{n+1}\omega_0+A_{n2} & \ldots & \omega_n\omega_{n-1}-\omega_{n+1}\omega_{n-2}+\ldots+(-1)^{n-1}\omega_{2n-1}\omega_0+A_{nn}\end{vmatrix}.\tag{23}$$

Our next problem is to prove that $D(\gamma)$ can differ from the following determinant only in sign:

$$\begin{vmatrix} \omega_1+B_{11} & \omega_2+B_{12} & \dots & \omega_n+B_{1n} \\ \omega_2+B_{21} & \omega_3+B_{22} & \dots & \omega_{n+1}+B_{2n} \\ \dots & \dots & \dots & \dots \\ \omega_n+B_{n1} & \omega_{n+1}+B_{n2} & \dots & \omega_{2n-1}+B_{nn} \end{vmatrix}, \tag{24}$$

where

$$B_{js}=\begin{cases} 0 & \text{for} \quad j>k, \\ (-1)^k\gamma\Delta_{j+s-2k-1} & \text{for} \quad j\leqslant k. \end{cases} \tag{24a}$$

Consider the determinant in which the first m columns coincide with the corresponding columns of (24) and the last $n-m$ columns coincide with the corresponding columns of (23):

$$\begin{vmatrix} \omega_1+B_{11} & \omega_2+B_{12} & \dots & \omega_m+B_{1m} & \omega_1\omega_m-\omega_2\omega_{m-1}+\dots+(-1)^{m-1}\omega_m\omega_1 \\ & & & & +(-1)^m\omega_{m+1}\omega_0+A_{1,\,m+1}\dots \\ \dots & \dots & \dots & \dots & \dots \\ \omega_\nu+B_{\nu 1} & \omega_{\nu+1}+B_{\nu 2} & \dots & \omega_{m+\nu-1}+B_{\nu m} & \omega_\nu\omega_m-\omega_{\nu+1}\omega_{m-1}+\dots \\ & & & & \dots+(-1)^{m-1}\omega_{m+\nu-1}\omega_1+(-1)^m\omega_{m+\nu}\omega_0+A_{\nu,\,m+1}\dots \\ \dots & \dots & \dots & \dots & \dots \\ \omega_n+B_{n1} & \omega_{n+1}+B_{n2} & \dots & \omega_{m+n-1}+B_{nm} & \omega_n\omega_m-\omega_{n+1}\omega_{m-1}+\dots \\ & & & & \dots+(-1)^{m-1}\omega_{m+n-1}\omega_1+(-1)^m\omega_{m+n}\omega_0+A_{n,\,m+1}\dots \end{vmatrix} \tag{25}$$

Multiply the first column by ω_m, the second by $-\omega_{m-1}$, and the mth by $(-1)^{m-1}\omega_1$, and subtract from the $(m+1)$th column. After multiplying the $(m+1)$th column by $(-1)^m$, we obtain a determinant in which the element in the νth row and $(m+1)$th column is

$$\omega_{m+\nu}+(-1)^m\,[A_{\nu,\,m+1}-\omega_m B_{\nu 1}+\omega_{m-1}B_{\nu 2}-\dots+(-1)^m\omega_1 B_{\nu m}]$$
$$=\omega_{m+\nu}+(-1)^m\left[A_{\nu,\,m+1}+\sum_{l=1}^{m}(-1)^l\omega_{m-l+1}B_{\nu l}\right]=\omega_{m+\nu}+B^*_{\nu,\,m+1}.$$

Let us compute $B^*_{\nu,m+1}$. For $\nu>k$ we have $B^*_{\nu j}=A_{\nu j}=0$ for $j=1,\dots,n$ (cf. (22) and (24)). Hence

$$B^*_{\nu,\,m+1}=B_{\nu,\,m+1}=0.$$

For $\nu\leqslant k$ we can write

$$B^*_{\nu,\,m+1}=(-1)^m A_{\nu,\,m+1}+(-1)^{k+m}\gamma\sum_{l=1}^{m}(-1)^l\omega_{m-l+1}\Delta_{\nu+l-2k-1}$$
$$=(-1)^{k+\nu+m}\gamma\Delta_{\nu+m-2k}\,\delta_{\nu+m,\,2k}+(-1)^{k+m}\gamma\sum_{s=1}^{m}(-1)^{m+1-s}\omega_s\Delta_{\nu+m-s-2k} \tag{26}$$
$$=(-1)^k\gamma\left\{\Delta_{\nu+m-2k}\delta_{\nu+m,\,2k}+\sum_{s=1}^{m}(-1)^{s+1}\omega_s\Delta_{\nu+m-s-2k}\right\}.$$

There are three cases to consider:

a) For $\nu+m-2k<0$, all terms in the curly brackets become zero, and consequently

$$B^*_{\nu,\,m+1}=B_{\nu,\,m+1}=0;$$

b) For $\nu + m - 2k = 0$, the first term in the curly brackets becomes 1 and all of the rest equal zero:

$$B^*_{\nu, m+1} = (-1)^k\gamma = B_{\nu, m+1};$$

c) For $\nu + m - 2k > 0$, the first term in the curly brackets becomes zero. Moreover, we have

$$\sum_{s=1}^{m} (-1)^{s+1}\omega_s\Delta_{\nu+m-s-2k} = \Delta_{\nu+m-2k}. \tag{27}$$

This relation is a special case of the relation $\Delta_p = \omega_1\Delta_{p-1} - \omega_2\Delta_{p-2} + \cdots + (-1)^{p-1}\omega_p\Delta_0$, obtained by expanding the determinant

$$\Delta_s = \begin{vmatrix} \omega_1 & \omega_0 & 0 & \dots & 0 \\ \omega_2 & \omega_1 & \omega_0 & \dots & 0 \\ \cdot & \cdot & \cdot & \cdot & \cdot \\ \omega_s & \omega_{s-1} & \omega_{s-2} & \dots & \omega_1 \end{vmatrix}$$

along its first column.

From (27) it follows that for $\nu + m - 2k > 0$

$$B^*_{\nu, m+1} = (-1)^k\gamma\Delta_{\nu+m-2k} = B_{\nu, m+1}.$$

Thus we have shown that $B^*_{\nu,m+1} = B_{\nu,m+1}$ for all $\nu = 1, \dots, n$. Consequently the absolute value of determinant (25) is not altered by substituting the $(m + 1)$th column of (24) in place of the $(m + 1)$th column of (25).

Thus after noting that determinants (23) and (24) have identical first columns ($\omega_0 = 1, A_{\nu 1} = B_{\nu 1} = 0$), we immediately obtain that (23) and (24) can differ only in sign.

If $\nu \geqslant k + 1$, then $B_{\nu m} = 0$ for $m = 1, \dots, n$. Consequently the rows from the $(k + 1)$th on in the determinant (24) have the form

$$\begin{vmatrix} \omega_{k+1} & \omega_{k+2} & \dots & \omega_n & 0 & \dots & 0 \\ \cdot & \cdot & \cdot & \cdot & \cdot & \cdot & \cdot \\ \omega_{n-1} & \omega_n & \dots & 0 & 0 & \dots & 0 \\ \omega_n & 0 & \dots & 0 & 0 & \dots & 0 \end{vmatrix}$$

This means that

$$D(\gamma) = \pm \begin{vmatrix} \omega_{n-k+1} + B_{1, n-k+1} & \omega_{n-k+2} + B_{1, n-k+2} & \dots & \omega_n + B_{1n} \\ \cdot & \cdot & \cdot & \cdot \\ \omega_n + B_{k, n-k+1} & \omega_{n+1} + B_{k, n-k+2} & \dots & \omega_{n+k-1} + B_{k,n} \end{vmatrix}.$$

If $\nu \leqslant k$, then $B_{\nu m} = (-1)^k\gamma\Delta_{m+\nu-2k-1}$ follows from (24a) ($\Delta_s = 0$ for $s < 0$). This means that

$$D(\gamma) = \pm \begin{vmatrix} \omega_{n-k+1} + (-1)^k\gamma\Delta_{n-3k+1} & \dots & \omega_n + (-1)^k\gamma\Delta_{n-2k} \\ \cdot & \cdot & \cdot \\ \omega_n + (-1)^k\gamma\Delta_{n-2k} & \dots & \omega_{n+k-1} + (-1)^k\gamma\Delta_{n-k-1} \end{vmatrix}.$$

Relation (20) allows us to simplify this formula:

$$D(\gamma) = \begin{vmatrix} \Delta_{n-k+1} & \Delta_{n-k+2} & \dots & \Delta_n \\ \Delta_{n-k+2} & \Delta_{n-k+3} & \dots & \Delta_{n+1} \\ \cdot & \cdot & \cdot & \cdot \\ \Delta_n & \Delta_{n+1} & \dots & \Delta_{n+k-1} \end{vmatrix}. \tag{28}$$

This concludes the transformation of the determinant $D(\gamma)$.

§4. An integral representation of Δ_s

In this section the quantities $\Delta_s(\gamma)$ will be expressed as integrals containing as a parameter only the variable γ.

Consider the function $\zeta^n p(1/\zeta) = \omega_n + \omega_{n-1}\zeta + \cdots + \omega_0\zeta^n$.

Let $\epsilon_1, \ldots, \epsilon_{2k}$ be the roots of 1 of degree $2k$. Then for $-(k-1) \leqslant j \leqslant k-1$

$$\varepsilon_\nu^{n-j}\zeta^n p\left(\frac{1}{\varepsilon_\nu\zeta}\right) = \omega_n\varepsilon_\nu^{-j} + \omega_{n-1}\varepsilon_\nu^{1-j}\zeta + \ldots + \omega_{n-j}\zeta^j + \ldots + \omega_0\varepsilon_\nu^{n-j}\zeta^n.$$

Thus

$$\frac{1}{2k\zeta^j}\sum_{\nu=1}^{2k}\varepsilon_\nu^{n-j}\zeta^n p\left(\frac{1}{\varepsilon_\nu\zeta}\right) = \omega_{n-j} + \omega_{n-j-2k}\zeta^{2k} + \cdots.$$

Consequently

$$\Delta_{n-j} = \frac{(i\gamma^{\frac{1}{2k}})^{n-j}}{2k}\sum_{\nu=1}^{2k}\varepsilon_\nu^{n-j} p\left(\frac{1}{i\varepsilon_\nu\gamma^{\frac{1}{2k}}}\right).$$

Or, for $-(k-1) \leqslant j \leqslant k-1$,

$$\Delta_{n-j}^* = \sum_{\nu=1}^{2k}\varepsilon_\nu^{n-j} p\left(\frac{1}{i\varepsilon_\nu\gamma^{\frac{1}{2k}}}\right), \tag{29}$$

where

$$\Delta_{n-j} = \frac{(i\gamma^{\frac{1}{2k}})^{n-j}}{2k}\Delta_{n-j}^*. \tag{29a}$$

The function $\ln p(z)$ is regular for Real $z \geqslant 0$. Consequently $(z = x + iy)$

$$\ln|p(z)| = \frac{x}{\pi}\int_{-\infty}^{\infty}\frac{\ln|p(i\eta)|}{x^2+(y-\eta)^2}d\eta$$

$$= \frac{x}{\pi}\int_0^\infty \ln|p(i\eta)|\left[\frac{1}{x^2+(y-\eta)^2} + \frac{1}{x^2+(y+\eta)^2}\right]d\eta.$$

But $p(i\eta)p(-i\eta) = q(i\eta) = \eta^{2n} - \gamma\eta^{2k} + 1$ and $p(-i\eta) = \overline{p(i\eta)}$. This means that

$$\ln|p(z)| = \frac{1}{\pi}\int_0^\infty \ln q(i\eta)\,\mathrm{Real}\,\frac{z}{z^2+\eta^2}d\eta,$$

and consequently

$$\ln p(z) = \frac{z}{\pi}\int_0^\infty\frac{\ln q(i\eta)}{z^2+\eta^2}d\eta + iC_0 \qquad (C_0 \text{ real}).$$

If in the right half-plane one chooses the branch for $\ln p(z)$ which is real on the real line, for Real $z > 0$ one will have

$$\ln p(z) = \frac{z}{\pi}\int_0^\infty\frac{\ln q(i\eta)}{z^2+\eta^2}d\eta.$$

We can rewrite the last formula in this form:

$$\ln p(z) = \frac{1}{2}\ln q(z) + \frac{z}{\pi}\int_0^\infty\frac{\ln q(i\eta) - \ln q(z)}{z^2+\eta^2}d\eta.$$

This is clearly valid for Real $z \geqslant 0$. Now set $z = 1/i\epsilon_j\gamma^{1/2k}$, where the ϵ_j are the roots of 1 of degree $2k$ which lie in the lower half-plane or on the real line. Then

$$\ln p\left(\frac{1}{i\varepsilon_j\gamma^{\frac{1}{2k}}}\right) = \frac{1}{2}\ln q\left(\frac{1}{i\varepsilon_j\gamma^{\frac{1}{2k}}}\right) + \frac{1}{\pi i\varepsilon_j\gamma^{\frac{1}{2k}}}\int_0^\infty \frac{\ln q(i\eta) - \ln q\left(\frac{1}{i\varepsilon_j\gamma^{\frac{1}{2k}}}\right)}{\eta^2 - \frac{1}{\varepsilon_j^2\gamma^{\frac{1}{k}}}}\,d\eta$$

$$= -\frac{n}{2k}\ln\gamma - n\ln\varepsilon_j + \frac{1}{i\varepsilon_j\pi\gamma^{\frac{1}{2k}}}\int_0^\infty \frac{\ln q(i\eta) + \ln(\gamma^{\frac{n}{k}})}{\eta^2 - \frac{1}{\varepsilon_j^2\gamma^{\frac{1}{k}}}}\,d\eta + \frac{2n\,\varepsilon_j}{i\varepsilon_j\pi\gamma^{\frac{1}{2k}}}\int_0^\infty \frac{d\eta}{\eta^2 - \frac{1}{\varepsilon_j^2\gamma^{\frac{1}{k}}}}.$$

Thus for $\operatorname{Im}\epsilon_j \leqslant 0$

$$\ln p\left(\frac{1}{i\varepsilon_j\gamma^{\frac{1}{2k}}}\right) = -\frac{n}{2k}\ln\gamma + \frac{\gamma^{\frac{1}{2k}}}{i\varepsilon_j\pi}\int_0^\infty \frac{\ln\left[\eta^{2n} - \gamma\eta^{2k} + 1\right]\gamma^{\frac{n}{k}}}{\gamma^{\frac{1}{k}}\eta^2 - \bar{\varepsilon}_j^2}\,d\eta$$

$$= -\frac{n}{2k}\ln\gamma + \frac{1}{i\varepsilon_j\pi}\int_0^\infty \frac{\ln\left[s^{2n} + (1 - s^{2k})\gamma^{\frac{n}{k}}\right]}{s^2 - \bar{\varepsilon}_j^2}\,ds.$$

Consequently for $\operatorname{Im}\epsilon_j \leqslant 0$

$$p\left(\frac{1}{i\varepsilon_j\gamma^{\frac{1}{2k}}}\right) = \gamma^{-\frac{n}{2k}}\exp\left\{\frac{1}{i\varepsilon_j\pi}\int_0^\infty \frac{\ln\left[s^{2n} + (1 - s^{2k})\gamma^{\frac{n}{k}}\right]}{s^2 - \bar{\varepsilon}_j^2}\,ds\right\}. \tag{30}$$

We have

$$\int_0^\infty \frac{\ln\left[s^{2n} + (1 - s^{2k})\gamma^{\frac{n}{k}}\right]}{s^2 - \bar{\varepsilon}_j^2}\,ds = 2n\int_0^\infty \frac{\ln s}{s^2 - \bar{\varepsilon}_j^2}\,ds$$

$$+ \frac{n}{k}\int_0^\gamma \eta^{\frac{n}{k}-1}\left(\int_0^\infty \frac{1 - s^{2k}}{(s^2 - \bar{\varepsilon}_j^2)\left[s^{2n} + (1 - s^{2k})\eta^{\frac{n}{k}}\right]}\,ds\right)d\eta.$$

If we assume that $\bar{\epsilon}_j = e^{i\theta_j}$ $(0 \leqslant \theta_j \leqslant \pi)$, then(1)

$$\frac{2}{\pi\varepsilon_j}\int_0^\infty \frac{\ln s}{s^2 - \bar{\varepsilon}_j^2}\,ds = \frac{\pi}{2} - \theta_j.$$

---

(1) This formula can be obtained as follows:

$$\frac{2}{\pi\varepsilon_j}\int_0^\infty \frac{\ln s}{s^2 - \bar{\varepsilon}_j^2}\,ds = \frac{2}{\pi\varepsilon_j}\int_{-\infty}^\infty \frac{te^t}{e^{2t} - \bar{\varepsilon}_j^2}\,dt = \frac{2}{\pi}\int_0^\infty \frac{te^t}{\varepsilon_j e^{2t} - \bar{\varepsilon}_j}\,dt$$

$$- \frac{2}{\pi}\int_0^\infty \frac{te^{-t}}{\varepsilon_j e^{-2t} - \bar{\varepsilon}_j}\,dt = \frac{4}{\pi}\operatorname{Real}\left(\frac{1}{\varepsilon_j}\int_0^\infty \frac{te^{-t}}{1 - \bar{\varepsilon}_j^2 e^{-2t}}\,dt\right)$$

$$= \frac{4}{\pi}\operatorname{Real}\sum_{\nu=0}^\infty \frac{\bar{\varepsilon}_j^{2\nu+1}}{(2\nu+1)^2} = \frac{\pi}{2} - \theta_j.$$

Then

$$\frac{1}{i\varepsilon_j\pi}\int_0^\infty \frac{\ln\left[s^{2n}+(1-s^{2k})\gamma^{\frac{n}{k}}\right]}{s^2-\bar{\varepsilon}_j^2}\,ds = -\frac{\pi}{2}ni + n\Theta_j i + \frac{n}{i\varepsilon_j\pi k}\int_0^\gamma \eta^{\frac{n}{k}-1}\Phi_j(\eta)\,d\eta,$$

where

$$\Phi_j(\eta) = \int_0^\infty \frac{1-s^{2k}}{(s^2-\bar{\varepsilon}_j^2)\left[s^{2n}+(1-s^{2k})\eta^{\frac{n}{k}}\right]}\,ds = \eta^{-\frac{n}{k}+\frac{1}{2k}}\int_0^\infty \frac{1-\eta t^{2k}}{(t^2\eta^{\frac{1}{k}}-\bar{\varepsilon}_j^2)[t^{2n}-\eta t^{2k}+1]}\,dt. \tag{31}$$

Hence for $\operatorname{Im}\epsilon_j \leqslant 0$ we have

$$p\left(\frac{1}{i\varepsilon_j\gamma^{\frac{1}{2k}}}\right) = \gamma^{-\frac{n}{2k}} i^{-n}\bar{\varepsilon}_j^n \exp\left[\frac{n}{i\varepsilon_j\pi k}\int_0^\gamma \eta^{\frac{n}{k}-1}\Phi_j(\eta)\,d\eta\right]. \tag{32}$$

In order to obtain the analogous formula for $\operatorname{Im}\epsilon_j \geqslant 0$, the relation $p(\zeta)p(-\zeta) = q(\zeta)$ (cf. (3)) is used:

$$p\left(\frac{1}{i\varepsilon_j\gamma^{\frac{1}{2k}}}\right) = \frac{q\left(\frac{1}{i\varepsilon_j\gamma^{\frac{1}{2k}}}\right)}{p\left(\frac{1}{i(-\varepsilon_j)\gamma^{\frac{1}{2k}}}\right)} = \frac{(-1)^n}{\varepsilon_j^{2n}\gamma^{\frac{n}{k}}}\gamma^{\frac{n}{2k}} i^n\bar{\varepsilon}_j^{-n}\exp\left[\frac{n}{i\varepsilon_j\pi k}\int_0^\gamma \eta^{\frac{n}{k}-1}\Phi_j(\eta)\,d\eta\right]$$

$$= \gamma^{-\frac{n}{2k}} i^{-n}\bar{\varepsilon}_j^n \exp\left[\frac{n}{i\pi\varepsilon_j k}\int_0^\gamma \eta^{\frac{n}{k}-1}\Phi_j(\eta)\,d\eta\right]. \tag{33}$$

Combining (29), (32), and (33), it is easy to show that

$$\Delta_{n-j}^* = \gamma^{-\frac{n}{2k}} i^{-n}\sum_{\nu=1} \bar{\varepsilon}_\nu^{-j}\exp\left[\frac{n}{i\varepsilon_\nu\pi k}\int_0^\gamma \eta^{\frac{n}{k}-1}\Phi_\nu(\eta)\,d\eta\right] = \gamma^{-\frac{n}{2k}} i^{-n}\widetilde{\Delta}_{n-j}, \tag{34}$$

where

$$\widetilde{\Delta}_{n-j} = \sum_{\nu=1}^{2k}\frac{1}{\varepsilon_\nu^j}\exp\left[\frac{n}{i\varepsilon_\nu\pi k}\int_0^\gamma \eta^{\frac{n}{k}-1}\Phi_\nu(\eta)\,d\eta\right]$$

$$= \sum_{\nu=1}^{2k}\frac{1}{\varepsilon_\nu^j}\exp\left[\frac{n}{i\varepsilon_\nu\pi k}\int_0^\gamma \eta^{\frac{1}{2k}-1}\left(\int_0^\infty \frac{1-\eta t^{2k}}{(t^2\eta^{\frac{1}{k}}-\bar{\varepsilon}_\nu^2)(t^{2n}-\eta t^{2k}+1)}\,dt\right)d\eta\right]. \tag{35}$$

From (29) and (34) we have

$$\Delta_{n-j} = \frac{1}{2k\,(i\gamma^{\frac{1}{2k}})^j}\widetilde{\Delta}_{n-j}.$$

It is easy to see that

$$\begin{vmatrix} \Delta_{n-k+1} & \Delta_{n-k+2} & \dots & \Delta_n \\ \Delta_{n-k+2} & \Delta_{n-k+3} & \dots & \Delta_{n+1} \\ \dots & \dots & \dots & \dots \\ \Delta_n & \Delta_{n+1} & \dots & \Delta_{n+k+1} \end{vmatrix} = \begin{vmatrix} \widetilde{\Delta}_{n-k+1} & \widetilde{\Delta}_{n-k+2} & \dots & \widetilde{\Delta}_n \\ \widetilde{\Delta}_{n-k+2} & \widetilde{\Delta}_{n-k+3} & \dots & \widetilde{\Delta}_{n+1} \\ \dots & \dots & \dots & \dots \\ \widetilde{\Delta}_n & \widetilde{\Delta}_{n+1} & \dots & \widetilde{\Delta}_{n+k+1} \end{vmatrix} \cdot \frac{1}{(2k)^k}$$

Thus we have proved

THEOREM 2. *The number γ_{nk} is the least positive solution of the equation*

$$D(\gamma) = \begin{vmatrix} \widetilde{\Delta}_{n-k+1} & \widetilde{\Delta}_{n-k+2} & \dots & \widetilde{\Delta}_{n} \\ \widetilde{\Delta}_{n-k+2} & \widetilde{\Delta}_{n-k+3} & \dots & \widetilde{\Delta}_{n+1} \\ \cdot & \cdot & \cdot & \cdot \\ \widetilde{\Delta}_{n} & \widetilde{\Delta}_{n+1} & \dots & \widetilde{\Delta}_{n+k-1} \end{vmatrix} = 0,$$

where

$$\widetilde{\Delta}_{n-j} = \sum_{\nu=1}^{2k} \varepsilon_\nu^{-j} \exp\left[\frac{n}{i\varepsilon_\nu \pi k} \int_0^\gamma \eta^{\frac{1}{2k}-1} \Psi_\nu(\eta)\, d\eta\right],$$

$$\Psi_\nu(\eta) = \int_0^\infty \frac{1 - s^{2k}\eta}{(s^2\eta^{\frac{1}{k}} - \varepsilon_\nu^2)(s^{2n} - \eta s^{2k} + 1)}\, ds,$$

and $\epsilon_1, \dots, \epsilon_{2k}$ are the roots of $+1$ of degree $2k$.

§5. Asymptotic formulas for γ_{n1}

In this section we shall consider the case $k = 1$.

It follows directly from Theorem 2 that γ_{n1} is the minimal positive root of the equation

$$\cos\left\{\frac{n}{\pi}\int_0^\gamma \frac{1}{\sqrt{\eta}}\left(\int_0^\infty \frac{ds}{s^{2n} - \eta s^2 + 1}\right) d\eta\right\} = 0.$$

or

$$\int_0^\gamma \frac{1}{\sqrt{\eta}}\left(\int_0^\infty \frac{ds}{s^{2n} - \eta s^2 + 1}\right) d\eta = \frac{\pi^2}{2n}. \tag{36}$$

Consequently

$$\int_0^{\gamma_{n1}} \frac{1}{\sqrt{\eta}}\left(\int_0^\infty \frac{ds}{s^{2n} + 1}\right) d\eta < \frac{\pi^2}{2n}$$

and

$$\sqrt{\gamma_{n1}} \leqslant \frac{\pi^2}{4n} \frac{2n \sin\frac{\pi}{2n}}{\pi} \leqslant \frac{\pi^2}{4n}. \tag{37}$$

From (36) it is easy to obtain

$$\sum_{\nu=1}^\infty \frac{\gamma^{\nu+\frac{1}{2}}}{2\nu+1} \int_0^\infty \frac{s^{2\nu}}{(1 + s^{2n})^{\nu+1}}\, ds = \frac{\pi^2}{4n}. \tag{38}$$

THEOREM 3. *The quantities γ_{n1} are the minimal positive roots of the equations*

$$\sum_{\nu=0}^\infty \frac{\gamma^{\nu+\frac{1}{2}}}{2\nu+1} \frac{\Gamma\left(\frac{2\nu+1}{2n}\right)\Gamma\left(1+\nu-\frac{2\nu+1}{2n}\right)}{\nu!} = \frac{\pi^2}{2}.$$

The proof follows immediately from (38). Set

$$g_\nu(n) = \frac{1}{2\nu+1} \int_0^\infty \frac{s^{2\nu}}{(1+s^{2n})^{\nu+1}}\, ds.$$

It is clear that all of the $g_\nu(n)$ are positive. It is equally easy to obtain the bound

$$g_\nu(n) \leqslant \frac{2}{(2\nu+1)^2}.$$

Taking this into account, we obtain from (38) that

$$\frac{\pi\sqrt{\gamma_{n1}}}{\sin\frac{\pi}{2n}} = \frac{\pi^2}{2} + \gamma_{n1}^{3/2}\varphi(\gamma_{n1}),$$

where

$$\varphi(\gamma_{n1}) = -\sum_{\nu=1}^{\infty} g_\nu(n)\,\gamma_{n1}^{\nu-1}.$$

All of the $\gamma_{n1} \leqslant 1$ (for $n = 2$ this is established by the corollary of Theorem 1, and for $n > 2$ it follows from (37)). Consequently

$$|\varphi(\gamma_{n1})| \leqslant 2\sum_{\nu=1}^{\infty} \frac{1}{(2\nu+1)^2} = \frac{\pi^2}{4} - 2.$$

This allows us to obtain

$$\sqrt{\gamma_{n1}} = \frac{\pi}{2}\sin\frac{\pi}{2n} + O\left(\frac{1}{n}\gamma_{n1}^{3/2}\right).$$

Combining this with (37), we conclude that

$$\sqrt{\gamma_{n1}} = \frac{\pi}{2}\sin\frac{\pi}{2n} + O\left(\frac{1}{n^4}\right) = \frac{\pi^2}{4n}\left[1 - \frac{\pi^2}{24n^2} + O\left(\frac{1}{n^3}\right)\right]$$

and

$$\gamma_{n1} = \frac{\pi^4}{16n^2}\left[\left(+1 - \frac{\pi^2}{24n^2}\right)^2 + O\left(\frac{1}{n^3}\right)\right] = \frac{\pi^4}{16n^2}\left[1 - \frac{\pi^2}{12n^2} + O\left(\frac{1}{n^3}\right)\right],$$

$$\frac{1}{\gamma_{n1}} = \frac{16n^2}{\pi^4} + \frac{4}{3\pi^2} + O\left(\frac{1}{n}\right).$$

Let us restrict ourselves to this asymptotic formula. One should keep in mind that (38) allows the possibility of obtaining bounds of the form

$$\frac{1}{\gamma_{n1}} = \frac{16n^2}{\pi^4} + \frac{4}{3\pi^2} + \frac{A_1}{n} + \frac{A_2}{n^2} + \dots + \frac{A_p}{n^p} + O\left(\frac{1}{n^{p+1}}\right).$$

However, the calculation of $A_1, A_2, \dots$ requires very tedious computations.

§6. Asymptotic formulas for γ_{nk} for fixed k

As a preliminary let us consider the following boundary value problem on the interval $[0, 1]$:

$$(-1)^k F^{(2k)}(t) = \lambda F(t); \tag{39}$$

$$\begin{gathered} F^{(k)}(0) = F^{(k+1)}(0) = \dots = F^{(2k-1)}(0) = 0, \\ F(1) = F'(1) = \dots = F^{k-1}(1) = 0. \end{gathered} \tag{40}$$

The operator corresponding to this problem is selfadjoint and positive, which can be verified without any difficulty. This means that the eigenvalues of this problem are real and positive.

Let $\epsilon_1, \ldots, \epsilon_{2k}$ be the roots of $+1$ of degree $2k$. The system of functions

$$F_0(t) = \frac{\lambda^{\frac{k+1}{2k}}}{2k} \sum_{j=1}^{2k} \frac{1}{(i\varepsilon_j)^{k-1}} e^{i\varepsilon_j \lambda^{\frac{1}{2k}} t},$$

$$\cdots\cdots\cdots\cdots\cdots\cdots$$

$$F_s(t) = \frac{\lambda^{\frac{s-k+1}{2k}}}{2k} \sum_{j=1}^{2k} \frac{1}{(i\varepsilon_j)^{k-s-1}} e^{i\varepsilon_j \lambda^{\frac{1}{2k}} t},$$

$$\cdots\cdots\cdots\cdots\cdots\cdots$$

$$F_{2k-1}(t) = \frac{\lambda^{\frac{k}{2k}}}{2k} \sum_{j=1}^{2k} \frac{1}{(i\varepsilon_j)^{k-2k}} e^{i\varepsilon_j \lambda^{\frac{1}{2k}} t}$$

is clearly a fundamental system of the solution to equation (39).

The following equations can be verified without difficulty:

a) If $0 \leqslant s \leqslant k-1$, then

$$F_s(0) = F_s'(0) = \ldots = F^{(k-s-2)}(0) = 0, \; F_s^{(k-s-1)}(0) = 1,$$

$$F_s^{(k-s)}(0) = \ldots = F_s^{(2k-1)}(0) = 0;$$

b) If $k \leqslant s \leqslant 2k-1$, then

$$F_s(0) \ldots = F^{(3k-s-2)}(0) \notin 0, \; F_s^{(3k-s-1)}(0) = (-1)^k \lambda,$$

$$F_s^{(3k-s)}(0) = \ldots = F_s^{(2k-1)}(0) = 0.$$

Let us find an equation for the eigenvalues of problem (39), (40). The general solution to (39) has the form

$$F(t) = \sum_{s=0}^{2k-1} A_s F_s(t).$$

The boundary conditions at the point $t = 0$ yield

$$F^{(j)}(0) = \sum_{s=0}^{k-1} A_s F_s^{(j)}(0) + \sum_{s=k}^{2k-1} A_s F_s^{(j)}(0) = 0$$

for $j = k, k+1, \ldots, 2k-1$. Consequently $F^{(j)}(0) = A_{3k-j-1} = 0$, i.e. $A_k = A_{k+1} = \cdots = A_{2k-1} = 0$. This means that each eigenfunction of the problem (39)–(40) necessarily has the form

$$F(t) = \sum_{s=0}^{k-1} A_s F_s(t).$$

The boundary conditions (40) at the point $t = 1$ yield

$$\sum_{s=0}^{k-1} A_s F_s^{(j)}(1) = 0 \tag{41}$$

for $j = 0, 1, \ldots, k-1$. Taking into account that not all of the A_s are equal to zero, we obtain from (41) an equation for studying the eigenvalues:

$$\begin{vmatrix} F_0(1) & F_1(1) & \dots & F_{k-1}(1) \\ \cdot & \cdot & \cdot & \cdot \\ F_0^{(k-1)}(1) & F_1^{(k-1)}(1) & \dots & F_{k-1}^{(k-1)}(1) \end{vmatrix} = 0. \tag{42}$$

If we set $\sqrt[2k]{\lambda} = \rho$, we can rewrite (42) in the form

$$\frac{1}{(2k)^k} d(\rho) = \begin{vmatrix} F_0(1) & F_1(1) & \dots & F_{k-1}(1) \\ \cdot & \cdot & \cdot & \cdot \\ F_0^{(k-1)}(1) & F_1^{(k-1)}(1) & \dots & F_{k-1}^{(k-1)}(1) \end{vmatrix}$$

$$= \frac{1}{(2k)^k} \begin{vmatrix} d_{n-k+1}(\rho) & d_{n-k+2}(\rho) & \dots & d_n(\rho) \\ d_{n-k+1}(\rho) & d_{n-k+3}(\rho) & \dots & d_{n+1}(\rho) \\ \cdot & \cdot & \cdot & \cdot \\ d_n(\rho) & d_{n+1}(\rho) & \dots & d_{n+k-1}(\rho) \end{vmatrix} = 0, \tag{42a}$$

where

$$d_{n-s}(\rho) = \sum_{\nu=1}^{2k} \frac{1}{\varepsilon_\nu^s} e^{i\varepsilon_\nu \rho}.$$

Thus if ρ_1 is the least positive solution of (42), then $\lambda_1 = \rho_1^{2k}$ is the least eigenvalue of the problem (39), (40), and conversely.

The function $d(\rho) \neq 0$ for $\rho = 0$ (zero is not an eigenvalue of problem (39), (40)). In what follows, we shall need the following property of the function $d(\rho)$:

$$\text{if} \quad d(\rho_1) = 0, \quad \text{then} \quad d'(\rho_1) \neq 0. \tag{43}$$

Assume the contrary assertion $d'(\rho_1) = 0$. Then

$$\begin{vmatrix} F_0(1) & F_1(1) & \dots & F_{k-1}(1) \\ \cdot & \cdot & \cdot & \cdot \\ F_0^{(k-2)}(1) & F_1^{(k-2)}(1) & \dots & F_{k-1}^{(k-2)}(1) \\ F_0^{(k)}(1) & F_1^{(k)}(1) & \dots & F_{k-1}^{(k)}(1) \end{vmatrix} = 0 \quad (\text{for } \rho = \rho_1) \tag{44}$$

since $(d/d\rho)F_s^{(j)}(1) = F_s^{(j+1)}(1)$. From (44) it follows that there exists a nontrivial function $\Phi(t) = \Sigma_0^{k-1} B_s F_s(t)$ satisfying the boundary conditions

$$\begin{gathered} \Phi(1) = \Phi'(1) = \dots = \Phi^{(k-2)}(1) = \Phi^{(k)}(1) = 0, \\ \Phi^{(k)}(0) = \Phi^{(k+1)}(0) = \dots = \Phi^{(2k-1)}(0) = 0. \end{gathered} \tag{45}$$

The function $\Phi(t)$ satisfies (39). Then

$$0 = \int_0^1 [F^{(2k)}\Phi - F\Phi^{(2k)}]\,dt = [F^{(2k-1)}\Phi - F^{(2k-2)}\Phi' + \dots \pm F^{(k+1)}\Phi^{(k-2)}$$

$$\mp F^{(k)}\Phi^{(k-1)} \pm F^{(k-1)}\Phi^{(k)} \mp \dots - F\Phi^{(2k-1)}]_0^1 = \mp F^{(k)}(1)\,\Phi^{(k-1)}(1).$$

This means that at least one of the quantities $F^{(k)}(1)$ or $\Phi^{(k-1)}(1)$ is equal to zero. Consequently there is a continuous function $\varphi(t)$ ($\varphi(t) = F(t)$ if $F^{(k)}(1) = 0$, or $\varphi(t) = \Phi(t)$ if $\Phi^{(k-1)}(1) = 0$) satisfying (39) and the boundary conditions

$$\begin{gathered} \varphi(1) = \varphi'(1) = \dots = \varphi^{(k-2)}(1) = \varphi^{(k-1)}(1) = \varphi^{(k)}(\hat{s}) = 0, \\ \varphi^{(k)}(0) = \varphi^{(k+1)}(0) = \dots = \varphi^{(2k-1)}(0) = 0. \end{gathered} \tag{46}$$

Now suppose that $\varphi^{(k-1)}(t)$ has N zeros on $(0, 1)$ (keep in mind everywhere here that the zeros are distinct). Taking the condition $\varphi^{(k-1)}(1) = 0$ into account, we obtain that $\varphi^{(k)}(t)$ has at least N zeros on $(0, 1)$. Keeping the equations $\varphi^{(k)}(0) = \varphi^{(k)}(1) = 0$ in mind, we conclude that $\varphi^{(k+1)}(t)$ has at least $N + 1$ zeros on $(0, 1)$. Then with regard to the condition $\varphi^{(k+1)}(0) = \cdots = \varphi^{(2k-1)}(0) = 0$, we easily obtain that each of the functions $\varphi^{(k+1)}(t), \ldots, \varphi^{(2k-1)}(t)$ has at least $N + 1$ zeros on $(0, 1)$. From the equation $\varphi(t) = (-1)^k \varphi^{(2k)}(t)/\lambda_1$ and the condition $\varphi(1) = 0$, we obtain that $\varphi(t)$ has $N + 1$ zeros on $(0, 1)$. Keeping the relations $\varphi'(0) = \cdots = \varphi^{(k-2)}(0) = 0$ in mind, we obtain without difficulty that $\varphi^{(k-1)}(t)$ has at least $N + 1$ zeros on $(0, 1)$. From this assertion it follows that $\varphi^{(k-1)}(t)$ has infinitely many zeros. Since the function $\varphi(t)$ is analytic, and by the boundary conditions, we obtain that $\varphi(t) \equiv 0$. But this contradicts the definition of the function $\varphi(t)$. Assertion (43) is established.

LEMMA 6. *Assume that a continuously differentiable function $d(\rho)$ on the interval $[0, a]$ satisfies the conditions $d(\rho) > 0$ for $0 \leqslant \rho < \rho_0 < a$, $d(\rho_0) = 0$ and $d'(\rho_0) \neq 0$, and suppose that a sequence $\{k_n\}$ of positive numbers satisfies $\lim_{n\to\infty} n/k_n^3 = 0$. In addition, assume that for some sequence of functions $\varphi_n(\rho)$, continuous on $[0, a/k_n]$, the inequality $|\varphi_n(\rho)| \leqslant C_0 n \rho^3$ holds on $[0, a/k_n]$, where C_0 does not depend on n or ρ. Then for each sufficiently large n there is a solution to the equation*

$$d(k_n\rho) = \varphi_n(\rho) \tag{47}$$

on the interval $[0, a/k_n]$, and for the least positive solution x_n of (47)

$$x_n = \frac{\rho_0}{k_n} + O\left(\frac{n}{k_n^4}\right). \tag{48}$$

PROOF. From the relations $d(\rho_0) = 0$ and $d'(\rho_0) < 0$ (the inequality $d'(\rho_0) > 0$ contradicts the requirements that $d(\rho) > 0$ for $0 \leqslant \rho < \rho_0$ and $d(\rho_0) = 0$), it follows that $d(\rho) < 0$ for $\rho \in (\rho_0, \rho_0 + h]$. We shall now consider $d(\rho)$ only on the interval $[0, \rho_0 + h]$. Choose an arbitrary positive number δ. Let $\mu_\delta = \min_{0\leqslant\rho\leqslant\rho_0-\delta} d(\rho)$ and choose δ' so that $\delta' < \delta$ and $\max_{\rho_0-\delta'\leqslant\rho\leqslant\rho_0+\delta'} |a(\rho)| < \mu_\delta$. Set

$$\min \{\max_{\rho_0-\delta'\leqslant\rho\leqslant\rho_0+\delta'} d(\rho), \max_{\rho_0-\delta'\leqslant\rho\leqslant\rho_0+\delta'} (-d(\rho))\} = \sigma.$$

It is clear that $0 < \sigma < \mu_\delta$. The function $\psi_n(\rho) = d(k_n\rho)$ is continuous on $[0, (\rho_0 + \delta')/k_n]$, and thus on this interval we have $\max \psi_n(\rho) \geqslant d(0) \geqslant \mu_\delta > \sigma$, and $\min \psi_n(\rho) \leqslant -\sigma$.

On the other hand, on $[0, a/k_n]$ for $k_n^3/n > C_0 a/\sigma$ we have the bound

$$|\varphi_n(\rho)| \leqslant C_0 n \frac{a}{k_n^3} < \sigma.$$

Consequently for $k_n^3/n > C_0 a/\sigma$ the equation

$$d(k_n\rho) = \varphi_n(\rho)$$

has at least one solution belonging to $[0, (\rho_0 + \delta')/k_n]$. It is easy to see that there is no solution to (47) on $[0, (\rho_0 - \delta)/k_n]$ for $k_n^3/n > C_0 a/\sigma$. Indeed,

$$\min_{0 \leqslant \rho \leqslant \frac{x_0 - \delta}{k_n}} \psi(\rho) = \min_{0 \leqslant \rho \leqslant x_0 - \delta} d(\rho) = \mu_\delta > \sigma$$

and

$$\max_{0 \leqslant \rho \leqslant \frac{x_0 - \delta}{k_n}} |\varphi_n(\rho)| < \sigma.$$

Thus, we have shown that:

a) for all sufficiently large n, solutions of (47) exist in $[0, a/k_n]$;

b) for each $\delta > 0$ and for all sufficiently large n, the least positive root of (47) belongs to the interval

$$\left[\frac{\rho_0 - \delta}{k_n}, \frac{\rho_0 + \delta}{k_n}\right].$$

Now it is not difficult to complete the proof. Let x_n be the least positive root of (47). We have

$$| d(k_n x_n) - d(\rho_0) | = | d(k_n x_n) | = | \varphi_n(x_n) | \leqslant C_0 n x_n^3. \tag{49}$$

Choose the quantity $\delta > 0$ so that $\min |d'(\rho)| = C_1 > 0$. For sufficiently large n, we have $k_n x_n \in [\rho_0 - \delta, \rho_0 + \delta]$. Then from (49) it follows by the mean value theorem that

$$C_1 | k_n x_n - \rho_0 | \leqslant C_0 n x_n^3.$$

Therefore

$$\left| x_n - \frac{\rho_0}{k_n} \right| \leqslant \frac{C_0}{C_1} \frac{n}{k_n} x_n^3 = O\left(\frac{n}{k_n^4}\right).$$

Relation (48) is established.

In the concluding part of this paper, let us turn to the derivation of asymptotic formulas for γ_{nk}.

THEOREM 4. *For fixed k, the following asymptotic formulas are valid for the quantities γ_{nk} as $n \to \infty$:*

$$\gamma_{nk} = \frac{\lambda_1(k) \pi^{2k}}{4^k n^{2k}} + O\left(\frac{1}{n^{2k+2}}\right),$$

$$\frac{1}{\gamma_{nk}} = \frac{4^k n^{2k}}{\lambda_1(k) \pi^{2k}} + O(n^{2k-2}),$$

where $\lambda_1(k)$ is the minimal eigenvalue of problem (39), (40).

PROOF. Consider the function $\Psi_\nu(\eta)$ from Theorem 2 (for $0 < \eta < 1/2$):

$$\Psi_\nu(\eta) = \int_0^\infty \frac{1 - s^{2k}\eta}{(s^2\eta^{\frac{1}{k}} - \bar{\varepsilon}_\nu^2)(s^{2n} - \eta s^{2k} + 1)} ds$$

$$= -\int_0^\infty \frac{\bar{\varepsilon}_\nu^{2k-2} + \bar{\varepsilon}_\nu^{2k-4} s^2 \eta^{\frac{1}{k}} + \ldots + s^{2k-2}\eta^{\frac{k-1}{k}}}{s^{2k} - \eta s^{2k} + 1} ds$$

$$= -\int_0^\infty \frac{\bar{\varepsilon}_\nu^{2k-2} + \bar{\varepsilon}_\nu^{2k-4}\eta^{\frac{1}{k}} s^2 + \ldots + s^{2k-2}\eta^{\frac{k-1}{k}}}{s^{2n}+1} ds - \eta \int_0^\infty \frac{[\bar{\varepsilon}_\nu^{2k-2} + \ldots + s^{2k-2}\eta^{\frac{k-1}{k}}]\, s^{2k}}{(s^{2n} - \eta s^{2k} + 1)(s^{2n}+1)} ds$$

$$= -\frac{1}{\bar{\varepsilon}_\nu^2}\int_0^\infty \frac{ds}{s^{2n}+1} - \eta^{\frac{1}{k}} \int_0^\infty \frac{\bar{\varepsilon}_\nu^{2k-4} s^2 + \ldots + s^{2k-2}\eta^{\frac{k-2}{k}}}{s^{2n}+1} ds \qquad (50)$$

$$- \eta \int_0^\infty \frac{[\bar{\varepsilon}_\nu^{2k-2} + \ldots + s^{2k-2}\eta^{\frac{k-1}{k}}]\, s^{2k}}{(s^{2n} - \eta s^{2k} + 1)(s^{2n}+1)} ds.$$

The following relations are evident:

$$\int_0^\infty \frac{s^{2j}}{s^{2n}+1} ds = O(1) \qquad \text{for} \qquad j = 1, 2, \ldots, n-1.$$

If $0 \leqslant \eta \leqslant ½$, then it is not difficult to verify that $s^{2n} - \eta s^{2k} + 1 \geqslant ½$ for $0 < s < \infty$ and $s^{2n} - \eta s^{2k} + 1 \geqslant s^{2n}/2 + 1$ for $1 \leqslant s < \infty$. From this it follows that

$$\int_0^\infty \frac{s^{2j}}{(s^{2n} - \eta s^{2k} + 1)(s^{2n}+1)} ds = O(1) \qquad \text{for} \qquad j = 1, \quad 2, \ldots, \quad 4k-2.$$

Consequently

$$\Psi_\nu(\eta) = -\frac{1}{\bar{\varepsilon}_\nu^2}\int_0^\infty \frac{ds}{s^{2n}+1} + O(\gamma^{\frac{1}{k}}) = -\frac{\pi}{\bar{\varepsilon}_\nu^2 2n \sin\frac{\pi}{2n}} + O(\gamma^{\frac{1}{k}}).$$

This means (cf. Theorem 2) that for $0 < \gamma < 1/2$

$$\tilde{\Delta}_{n-j} = \sum_{\nu=1}^{2k} \frac{1}{\varepsilon_\nu^j} \exp\left\{\frac{n}{i\varepsilon_\nu \pi k}\left[-\frac{\pi 2k\gamma^{\frac{1}{2k}}}{\bar{\varepsilon}_\nu^2 2n \sin\frac{\pi}{2n}} + O(\gamma^{\frac{3}{2k}})\right]\right\}$$

$$= \sum_{\nu=1}^{2k} \frac{1}{\varepsilon_\nu^j} \exp\left[i\varepsilon_\nu \frac{\gamma^{\frac{1}{2k}}}{\sin\frac{\pi}{2n}} + O(n\gamma^{\frac{3}{2k}})\right] = \sum_{\nu=1}^{2k} \frac{1}{\varepsilon_\nu^j} \exp(i\varepsilon_\nu k_n \rho)\,[1 + O(n\rho^3)],$$

where $k_n = 1/\sin(\pi/2n)$ and $\rho = \gamma^{1/2k}$. Consequently if $k_n\rho \leqslant a$ (where $a = 2\lambda_1^{1/2k}(k)$), then one can represent $\tilde{D}(\gamma)$ (cf. Theorem 2 and (42a)) in the form

$$\tilde{D}(\gamma) = d(k_n\rho) - \varphi_n(\rho).$$

The function $d(\rho)$ satisfies the requirements of Lemma 6 on the interval $[0, 2\lambda_1^{1/2k}(k)]$, and $\varphi_n(\rho)$ satisfies on $[0, a/k_n]$ the inequality

$$|\varphi_n(\rho)| \leqslant C_0 n\rho^3.$$

From this we obtain for the least positive solution γ_{nk} of the equation for $\widetilde{D}(\gamma)$ that

$$\gamma_{nk}^{\frac{1}{2k}} = \frac{\lambda_1^{\frac{1}{2k}}(k)}{k_n} + O\left(\frac{n}{k_n^4}\right) = \frac{\lambda_1^{\frac{1}{2k}}(k)}{k_n} + O\left(\frac{1}{n^3}\right).$$

or

$$\gamma_{nk} = \frac{\lambda_1(k)}{k_n^{2k}} + O\left(\frac{1}{n^{2k-2}}\right),$$

$$\frac{1}{\gamma_{nk}} = \frac{k_n^{2k}}{\lambda_1(k)} + O(n^{2k-2}).$$

The assertions of the theorem are obtained from this without difficulty if one recalls that $1/k_n = \sin(\pi/2n)$.

BIBLIOGRAPHY

1. G. H. Hardy, J. E. Littlewood and G. Pólya, *Inequalities*, Cambridge Univ. Press, New York, 1934; Russian transl., IL, Moscow, 1948.

2. A. N. Kolmogorov, *On inequalities between the upper bounds of the successive derivatives of an arbitrary function on an infinite interval*, Učen. Zap. Moskov. Gos. Univ. Mat. **30** (1939), 3–13; English transl., Amer. Math. Soc. Transl. (1) **2** (1962), 233–243. MR **1**, 298.

3. Elias M. Stein, *Functions of exponential type*, Ann. of Math. (2) **65** (1957), 582–592. MR **19**, 23.

4. L. V. Taĭkov, *Inequalities of Kolmogorov type and best formulas for numerical differentiation*, Mat. Zametki 4 (1968), 233–238 = Math. Notes **4** (1968), 631–634. MR **38** # 1459.

5. Ju. N. Subbotin and L. V. Taĭkov, *Best approximation of a differentiation operator in the space* L_2, Mat. Zametki 3 (1968), 157–164 = Math. Notes 3 (1968), 100–105. MR **37** #4471.

6. S. B. Stečkin, *Inequalities between norms of derivatives of arbitrary functions*, Acta Sci. Math. (Szeged) **26** (1965), 225–230. (Russian) MR **32** # 2534.

7. ———, *Best approximation of linear operators*, Mat. Zametki **1** (1967), 137–148 = Math. Notes **1** (1967), 91–99. MR **35** # 2051.

8. V. V. Arestov, *Sharp inequalities between the norms of functions and their derivatives*, Acta Sci. Math. (Szeged) **33** (1972), 243–267. (Russian) MR **47** # 9263.

9. V. N. Gabušin, *Best approximation of a differentiation operator on the half-line*, Mat. Zametki **6** (1969), 573–582 = Math. Notes **6** (1969), 804–810. MR **41** # 4080.

Translated by W. L. BYNUM

Trudy Mat. Inst. Steklov.
138 (1975)

Proc. Steklov Inst. Math.
138 (1975)

EXTREMAL PROBLEMS OF FUNCTIONAL INTERPOLATION AND INTERPOLATION-IN-THE-MEAN SPLINES

UDC 517: 518.12

Ju. N. SUBBOTIN

ABSTRACT. The problem of optimal functional interpolation by interpolation-in-the-mean splines is investigated in the metric of L_p. Let n be a natural number, let $Y = \{y_k\}$ $(k = 0, \pm 1, \pm 2, \ldots)$ be a sequence of real numbers, and let

$$\|Y\|_{l_p^{(n)}} = \left\{ \sum_{k=-\infty}^{\infty} |\Delta^n y_k|^p \right\}^{1/p} \quad \text{for} \quad 1 \leqslant p < \infty,$$

$$\|Y\|_{l_\infty^{(n)}} = \sup_k |\Delta^n y_k|.$$

Let $L_p^{(n)}$ denote the totality of functions defined on the real line that have a locally absolutely continuous $(n-1)$th derivative and an nth derivative in $L_p(-\infty, \infty)$ $(1 \leqslant p \leqslant \infty)$, let $\|f\|_{L_p^{(n)}} = \|f^{(n)}\|_{L_p}$, and let

$$F_k(f) = \int_{-\infty}^{\infty} f(x + kh)\,dg(x) \quad (0 < h < \infty,\ k = 0; \pm 1, \pm 2, \ldots),$$

where $g(x)$ is a function of bounded variation. One wishes to find

$$\sup_{\|Y\|_{l_p^{(n)}} \leqslant M} \ \inf_{\substack{f \in L_p^{(n)} \\ F_k(f) = y_k}} \|f\|_{L_p^{(n)}}.$$

The question of finiteness of this quantity is studied, and upper and lower bounds are given. Exact solutions are given for a specific choice of $g(x)$. The approximative properties of interpolation-in-the-mean splines are investigated.

Bibliography: 43 titles.

Introduction

The paper consists of two parts. In the first part we consider extremal problems of functional interpolation. The first three sections are devoted to these investigations. The first is introductory; in it are presented, in particular, general upper and lower bounds for the corresponding extremal quantities. In §2 we study the properties of the zeros of some algebraic polynomials. §3 is the main section; here we give exact solutions of extremal problems for special functionals. We note that in some cases the extremal functions are polynomial interpolation-in-the-mean splines. The fourth section is devoted to questions of existence and convergence in the metric of L_p $(1 \leqslant p \leqslant \infty)$ for such splines. In the fifth section we make use of the jumps of the highest nonzero derivative of an interpolation spline with equidistant knots to obtain approximations of higher derivatives of the function being interpolated.

AMS (MOS) subject classifications (1970). Primary 41A05; Secondary 41A15, 41A30.

§1. Statement of the problem. General upper and lower bounds

Let n be a fixed natural number and $Y = \{y_k\}$ $(k = 0, \pm 1, \pm 2, \ldots)$ be a sequence of real numbers. We put $\Delta^0 y_k = y_k$, $\Delta y_k = y_{k+1} - y_k$ and $\Delta^n y_k = \Delta\Delta^{n-1} y_k$ for $k = 0, \pm 1, \pm 2, \ldots$ and $n = 0, 1, 2, \ldots$. It will be assumed that the sequence $\{\Delta^n y_k\} \in l_p$. Also, suppose given a nonnegative number M and suppose $\| Y\|_{l_p^{(n)}} = \|\Delta^n y_k\|_{l_p} \leqslant M$, where

$$\|\Delta^n y_k\|_{l_p} = \begin{cases} \left\{\sum\limits_{k=-\infty}^{\infty} |\Delta^n y_k|^p\right\}^{1/p}, & 1 \leqslant p < \infty, \\ \sup\limits_k |\Delta^n y_k|, & p = \infty. \end{cases}$$

Let $L_p^{(n)}$ denote the totality of functions defined on $I = (-\infty, \infty)$ that have a locally absolutely continuous $(n-1)$th derivative and a p-power $(p \geqslant 1)$ summable nth derivative. We introduce in $L_p^{(n)}$ the seminorm

$$\|f\|_{L_p^{(n)}} = \|f^{(n)}(x)\|_{L_p(I)}, \tag{1.1}$$

where

$$\|f^{(n)}(x)\|_{L_p(I)} = \begin{cases} \left\{\int\limits_I |f^{(n)}(x)|^p \, dx\right\}^{1/p}, & 1 \leqslant p < \infty, \\ \operatorname{ess\,sup}\limits_{x \in I} |f^{(n)}(x)|, & p = \infty. \end{cases} \tag{1.2}$$

We define a sequence $\{F_k\}$ of linear functionals by first letting

$$F(f) = \int\limits_{-\infty}^{\infty} f(x)\, dg(x), \tag{1.3}$$

where $g(x)$ is (i) a function of bounded variation if $n \geqslant 1$ or $n = 0$, $p = \infty$, and (ii) locally absolutely continuous with

$$g'(x) \in L_q(I) \left(\frac{1}{p} + \frac{1}{q} = 1,\ 1 < q \leqslant \infty\right)$$

if $n = 0$, and then letting

$$F_k(f) = \int\limits_{-\infty}^{\infty} f(x + kh)\, dg(x),\quad 0 < h < \infty,\quad k = 0, \pm 1, \pm 2, \ldots. \tag{1.4}$$

The problems considered below have the following general statements. Suppose

$$F_k(f) = y_k \quad (k = 0, \pm 1, \pm 2, \ldots) \tag{1.5}$$

and

$$\|\Delta^n y_k\|_{l_p} \leqslant M.$$

1. When does there exist, for any sequence Y with $\|\Delta^n y_k\|_{l_p} \leqslant M$, a function $f(x) \in L_p^{(n)}$ satisfying (1.5)?

2. What is the value of the norm of the best function for the worst sequence? In more formal language, we wish to find the value of the quantity

$$\sup_{\|Y\|_{l_p^{(n)}}\leqslant M}\ \inf_{\substack{f\in L_p^{(n)}\\ F_k(f)=y_k}} \|f\|_{L_p^{(n)}} = A_{n,p}. \tag{1.6}$$

3. When does there exist a bounded linear operator B from $l_p^{(n)}$ into $L_p^{(n)}$ that satisfies (1.5)?

4. What is the infimum of the norms of the linear operators from $l_p^{(n)}$ into $L_p^{(n)}$ that satisfy (1.5)?

We note that problems 1 and 3 were considered for

$$g(x) = \begin{cases} 1, & x \geqslant 0, \\ 0, & x < 0, \end{cases} \tag{1.7}$$

by Rjaben'kiĭ [1] (when $p = \infty$) (see also Rjaben'kiĭ and Filippov [2]) and Sobolev [3], who showed that the Rjaben'kiĭ process also yields finiteness when $1 \leqslant p < \infty$. The investigations of Holladay, Ahlberg, Nilson, Walsh and others (see, for example, [4]) connected with the minimal norm property also yield a positive answer to question 1 when $p = 2$, but in the case of a finite interval; the function $g(x)$ in this connection also has the form (1.7). In these cases the proof of finiteness is carried out with the use of a linear operator, so that the answer to question 3 is also positive.

In [5] and [6] the author found the infimum of the norms connected with problems 2 and 3 for the functional (1.7) and $1 \leqslant p \leqslant \infty$. More recently, this case was also considered for $p = 1, 2, \infty$ by Schoenberg [7], who found for $p = 2$ the best extension for an individual sequence Y.

Below we give a general lower bound for the quantity (1.6) and discuss the problem of obtaining upper bounds.

Suppose $f(x) \in L_p^{(n)}$ satisfies (1.5). Then

$$\Delta^n y_k = \int_{-\infty}^{\infty} \Delta^n f(kh+x)\,dg(x) = \int_{-\infty}^{\infty}\int_{-\infty}^{\infty} \varphi_n(kh-t) f^{(n)}(x+t)\,dt\,dg(x)$$
$$= \int_{-\infty}^{\infty} f^{(n)}(kh+t)\,dt \int_{-\infty}^{\infty} \varphi_n(x-t)\,dg(x), \tag{1.8}$$

where

$$\varphi_n(x-t) = \frac{1}{(n-1)!}\sum_{s=0}^{n} (-1)^{n-s} C_n^s (x+sh-t)_+^{n-1}, \quad -\infty < t < \infty, \tag{1.9}$$

and $x_+^m = \max(x^m, 0)$.

We will indicate a necessary condition for (i) the existence of a function $f(x) \in L_p^{(n)}$ satisfying condition (1.5) for any sequence Y with $\|Y\|_{l_p^{(n)}} \leqslant M$, and (ii) the uniform boundedness of the seminorms of such functions. We take an arbitrary sequence $\{\xi_k\} \in L_q$ $(p^{-1} + q^{-1} = 1)$ and form the sum

$$\sum_{k=-\infty}^{\infty} \xi_k \Delta^n y_k = \int_{-\infty}^{\infty} f^{(n)}(t) \sum_{k=-\infty}^{\infty} \xi_k \psi_k(t)\,dt, \tag{1.10}$$

where

$$\psi_k(t) = \int_{-\infty}^{\infty} \varphi_n(kh - t + x)\, dg(x), \qquad n \geqslant 1, \tag{1.11}$$

and $\psi_k(t) = dg(x - kh)$; $n = 0$. In the latter case it is also assumed when $1 \leqslant p < \infty$ that $g(x)$ is locally absolutely continuous. From (1.10), applying Hölder's inequality, we get

$$\Big| \sum_{k=-\infty}^{\infty} \xi_k \Delta^n y_k \Big| \leqslant \| f^{(n)} \|_{L_p(I)} \Big\| \sum_{k=-\infty}^{\infty} \xi_k \psi_k(t) \Big\|_{L_q(I)} \tag{1.12}$$

for a fixed sequence $\xi = \{\xi_k\}$ and an arbitrary $Y \in l_p^{(n)}$. It will be assumed in the sequel that $\|\xi_k\|_{l_q} = 1$. We put $\Delta^n y_k = |\xi_k|^{q-1} \operatorname{sgn} \xi_k$ and

$$\Phi(\xi, t) = \sum_{k=-\infty}^{\infty} \xi_k \psi_k(t). \tag{1.13}$$

Then

$$\| f^{(n)}(x) \|_{L_p(I)} \geqslant \| \Phi(\xi, t) \|_{L_q(I)}^{-1}$$

for any sequence $\xi = \{\xi_k\} \in l_q$ with $\|\xi\|_{l_q} = 1$; in particular,

$$\| f^{(n)}(x) \|_{L_p(I)} \geqslant \{\inf_{\|\xi\|_{l_q}=1} \| \Phi(\xi, t) \|_{L_q(I)}\}^{-1} \left(\frac{1}{p} + \frac{1}{q} = 1 \right). \tag{1.14}$$

A sharp lower bound for the quantity $A_{n,p}$ will be found by this method in the sequel for certain special cases.

We cite some sufficient conditions for the finiteness of $A_{n,p}$. It will be assumed for the sake of simplicity that $h = 1$. It is evident that this does not lead to a loss of generality, since we can always replace x by hu and put $g_1(u) = g(hu)$ and $f_1(u) = f(uh)$.

Suppose

$$\inf_{\|\xi\|_{l_q}=1} \| \Phi(\xi, t) \|_{L_q(I)} = \sigma > 0.$$

Two cases are possible: 1) the infimum is achieved at some element $\xi = \bar{\xi}$, or 2) an extremal element does not exist. In the first case we put $\bar{\Phi}(t) = \Phi(\bar{\xi}, t)$. In the second, we have a sequence of elements $\xi^{(m)}$ for which

$$\| \Phi(\xi^{(m)}, t) \|_{L_q(I)} \to \sigma$$

as $m \longrightarrow \infty$, so that for any $\epsilon \in (0, \sigma)$ there exists an m_0 such that $\|\Phi(\xi^{(m_0)}, t)\|_{L_q(I)} \geqslant \sigma - \epsilon$. In this case we put $\bar{\Phi}(t) = \Phi(\xi^{(m_0)}, t)$.

We will seek a function $f(x) \in L_p^{(n)}$ satisfying conditions (1.5) in the form

$$f^{(n)}(m + t + \alpha) = Z_m |\bar{\Phi}(m + t + \alpha)|^{q-1}, \qquad 0 \leqslant t < 1, \tag{1.15}$$

where the Z_m are unknown constants, α is a fixed number such that $0 \leqslant \alpha < 1$, and $m = 0, \pm 1, \pm 2, \ldots$. We substitute these values of $f^{(n)}(k + t + \alpha)$ in the right side of (1.8) for $h = 1$:

$$\Delta^n y_k = \sum_{m=-\infty}^{\infty} \int_{m+\alpha}^{m+1+\alpha} f^{(n)}(t) \int_{-\infty}^{\infty} \varphi_n(k-t+x)\, dg(x)\, dt$$

$$= \sum_{m=-\infty}^{\infty} \int_0^1 f^{(n)}(m+\alpha+t) \int_{-\infty}^{\infty} \varphi_n(k-m+\alpha-t+x)\, dg(x)\, dt$$

$$= \sum_{m=-\infty}^{\infty} Z_m \int_0^1 |\overline{\Phi}(m+t+\alpha)|^{q-1} \int_{-\infty}^{\infty} \varphi_n(k-m+\alpha-t+x)\, dg(x)\, dt.$$

Hence

$$\sum_{m=-\infty}^{\infty} a_{m,k} Z_m = \Delta^n y_k, \qquad k=0, \quad \pm 1, \quad \pm 2, \ldots, \tag{1.16}$$

where

$$a_{m,k} = \int_0^1 |\overline{\Phi}(m+t+\alpha)|^{q-1} \int_{-\infty}^{\infty} \varphi_n(k-m-\alpha+x-t)\, dg(x)\, dt. \tag{1.17}$$

If, for some α_0, the matrix of system (1.16) has a bounded inverse (from l_p into l_p) and

$$\sup_m \int_0^1 |\overline{\Phi}(m+t+\alpha_0)|^q\, dt < \infty,$$

then system (1.16) has a solution in l_p and the resultant function (1.15) belongs to $L_p^{(n)}$.

This is true, in particular, if

$$|a_{k,k}| - \sum_{m\neq k} |a_{m,k}| \geqslant \delta > 0$$

(see, for example, [8]).

One can hope to find in this way a best upper bound for the quantity $A_{n,p}$. A sharp upper bound of $A_{n,p}$ will in fact be obtained in the sequel by this approach for certain special functionals. In these cases $\overline{\Phi}(t)$ is 1-periodic and system (1.16) is simplified. As a result, the problem reduces to a discrete Wiener-Hopf problem; more precisely, to the solution of an infinite difference equation.

Let us show how it is possible in this way to obtain an upper bound for $A_{n,p}$ We put

$$f^{(n)}(m+\alpha+t) = Z_m \psi(t), \quad 0 \leqslant t < 1, \quad m=0, \ \pm 1, \ \pm 2, \ \ldots, \tag{1.18}$$

where α is a fixed number such that $0 \leqslant \alpha < 1$ and $\psi(t)$ is a measurable and q-power integrable function. In this case we have in place of (1.16) the difference equation

$$\sum_{m=-\infty}^{\infty} a_{m+k} Z_m = \Delta^n y_k \quad (k=0, \pm 1, \pm 2, \ldots), \tag{1.19}$$

where

$$a_m = a_m(\alpha) = \int_0^1 \psi(t) \int_{-\infty}^{\infty} \varphi_n(x-m-\alpha-t)\, dg(x)\, dt. \tag{1.20}$$

The following theorem holds.

THEOREM A. *If there exists an* $\alpha \in [0, 1)$ *such that* $\Sigma_{-\infty}^{\infty} a_m$ *is absolutely convergent and the function* $\eta(z) = \Sigma_{-\infty}^{\infty} a_m Z^m$ *does not have any zeros on the unit circle, then there exists a function* $f(x) \in L_p^{(n)}$ *satisfying* (1.5) *and the condition* $\|f^{(n)}\|_{L_p(I)} \leqslant K \|Y\|_{l_p^{(n)}}$, *where* K *is a constant not depending on* f *or* Y.

This theorem is a consequence of the following result.

THEOREM B [9]. *If the series of the* a_m *absolutely converges:* $\Sigma_{-\infty}^{\infty} |a_m| < \infty$, *and the function* $\eta(z) = \Sigma_{-\infty}^{\infty} a_m Z^m$ *does not vanish on the unit circle* $|Z| = 1$, *then equation* (1.19) *has the solution*

$$Z_m = \sum_{s=-\infty}^{\infty} \beta_{-s-m} \Delta^n y_s \ (m = 0, \pm 1, \pm 2, \ldots), \tag{1.21}$$

where $1/\eta(z) = \Sigma_{-\infty}^{\infty} \beta_k z^k$; *in this connection,*

$$\|Z_m\|_{l_p} \leqslant \sum_{k=-\infty}^{\infty} |\beta_k| \|\Delta^n y_k\|_{l_p}. \tag{1.22}$$

The absolute convergence of $\Sigma_{-\infty}^{\infty} \beta_k$ follows from a theorem of Wiener [10].

PROOF. We substitute (1.21) into (1.19):

$$\sum_{m=-\infty}^{\infty} a_{m+k} \sum_{s=-\infty}^{\infty} \beta_{-s-m} \Delta^n y_s = \sum_{s=-\infty}^{\infty} \Delta^n y_s \sum_{m=-\infty}^{\infty} a_{m+k} \beta_{-m-s} = \sum_{s=-\infty}^{\infty} \delta_{k,s} \Delta^n y_s = \Delta^n y_k,$$

where $\delta_{k,s}$ is the Kronecker delta. We have made use of the identity

$$1 = \eta(z) \frac{1}{\eta(z)} = \sum_{m=-\infty}^{\infty} a_m Z^m \sum_{k=-\infty}^{\infty} \beta_k Z^k = \sum_{m=-\infty}^{\infty} Z^m \sum_{s=-\infty}^{\infty} a_s \beta_{m-s}.$$

The multiplication of series is possible since both series converge absolutely. Further,

$$\|Z_m\|_{l_p} = \Big\{ \sum_{m=-\infty}^{\infty} \Big| \sum_{s=-\infty}^{\infty} \beta_{-m-s} \Delta^n y_s \Big|^p \Big\}^{1/p} = \Big\{ \sum_{m=-\infty}^{\infty} \Big| \sum_{s=-\infty}^{\infty} \beta_{-s} \Delta^n y_{s-m} \Big|^p \Big\}^{1/p}$$

$$\leqslant \sum_{s=-\infty}^{\infty} |\beta_{-s}| \Big\{ \sum_{m=-\infty}^{\infty} |\Delta^n y_{s-m}|^p \Big\}^{1/p} = \sum_{s=-\infty}^{\infty} |\beta_s| \|\Delta^n y_m\|_{l_p}.$$

REMARK 1.1. If $\{\Delta^n y_s\} \in l_p$ and $|\beta_s| = (-1)^s \beta_s \epsilon$, $s = 0, \pm 1, \pm 2, \ldots$, where ϵ is equal to 1 or -1, then the estimate

$$\|Z_m\|_{l_p} \leqslant \sum_{s=-\infty}^{\infty} |\beta_s| \|\Delta^n y_m\|_{l_p}$$

is unimprovable, i.e.

$$\sup_{\|\Delta^n y_m\|_{l_p}=1} \|Z_m\|_{l_p} = \sum_{s=-\infty}^{\infty} |\beta_s|. \tag{1.23}$$

PROOF. We put

$$\Delta^n y_s^* = \begin{cases} (-1)^s (2N+1)^{-1/p}, & |s| \leqslant N, \\ 0, & |s| > N, \quad N = r^2. \end{cases}$$

Then $\|\Delta^n y_s^*\|_{l_p} = 1$ and

$$\|Z_m\|_{l_p} = \Big\{ \sum_{m=-\infty}^{\infty} \Big| \sum_{s=-m-N}^{-m+N} \beta_s(-1)^{s+m}(2N+1)^{-\frac{1}{p}} \Big|^p \Big\}^{1/p}$$

$$\geqslant (2N+1)^{-1/p} \Big\{ \sum_{m=-N+\sqrt{N}}^{N-\sqrt{N}} \Big| \sum_{s=-N}^{N} \beta_{s-m}(-1)^s \Big|^p \Big\}^{1/p}$$

$$\geqslant (2N+1)^{-1/p} \Big\{ \Big[\sum_{m=-N+\sqrt{N}}^{N-\sqrt{N}} \Big(\sum_{s=-\infty}^{\infty} |\beta_{s-m}|^p \Big]^{1/p}$$

$$- \Big[\sum_{m=-N+\sqrt{N}}^{N-\sqrt{N}} \Big(\sum_{|s|>N} |\beta_{s-m}| \Big)^p \Big]^{1/p} \Big\}$$

$$\geqslant \left(\frac{2N-2\sqrt{N}+1}{2N+1} \right)^{1/p} \Big\{ \sum_{s=-\infty}^{\infty} |\beta_s| - \sum_{|s|>\sqrt{N}} |\beta_s| \Big\}.$$

Since the right side of the inequality tends to $\Sigma_{-\infty}^{\infty} |\beta_s|$ as $N \to \infty$, equality (1.23) is proved. Clearly, (1.23) also holds when β_s is constant in sign.

In the sequel, we will frequently deal with difference equations of the form

$$\sum_{k=0}^{n} b_k Z_{m+k} = C_m, \qquad m = 0, \quad \pm 1, \quad \pm 2, \ldots, \tag{1.24}$$

where the b_k and the C_m are known, while the Z_m are unknown quantities. It will be assumed that $\{C_m\} \in l_p$ $(1 \leqslant p \leqslant \infty)$.

By virtue of Theorem B, if the characteristic polynomial

$$P_n(x) = \sum_{k=0}^{n} b_k x^k \tag{1.25}$$

does not vanish on the unit circle, then the difference equation has a unique solution $\{Z_m^0\}$ in l_p.

LEMMA 1.1. *If all of the zeros* $\{x_s\}$ $(x_s \neq -1)$ *of the polynomial* (1.25) *are negative and mutually distinct, then*

$$P_n^{-1}(Z) = \sum_{s=-\infty}^{\infty} C_s Z^s, \tag{1.26}$$

$$C_0 = P_n^{-1}(0) + \sum_{s=1}^{k} \frac{x_s^{-1}}{P'_n(x_s)},$$

$$C_m = \sum_{s=1}^{k} \frac{x_s^{-m-1}}{P'_n(x_s)}, \qquad m = -1, -2, \ldots,$$

$$C_m = -\sum_{s=k+1}^{n} \frac{x_s^{-m-1}}{P'_n(x_s)}, \qquad m = 1, 2, \ldots, \tag{1.27}$$

where $-\infty < x_n < \cdots < x_{k+1} < -1 < x_k < \cdots < x_2 < x_1 < 0$ *are the zeros of* $P_n(x)$. *In this connection, if* $b_n > 0$, *then*

$$|C_m| = C_m(-1)^{m-k}, \qquad m = 0, \quad \pm 1, \pm 2, \ldots. \tag{1.28}$$

PROOF. In fact,

$$C_{-m-1} = \frac{1}{2\pi i} \int_{L^+} \frac{Z^m\, dZ}{P_n(Z)}, \qquad m \geqslant -1.$$

$$C_m = -\frac{1}{2\pi i} \int_{L^-} \frac{Z^{-m-1}}{P_n(Z)}\, dZ, \qquad m \geqslant 1,$$

where L^+ (L^-) is the unit circle oriented in the positive (negative) sense.

The equalities (1.27) follow upon calculating the residues of the corresponding integrands at the poles situated inside the contour. When $m = -1$ there is a residue at $Z = 0$ as well as at the zeros of the polynomial.

For a proof of (1.28) we note that

$$\frac{1}{P_n(-Z)} = \frac{(-1)^n}{b_n(Z + x_1)\dots(Z + x_n)}$$

$$= \frac{(-1)^n (x_{k+1}\dots x_n)^{-1}}{b_n Z^k \left(1 + \frac{x_1}{Z}\right)\dots\left(1 + \frac{x_k}{Z}\right)\left(1 + \frac{Z}{x_{k+1}}\right)\dots\left(1 + \frac{Z}{x_n}\right)}$$

$$= (-1)^k b_n^{-1} |x_{k+1}\dots x_n|^{-1} Z^{-k} \prod_{i=1}^{k} \sum_{s=0}^{\infty} \left(-\frac{x_i}{z}\right)^s \cdot \prod_{i=k+1}^{n} \sum_{s=0}^{\infty} \left(-\frac{Z}{x_i}\right)^s$$

$$= (-1)^k b_n^{-1} |x_{k+1}\dots x_n|^{-1} Z^{-k} \prod_{i=1}^{k} \sum_{s=0}^{\infty} |x_i|^s Z^{-s} \prod_{i=k+1}^{n} \sum_{s=0}^{\infty} |x_i|^{-s} Z^s$$

$$= (-1)^k \sum_{s=-\infty}^{\infty} \bar{C}_s Z^s,$$

where, clearly, $\bar{C}_s \geqslant 0$. Hence

$$\frac{1}{P_n(z)} = (-1)^k \sum_{s=-\infty}^{\infty} \bar{C}_s (-1)^s Z^s, \tag{1.29}$$

and the second part of the assertion of the lemma follows from the uniqueness of the Laurent expansion.

REMARK 1.2. If the polynomial $P_n(x)$ of the difference equation (1.24) satisfies the condition of the lemma, then Remark 1.1 and (1.29) imply

$$\|Z_m\|_{l_p} \leqslant \frac{\|C_m\|_{l_p}}{|P_n(-1)|}, \tag{1.30}$$

where $|P_n(-1)|^{-1}$ cannot be replaced by a smaller constant. For a narrower class of difference equations, (1.27) can be replaced by more suitable formulas. We formulate a corresponding result due to the author [6].

THEOREM C. *If the polynomial $P_{2n}(Z) = \Sigma_0^{2n} b_l Z^l$ satisfies the condition*

$$P_{2n}(Z) \equiv Z^{2n} P_{2n}\left(\frac{1}{Z}\right) \tag{1.31}$$

and all of its zeros are mutually distinct and do not lie on the unit circle while $\{C_m\} \in l_p$, then the difference equation

$$\sum_{l=0}^{2n} b_l Z_{m+l} = C_m \quad (m = 0, \pm 1, \pm 2, \dots) \tag{1.32}$$

has a unique solution $\{Z_m^0\} \in l_p$, *which is representable in the form*

$$Z_m^0 = \sum_{k=-\infty}^{\infty} \sum_{s=1}^{n} \frac{x_s^{n-1+|k|}}{P_{2n}^1(x_s)} C_{m-n+k}, \tag{1.33}$$

where the x_s $(s = 1, \ldots, n)$ *are the zeros of* $P_{2n}(x)$ *lying inside the disk* $|Z| < 1$. *If, in addition, all of these zeros are negative, then*

$$\sup_{\|c_m\|_{l_p} \leqslant 1} \|Z_m\|_{l_p} = |P_{2n}(-1)|^{-1}. \tag{1.34}$$

The existence of a bounded solution follows from Theorem B. We cite an example of the application of Theorem A.

EXAMPLE 1. Suppose $n = 0$,

$$g(x) = \begin{cases} -e^{-x} + 1, & x \geqslant 0, \\ e^{x} - 1, & x < 0 \end{cases} \tag{1.35}$$

and $\{y_k\} \in l_p$. Does there exist a function $f(x) \in L_p(-\infty, \infty)$ satisfying the conditions

$$y_k = \int_{-\infty}^{\infty} f(kh + x)\, dg(x) = \int_{-\infty}^{\infty} e^{-|x|} f(kh + x)\, dx, \ k = 0, \pm 1, \pm 2, \ldots, h > 0?$$

We put $\alpha = h/2$ and

$$f\left(h_s - \frac{h}{2} + th\right) = Z_s \, (s = 0, \pm 1, \ldots), \qquad 0 \leqslant t < 1.$$

We then get

$$Z_k 2(1 - e^{-h/2}) + \sum_{s=1}^{\infty} (Z_{k+s} + Z_{k-s})\, e^{(h/2)-sh}(1 - e^{-h}) = y_k,$$

and all of the conditions of Theorem A are satisfied.

In fact, the series

$$2(1 - e^{-h/2}) + \sum_{s=1}^{\infty} 2e^{(h/2)-sh}(1 - e^{-h})$$

is absolutely convergent, and the function

$$\eta(Z) = 2(1 - e^{-h/2}) + \sum_{s=1}^{\infty} (Z^s + Z^{-s})\, e^{(h/2)-sh}(1 - e^{-h})$$

$$= \frac{(1 - e^{-h/2})\,[e^{-h/2}(1 - e^{-h/2}) + 2Z(1 - e^{-3h/2}) + Z^2 e^{-h/2}(1 - e^{-h/2})]}{(1 - Ze^{-h})(Z - e^{-h})}$$

does not have any zeros on the unit circle. Moreover, according to the same theorem,

$$\|f\|_{L_p(I)} = h^{1/p} \|Z_m\|_{l_p} \leqslant \sum_{m=-\infty}^{\infty} |\beta_m| \, \|y_k\|_{l_p},$$

where the β_m are the Fourier coefficients of the function $\eta^{-1}(e^{i\theta})$. One cannot put $\alpha = 0$ in this example, since then $\eta(-1) = 0$.

The main goal of this paper is an exact solution of problems 2 and 4 for certain special functionals. The method of solution is the following. We obtain lower bounds with the use of (1.14), and deduce upper bounds by making use of a difference equation of form (1.32) and Theorem A. In this connection, we must show that it is possible to satisfy the conditions of Theorem C for the characteristic polynomials arising in the solution of the corresponding functional interpolation problems.

These auxiliary results are formulated and proved in the next section.

§2. On some properties of polynomials

Let

$$Q_n^*(t) = n! \sum_{l=0}^{n+1} (-1)^l \varphi_{n+1}(1-l-t), \tag{2.1}$$

where $\varphi_n(x-t)$ is defined by (1.9) for $h = 1$. Then for $0 \leqslant t \leqslant 1$

$$\begin{aligned} Q_n^*(t) &= \sum_{l=0}^{n+1} (-1)^l \sum_{s=l}^{n+1} (-1)^{n-s+1} C_{n+1}^s (s+1-l-t)^n \\ &= \sum_{l=0}^{n+1} \sum_{s=0}^{n+1-l} (-1)^{n-s+1} C_{n+1}^{s+l} (s+1-t)^n. \end{aligned} \tag{2.2}$$

The following lemma is valid.

LEMMA 2.1. *The polynomial* (2.1) *satisfies the following conditions*:

1) $$Q_n^{*\prime}(t) = 2\,n\,Q_{n-1}^*(t);$$

2) *it is even* (*odd*) *relative to the point* $t = 1/2$ *for* n *even* (*odd*);

3) $$|Q_n^*(t)| = (-1)^{[n/2]} Q_n^*(t), \quad \frac{1}{2} \leqslant t \leqslant 1;$$

4) $$Q_{2s}^*(1) = 0, \quad s \geqslant 1.$$

PROOF. Let $Q_n(x) = Q_n^*((1+x)/2)$. Then the following assertions are valid ([5], Lemma 4):

1) $$Q_n'(x) = nQ_{n-1}(x);$$

2) the polynomial $Q_n(x)$ is even (odd) for n even (odd);

3) $$|Q_n(x)| = (-1)^{[n/2]} Q_n(x), \quad 0 \leqslant x \leqslant 1.$$

Properties 1)–3) of Lemma 2.1 readily follow from these assertions. Property 4) follows from equality (46) in [5] and the fact that $B_{2s+1}(1/2) = B_{2s+1}(0)$ for $s \geqslant 1$, where the $B_n(x)$ are the Bernoulli polynomials (see, for example, [11] or [12]).

Let

$$\bar{g}_n(t-u, x) = \sum_{l=0}^{\infty} (l-1+t-u)_+^n x^l, \tag{2.3}$$

where, as before, $x_+^m = \max(x^m, 0)$ and

$$\begin{aligned} \bar{P}_{n+1}(t-u, x) &= (1-x)^{n+1} \bar{g}_n(t-u, x) \\ &= \sum_{l=0}^{n+2} \sum_{s=0}^{n+1} (-1)^s C_{n+1}^s (l-1-s+t-u)_+^n x^l. \end{aligned} \tag{2.4}$$

The following theorem on the zeros of algebraic polynomials is valid.

THEOREM 2.1. *If an algebraic polynomial of even degree is representable in either the form*

$$
\begin{aligned}
P_{2s+2}(x) &= \int_0^1 \varphi(t)\frac{1}{h}\int_{-h/2}^{h/2} \overline{P}_{2s+1}(t-u,x)\,du\,dt \\
&= \sum_{r=0}^{2s+2} x^r \int_0^1 \varphi(t)\frac{1}{h}\int_{-h/2}^{h/2} \sum_{k=0}^{2s+1}(-1)^k C_{2s+1}^k (t+r-k-1-u)_+^{2s}\,du\,dt
\end{aligned}
\tag{2.5}
$$

or the form

$$
\begin{aligned}
P_{2s}^*(x) &= \int_0^{1/2} \varphi^*(t)\,dt\frac{1}{h}\int_{-h/2}^{h/2} \overline{P}_{2s}(1+t-u,x)\,du \\
&+ \int_{1/2}^1 \varphi^*(t)\,dt\frac{1}{h}\int_{-h/2}^{h/2} x\overline{P}_{2s}(t-u,x)\,du \\
&= \sum_{r=0}^{2s} x^r \Big\{\int_0^{1/2} \varphi^*(t)\,dt\frac{1}{h}\int_{-h/2}^{h/2} \sum_{k=0}^{2s}(-1)^k C_{2s}^k (t+r-k-u)_+^{2s-1}\,du_+ \\
&+ \int_{1/2}^1 \varphi^*(t)\,dt\frac{1}{h}\int_{-h/2}^{h/2} \sum_{k=0}^{2s}(-1)^k C_{2s}^k (t+r-k-1-u)_+^{2s-1}\,du\Big\},
\end{aligned}
\tag{2.6}
$$

where $0 \leqslant h \leqslant 1$ *and* $\varphi(t)$ $(\varphi^*(t))$ *is nondecreasing (nonincreasing) on the interval* $[0, 1/2]$, *integrable, positive a.e. and such that* $\varphi(t) = \varphi(1-t)$ $(\varphi^*(t) = \varphi^*(1-t))$, *then it is symmetric, i.e.*

$$
P_{2s+2}(x) \equiv x^{2s+2}P_{2s+2}\left(\frac{1}{x}\right), \quad P_{2s}^*(x) \equiv x^{2s}P_{2s}^*\left(\frac{1}{x}\right), \tag{2.7}
$$

and its zeros are negative, mutually distinct and do not include the point $x = -1$.

This theorem was proved for $h = 0$ by the author in [6]. Analogous properties of the polynomial (2.5) for $h = 0$ and $\varphi(t) \equiv 1$ were studied still earlier by Peyerimhoff [13] and Hille [14] (see also Ahlberg, Nilson and Walsh [4], and Schoenberg [7]).

The proof of this theorem is prefaced by several lemmas. We first transform the polynomial (2.5).

From (2.4) and (2.3) we find that

$$
\overline{P}_n(t,x) = \frac{1}{(n+1)(1-x)}\frac{\partial}{\partial t}\overline{P}_{n+1}(t,x). \tag{2.8}
$$

Thus

$$P_{2s+2}(x)=\int_0^1 \varphi(t)\,dt\,\frac{1}{h}\int_{-h/2}^{h/2}\bar P_{2s+1}(t-u,x)\,du$$

$$=\frac{1}{(2s+2)(1-x)h}\int_0^1 \varphi(t)\left[\bar P_{2s+2}\left(t+\frac{h}{2},x\right)-\bar P_{2s+2}\left(t-\frac{h}{2},x\right)\right]dt$$

$$=\frac{1}{(2s+2)(1-x)h}\left\{\int_0^{1-(h/2)}\varphi(t)\bar P_{2s+2}\left(t+\frac{h}{2},x\right)dt\right. \tag{2.9}$$

$$+\int_{1-h/2}^{1}\varphi(t)\bar P_{2s+2}\left(t+\frac{h}{2},x\right)dt-\int_0^{h/2}\varphi(t)\bar P_{2s+2}\left(t-\frac{h}{2},x\right)dt$$

$$\left.-\int_{h/2}^{1}\varphi(t)\bar P_{2s+2}\left(t-\frac{h}{2},x\right)dt\right\}.$$

Let

$$P_n(t,x)=\sum_{l=0}^{n}x^l\sum_{k=0}^{l}(-1)^k C_{n+1}^k(l+t-k)^n \tag{2.10}$$

and

$$g_n(t,x)=\sum_{l=0}^{\infty}(l+t)^n x^l. \tag{2.11}$$

It is not difficult to verify [6] that

$$P_n(t,x)=(1-x)^{n+1}g_n(t,x). \tag{2.12}$$

From (2.10)–(2.12), (2.3) and (2.4) we find that:

a) for $0\leqslant t\leqslant 1-h/2$

$$\bar P_{2s+2}\left(t+\frac{h}{2},x\right)=(1-x)^{2s+2}\sum_{l=0}^{\infty}\left(l-1+t+\frac{h}{2}\right)_+^{2s+1}x^l$$

$$=(1-x)^{2s+2}x\sum_{l\geq 1}^{\infty}\left(l-1+t+\frac{h}{2}\right)^{2s+1}x^{l-1}=xP_{2s+1}\left(t+\frac{h}{2},x\right);$$

b) for $1-h/2\leqslant t\leqslant 1$

$$\bar P_{2s+2}\left(t+\frac{h}{2},x\right)=P_{2s+1}\left(t+\frac{h}{2}-1,x\right);$$

c) for $0\leqslant t\leqslant h/2$

$$\bar P_{2s+2}\left(t-\frac{h}{2},x\right)=x^2P_{2s+1}\left(1+t-\frac{h}{2},x\right);$$

d) for $h/2\leqslant t\leqslant 1$

$$\bar P_{2s+2}\left(t-\frac{h}{2},x\right)=xP_{2s+1}\left(t-\frac{h}{2},x\right).$$

We substitute these values of $\bar{P}_{2s+2}(u, x)$ into (2.9):

$$(2s+2)(1-x)hP_{2s+2}(x) = \int_0^{1-h/2} \varphi(t)\, xP_{2s+1}\left(t+\frac{h}{2}, x\right) dt$$

$$+ \int_{1-(h/2)}^{1} \varphi(t)\, P_{2s+1}\left(t+\frac{h}{2}-1, x\right) dt - \int_0^{h/2} \varphi(t)\, x^2 P_{2s+1}\left(1+t-\frac{h}{2}, x\right) dt$$

$$- \int_{h/2}^{1} \varphi(t)\, xP_{2s+1}\left(t-\frac{h}{2}, x\right) dt$$

$$= \int_0^{1/2} + \int_{1/2}^{1-(h/2)} + \int_{1-(h/2)}^{1} - \int_{0}^{h/2} - \int_{h/2}^{1/2} - \int_{1/2}^{1}, \quad 0 \leqslant h \leqslant 1.$$

By assumption, $\varphi((1+t)/2) \equiv \varphi((1-t)/2)$. We put

$$\psi(t) \equiv \frac{1}{2}\varphi\left(\frac{1+t}{2}\right) \equiv \frac{1}{2}\varphi\left(\frac{1-t}{2}\right)$$

and replace t by $(1-t)/2$ ($(1+t)/2$) in the first, fourth and fifth (second, third and sixth) integrals. We then get

$$(2s+2)(1-x)hP_{2s+2}(x) = \int_0^1 \psi(t)\, xP_{2s+1}\left(\frac{1-t+h}{2}, x\right) dt$$

$$+ \int_0^{1-h} \psi(t)\, xP_{2s+1}\left(\frac{1+t+h}{2}, x\right) dt + \int_{1-h}^{1} \psi(t)\, P_{2s+1}\left(\frac{t+h-1}{2}, x\right) dt$$

$$- \int_{1-h}^{1} \psi(t)\, x^2 P_{2s+2}\left(\frac{3-t-h}{2}, x\right) dt - \int_0^{1-h} \psi(t)\, xP_{2s+1}\left(\frac{1-t-h}{2}, x\right) dt$$

$$- \int_0^1 \psi(t)\, xP_{2s+1}\left(\frac{1+t-h}{2}, x\right) dt.$$

We replace t by $t-h$ (by $2-h-t$) in the second and fifth (third and fourth) integrals. Also, we split the first and last integrals into $\int_0^h$ and $\int_h^1$. In the integral $\int_0^h$ ($\int_h^1$) we replace t by $h-t$ (by $h+t$). Then

$$
\begin{aligned}
(2s+2)(1-x)hP_{2s+2}(x) &= \int_0^h \psi(h-t)\,xP_{2s+1}\left(\frac{1+t}{2}, x\right)dt \\
&+ \int_0^{1-h} \psi(t+h)\,xP_{2s+1}\left(\frac{1-t}{2}, x\right)dt + \int_h^1 \psi(t-h)\,xP_{2s+1}\left(\frac{1+t}{2}, x\right)dt \\
&+ \int_{1-h}^1 \psi(2-h-t)\,P_{2s+1}\left(\frac{1-t}{2}, x\right)dt \\
&- \int_{1-h}^1 \psi(2-h-t)\,x^2P_{2s+1}\left(\frac{1+t}{2}, x\right)dt \qquad (2.13)\\
&- \int_h^1 \psi(t-h)\,xP_{2s+1}\left(\frac{1-t}{2}, x\right)dt - \int_0^h \psi(h-t)\,xP_{2s+1}\left(\frac{1-t}{2}, x\right)dt \\
&- \int_0^{1-h} \psi(h+t)\,xP_{2s+1}\left(\frac{1+t}{2}, x\right)dt.
\end{aligned}
$$

We will prove a series of properties of the polynomials $P_n(t, x)$ defined by equalities (2.10)–(2.12).

LEMMA 2.2. *The polynomials $P_n(t, x)$ satisfy the recurrence relation*

$$P_{n+1}(t, x) = x(1-x)P_n'(t, x) + [t + (n+1-t)x]\,P_n(t, x) \qquad (2.14)$$

relative to x.

PROOF. In fact,

$$
\begin{aligned}
P_{n+1}(t, x) &= (1-x)^{n+2}\sum_{l=0}^{\infty}(l+t)^{n+1}x^l = (1-x)^{n+2}x\sum_{l=0}^{\infty} l(l+t)^n x^{l-1} \\
&+ (1-x)^{n+2}t\sum_{l=0}^{\infty}(l+t)^n x^l = x(1-x)^{n+2}g_n'(t, x) + (1-x)tP_n(t, x). \qquad (2.15)
\end{aligned}
$$

Here and below the differentiation is carried out with respect to x. We differentiate (2.12) with respect to x:

$$P_n'(t, x) = -(n+1)(1-x)^n g_n(t, x) + (1-x)^{n+1}g_n'(t, x).$$

Thus

$$(1-x)^{n+2}g_n'(t, x) = (1-x)P_n^1(t, x) + (n+1)P_n(t, x).$$

From this and (2.15) we get (2.14), and the lemma is proved.

LEMMA 2.3. *The polynomial $P_n(t, x)$ has n mutually distinct negative zeros relative to x for any $t \in (0, 1)$.*

The proof is carried out by induction. From the definition of $P_n(t, x)$ it follows that $P_0(t, x) \equiv 1$. Further, from the recurrence relation (2.14) we have

$$P_1(t, x) = t + (1 - t)\, x,$$

and the lemma is correct for $n = 1$. Suppose it is correct for $n = k$, i.e. the polynomial $P_k(t, x)$ has k distinct negative zeros $-\infty < x_k < \cdots < x_2 < x_1 < 0$. Since all of the zeros of $P_k(t, x)$ are simple and (2.11), (2.12) imply $P_k(t, 0) = t^k$ for any k, it follows that

$$\operatorname{sgn} P_k'(t, x_i) = (-1)^{i-1} \qquad (i = 1, 2, \ldots, k).$$

We will prove that $P_{k+1}(t, x)$ then has $k + 1$ distinct negative zeros. Since the x_i are the zeros of $P_k(t, x)$, from (2.14) we have

$$\operatorname{sgn} P_{k+1}(t, x_i) = (-1)^i \qquad (i = 1, 2, \ldots, k).$$

Also, as was already noted, $P_{k+1}(t, 0) = t^{k+1} > 0$. Let a_k denote the coefficient of the highest power of x in $P_k(t, x)$. Then the recurrence relation (2.14) implies $a_{k+1} = -ka_k + (k + 1 - t)a_k = (1 - t)a_k$, and hence $a_{k+1} = (1 - t)^{k+1}$ since $a_0 = 1$. There consequently exists a point $x_{k+1} < x_k$ such that

$$\operatorname{sgn} P_{k+1}(t, x_{k+1}) = (-1)^{k+1}.$$

Thus $P_{k+1}(t, x)$ has $k + 2$ variations in sign on $(-\infty, 0)$, and hence $k + 1$ distinct negative zeros. The lemma is proved.

REMARK 2.1. It follows from the proof of the lemma that the polynomials $P_n(t, x)$ and $P_{n+1}(t, x)$ have alternating zeros for any fixed $t \in (0, 1)$. This holds, in particular, for the polynomials $P_n(1/2, x)$ and $P_{n+1}(1/2, x)$.

LEMMA 2.4. *The zeros of the polynomials $P_n(t, x)$ strictly decrease with increasing t $(0 \leqslant t \leqslant 1)$.*

PROOF. From Lemma 2.3 it follows that $P_n(t, x)$ has n distinct negative zeros x_i^0 $(i = 1, \ldots, n)$ for each fixed $t_0 \in (0, 1)$. Since $\partial P_n(t, x)/\partial x$ and $\partial P_n(t, x)/\partial t$ are continuous, and since $\partial P_n(t_0, x_i^0)/\partial x \neq 0$ and $P_n(t_0, x_i^0) = 0$, we see that the equation $P_n(t, x) = 0$ determines x_i as an implicit function of t: $x_i = x_i(t)$, in a neighborhood of the point (t_0, x_i^0). In this connection,

$$x_i'(t)_t = -\frac{\partial P_n(t, x_i)/\partial t}{\partial P_n(t, x_i)/\partial x_i}.$$

We note that

$$\frac{\partial P_n(t, x)}{\partial t} = n\,(1 - x)\, P_{n-1}(t, x).$$

Further, from the recurrence relation (2.14) we have

$$\left.\frac{\partial P_n(t, x)}{\partial x}\right|_{x = x_i} = \frac{P_{n+1}(t, x_i)}{x_i\,(1 - x_i)}.$$

Thus

$$\frac{dx_i(t)}{dt} = -\frac{n x_i\,(1 - x_i)^2\, P_{n-1}(t, x_i)}{P_{n+1}(t, x_i)}. \tag{2.16}$$

As was already noted, $P_k(t, 0) > 0$ and, according to Remark 2.1, the polynomials $P_{n-1}(t, x)$ and $P_n(t, x)$ have alternating zeros. Therefore

$$\operatorname{sgn} P_{n-1}(t, x_i) = (-1)^{i-1}, \qquad i = 1, 2, \ldots, n. \tag{2.17}$$

Further, the recurrence relation (2.14) implies $P_{n+1}(t, x_i)$ and $P_n'(t, x_i)$ have opposite signs. But $\operatorname{sgn} P_n'(t, x_i) = (-1)^{i-1}$, $i = 1, 2, \ldots, n$, and consequently

$$\operatorname{sgn} P_{n+1}(t, x_i) = (-1)^i, \; i = 1, 2, \ldots, n.$$

From this result, (2.17) and (2.16) it follows that $dx_i(t)/dt < 0$ and hence that the zeros $x_i(t)$ strictly decrease with increasing t. Lemma 2.4 is proved.

We will study some properties of the polynomials $P_n(0, x)$ and $P_n(1, x)$. From (2.14) we have

$$P_{n+1}(0, x) = x(1-x) P_n'(0, x) + (n+1) x P_n(0, x), \tag{2.18}$$

$$P_{n+1}(1, x) = x(1-x) P_n'(1, x) + (1+nx) P_n(1, x). \tag{2.19}$$

It follows from (2.18) that $P_n(0, 0) = 0$ for $n \geqslant 1$. Let $R_n(x) = x^{-1} P_{n-1}(0, x)$. Then

$$R_{n-1}'(x) = -\frac{1}{x^2} P_n(0, x) + \frac{1}{x} P_n'(0, x)$$

or

$$\frac{1}{x} P_n'(0, x) = R_{n-1}'(x) + \frac{1}{x^2} P_n(0, x),$$

which together with (2.18) implies

$$R_n(x) = x(1-x) R_{n-1}'(x) + (1+nx) R_{n-1}(x), \; n \geqslant 1.$$

In this connection, $R_0(x) \equiv 1$. We further note that $P_{n+1}(1, x)$ is a polynomial of degree n. In fact, since $P_0(1, x) \equiv 1$, it follows from (2.19) that $P_1(1, x) \equiv 1$, and the assertion is proved. By virtue of the fact that $R_n(x)$ and $P_{n+1}(1, x)$ satisfy one and the same recurrence relation and $P_1(1, x) \equiv R_0(x) \equiv 1$, we have $R_n(x) \equiv P_{n+1}(1, x)$, $n \geqslant 0$, i.e.

$$P_{n+1}(1, x) \equiv \frac{1}{x} P_{n+1}(0, x), \; n \geqslant 0. \tag{2.20}$$

In the same way as in the proof of Lemma 2.3, we establish that $R_n(x)$ has n distinct negative zeros.

REMARK 2.2. It follows from the latter fact and (2.20) that the negative zeros of $P_{n+1}(1, x)$ and $P_{n+1}(0, x)$ coincide, are mutually distinct and are equal to n in number. Moreover, the polynomial $P_{n+1}(1, x)$ does not have any other zeros since its degree is equal to n. The polynomial $P_{n+1}(0, x)$ has one additional zero at $x = 0$ $(n \geqslant 0)$.

LEMMA 2.5. *The polynomials $P_{n+1}(1, x)$ and $P_{n+1}(1/2, x)$ have alternating zeros for $n \geqslant 1$.*

The proof is carried out by induction. Since $P_0(t, x) \equiv 1$, we have from (2.14) that

$$P_2(1,x) = 1 + x,\ P_2(1/2, x) = \frac{x^2+6x+1}{4},$$

and the assertion is proved for $n = 1$. Suppose $P_n(1, x)$ and $P_n(1/2, x)$ have alternating zeros. Then

$$\frac{P_n(1,x)}{P_n\left(\frac{1}{2}, x\right)} = \frac{A_1}{x-x_1} + \ldots + \frac{A_n}{x-x_n}, \tag{2.21}$$

where all of the A_i have the same sign and the x_i are the zeros of the polynomial $P_n(1/2, x)$ $(i = 1, \ldots, n)$ (see [15], p. 18).

We will show that there is a zero of $P_{n+1}(1, x)$ between any two adjacent zeros $\bar{x}_1$ and $\bar{x}_2$ of $P_{n+1}(1/2, x)$. The polynomial $P_{n+1}(1/2, x)$ has $n + 1$ distinct negative zeros by virtue of Lemma 2.3, while $P_{n+1}(1, x)$, which is of degree n, has n distinct negative zeros according to Remark 2.2. By virtue of (2.14),

$$P_{n+1}(1/2, x) = x(1-x)P_n'(1/2, x) + [1/2 + (n+1/2)x]\, P_n(1/2, x). \tag{2.22}$$

We multiply (2.22) by $P_n(1, x)$, and (2.19) by $P_n(1/2, x)$. Subtracting the first product from the second and making use of (2.21), we get

$$\begin{aligned} P_{n+1}(1,x)P_n(1/2,x) - P_n(1,x)P_{n+1}(1/2,x) &= x(1-x)\Big[P_n'(1,x)P_n\left(\frac{1}{2},x\right) \\ &\quad - P_n'\left(\frac{1}{2},x\right)P_n(1,x)\Big] + \frac{1-x}{2}P_n(1,x)P_n(1/2,x) \\ &= \frac{1-x}{2}P_n^2(1/2,x)\left\{2x\left[\frac{P_n(1,x)}{P_n(1/2,x)}\right]' + \frac{P_n(1,x)}{P_n(1/2,x)}\right\} \\ &= \frac{1-x}{2}P_n^2(1/2,x)\left[\sum_{i=1}^n \frac{A_i}{x-x_i} - \sum_{i=1}^n \frac{2xA_i}{(x-x_i)^2}\right] \\ &= -\frac{1-x}{2}P_n^2(1/2,x)\sum_{i=1}^n \frac{A_i(x+x_i)}{(x-x_i)^2}. \end{aligned} \tag{2.23}$$

The right side of (2.23) does not change in sign for $x < 0$ $(x \neq x_i)$ since all of the A_i have the same sign while $x_i < 0$ $(i = 1, \ldots, n)$. We put $x = \bar{x}_k$ $(k = 1, 2)$ in (2.23). Then

$$P_{n+1}(1, \bar{x}_k) = \frac{1}{P_n(1/2, \bar{x}_k)}\left[-\frac{(1-\bar{x}_k)P_n^2(1/2,\bar{x}_k)}{2}\sum_{i=1}^n \frac{A_i(x_i+\bar{x}_k)}{(\bar{x}_k - x_i)^2}\right].$$

According to what was proved above, the expression in square brackets does not change in sign, while $P_n(1/2, \bar{x}_1)$ and $P_n(1/2, \bar{x}_2)$ have opposite signs since $P_n(1/2, x)$ and $P_{n+1}(1/2, x)$ have alternating zeros by virtue of Remark 2.1. It also follows from the fact that these polynomials have alternating zeros that $P_n(1/2, \bar{x}_k) \neq 0$ $(k = 1, 2)$. Thus $P_{n+1}(1, \bar{x}_1)$ and $P_{n+1}(1, \bar{x}_2)$ have alternating signs, and hence $P_{n+1}(1, x)$ has a zero in the interval $(\bar{x}_1, \bar{x}_2)$. Since $\bar{x}_1$ and $\bar{x}_2$ are any two adjacent zeros of $P_{n+1}(1/2, x)$, the lemma follows.

We now formulate a fundamental lemma.

LEMMA 2.6. *Let x_i $(i = 1, \ldots, 2s+1)$ be the mutually distinct zeros of the* *polynomial* $P_{2s+1}(1/2, x)$. *Then for* $0 < t < 1$

$$\operatorname{sgn} P_{2s+1}\left(\frac{1+t}{2}, x_i\right) = - \operatorname{sgn} P_{2s+1}\left(\frac{1-t}{2}, x_i\right) = (-1)^i, \quad -\infty < x_{2s+1} < \ldots < x_2 < x_1 < 0. \tag{2.24}$$

PROOF. The fact that $P_{2s+1}(1/2, x)$ has $2s + 1$ distinct negative zeros follows from Lemma 2.3, which also implies this property for the polynomials $P_{2s+1}((1 + t)/2, x)$ and $P_{2s+1}((1 - t)/2, x)$ when $t \in (0, 1)$. Let the zeros of the polynomial $P_{2s+1}((1 + t)/2, x)$ ($P_{2s+1}((1 - t)/2, x)$) be denoted by $z_i = z_i(t)$ ($\bar{z}_i = \bar{z}_i(t)$), $i = 1, \ldots, 2s + 1$, $-\infty < z_{2s+1} < \cdots < z_2 < z_1 < 0$ ($-\infty < \bar{z}_{2s+1} < \cdots < \bar{z}_2 < \bar{z}_1 < 0$). Clearly, $z_i(0) = \bar{z}_i(0) = x_i$. Further, it was proved in Lemma 2.4 that the zeros of $P_{2s+1}(t, x)$ strictly decrease with increasing t. Consequently, $z_i(t)$ ($\bar{z}_i(t)$) strictly decreases (increases) with increasing t, i.e. for $0 < t < 1$

$$z_i(t) < x_i < \bar{z}_i(t), \quad i = 1, 2, \ldots, 2s + 1. \tag{2.25}$$

The negative zeros of $P_{2s+1}(0, x)$ and $P_{2s+1}(1, x)$ coincide by virtue of Remark 2.2. We denote them by y_i, $-\infty < y_{2s} < \cdots < y_2 < y_1 < 0$. In addition, $P_{2s+1}(0, 0) = 0$ and $P_{2s+1}(1, x)$ is a polynomial of degree $2s$. From Lemma 2.4 and Remark 2.2 it follows, in particular, that

$$z_i(t) \to y_i, \; \bar{z}_{i+1}(t) \to y_i \; (i = 1, 2, \ldots, 2s), \quad \bar{z}_1(t) \to 0, \; z_{2s+1}(t) \to -\infty \; \text{ for } t \to 1. \tag{2.26}$$

It follows from Lemma 2.5 that $P_{2s+1}(1, x)$ and $P_{2s+1}(1/2, x)$ have alternating zeros, i.e.

$$-\infty < x_{2s+1} < y_{2s} < x_{2s} < \ldots < y_2 < x_2 < y_1 < x_1 < 0. \tag{2.27}$$

Comparing (2.25) and (2.27), as well as (2.26), we find that

$$-\infty < z_{2s+1}(t) < x_{2s+1} < \bar{z}_{2s+1}(t) < y_{2s} < z_{2s}(t) < x_{2s} < \bar{z}_{2s}(t) < \ldots < z_2(t) < x_2 < \bar{z}_2(t) < y_1 < z_1(t) < x_1 < \bar{z}_1(t) < 0. \tag{2.28}$$

Since $P_{2s+1}(t, x)$ has only simple zeros relative to x for $t \in [0, 1]$ while $P_{2s+1}(t, 0) > 0$ for $t \in (0, 1)$, the validity of Lemma 2.6 follows from (2.28).

Let us complete the proof of the theorem for the polynomial (2.5). We assume initially that $h \geqslant 1/2$. Then from (2.13) for $x = x_i$ we have

$$
\begin{aligned}
(2s+2)(1-x_i)hP_{2s+2}(x_i) &= \int_0^{1-h} \psi(h-t)\,x_iP_{2s+1}\left(\frac{1+t}{2}, x_i\right)dt \\
&+ \int_{1-h}^{h} \psi(h-t)\,x_iP_{2s+1}\left(\frac{1+t}{2}, x_i\right)dt + \int_0^{1-h} \psi(h+t)\,x_iP_{2s+1}\left(\frac{1-t}{2}, x_i\right)dt \\
&+ \int_h^1 \psi(t-h)\,x_i\left[P_{2s+1}\left(\frac{1+t}{2}, x_i\right) - P_{2s+1}\left(\frac{1-t}{2}, x_i\right)\right]dt \\
&+ \int_{1-h}^{1} \psi(2-h-t)\left[P_{2s+1}\left(\frac{1-t}{2}, x_i\right) - x_i^2P_{2s+1}\left(\frac{1+t}{2}, x_i\right)\right]dt \\
&- \int_0^{1-h} \psi(h-t)\,x_iP_{2s+1}\left(\frac{1-t}{2}, x_i\right)dt - \int_{1-h}^{h} \psi(h-t)\,x_iP_{2s+1}\left(\frac{1-t}{2}, x_i\right)dt \\
&- \int_0^{1-h} \psi(h+t)\,x_iP_{2s+1}\left(\frac{1+t}{2}, x_i\right)dt \qquad (2.29) \\
&= \int_0^{1-h} [\psi(h-t)-\psi(h+t)]\,x_i\left[P_{2s+1}\left(\frac{1+t}{2}, x_i\right) - P_{2s+1}\left(\frac{1-t}{2}, x_i\right)\right]dt \\
&+ \int_{1-h}^{h} \psi(h-t)\,x_i\left[P_{2s+1}\left(\frac{1+t}{2}, x_i\right) - P_{2s+1}\left(\frac{1-t}{2}, x_i\right)\right]dt \\
&+ \int_h^1 \psi(t-h)\,x_i\left[P_{2s+1}\left(\frac{1+t}{2}, x_i\right) - P_{2s+1}\left(\frac{1-t}{2}, x_i\right)\right]dt \\
&+ \int_{1-h}^{1} \psi(2-h-t)\left[P_{2s+1}\left(\frac{1-t}{2}, x_i\right) - x_i^2P_{2s+1}\left(\frac{1+t}{2}, x_i\right)\right]dt.
\end{aligned}
$$

Since, by assumption of the theorem, $\psi(t) = \varphi((1-t)/2) = \varphi((1+t)/2) \geqslant 0$ for $t \in [0, 1]$, with the strict inequality holding on a subset of $[0, 1]$ of positive measure, and $\psi(h-t) - \psi(h+t) \geqslant 0$ for $t \in [0, 1-h]$ (and $h \in [1/2, 1]$), it follows from Lemma 2.6 that the integrands in (2.29) do not change in sign, and hence

$$\operatorname{sgn} P_{2s+2}(x_i) = (-1)^i, \quad i = 1, 2, \ldots, \quad 2s+1. \qquad (2.30)$$

Further, as was noted in the proof of Lemma 2.3, $P_{2s+1}(t, 0) = t^{2s+1}$ for $t \in (0, 1)$, and therefore (2.29) with x_i replaced by 0 implies ($\varphi(t)$ is nondecreasing on $(0, 1/2)$)

$$(2s+2)hP_{2s+2}(0) = \int_{1-h}^{1} \psi(2-h-t)\,P_{2s+1}\left(\frac{1-t}{2}, 0\right)dt > 0. \qquad (2.31)$$

We prove in addition that $P_{2s+2}(x)$ satisfies (2.7), i.e. $b_{2s+2-r} = b_r$ $(r = 0, 1, \ldots, s)$, where b_r is the coefficient of x^r in $P_{2s+2}(x)$. In fact,

$$b_{2s+2-r} = \int_0^1 \varphi(t)\frac{1}{h}\int_{-h/2}^{h/2} \sum_{k=0}^{2s+1} (-1)^k C_{2s+1}^k (t+2s+1-r-k-u)_+^{2s}\,du\,dt.$$

Here we replace k by $2s+1-k$, t by $1-t$ and u by $-u$ and make use of the fact that $\varphi(1-t) = \varphi(t)$. We then get

$$b_{2s+2-r} = -\int_0^1 \varphi(t)\frac{1}{h}\int_{-h/2}^{h/2}\sum_{k=0}^{2s+1}(-1)^k C_{2s+1}^k (1-t-r+k+u)_+^{2s}\,du\,dt$$

or

$$b_r - b_{2s-2-r} = \int_0^1 \varphi(t)\frac{1}{h}\int_{-h/2}^{h/2}\sum_{k=0}^{2s+1}(-1)^k C_{2s+1}^k [(t+\bar{u}_k)_+^{2s} + (-t-\bar{u}_k)_+^{2s}]\,du\,dt,$$

where $\bar{u}_k = r - k - 1 - u$. But

$$(t+\bar{u}_k)_+^{2s} + (-t-\bar{u}_k)_+^{2s} = (t+\bar{u}_k)^{2s}.$$

Therefore

$$\sum_{k=0}^{2s+1}(-1)^k C_{2s+1}^k [(t+\bar{u}_k)_+^{2s} + (-t-\bar{u}_k)_+^{2s}] \equiv 0 \tag{2.32}$$

and hence $b_r = b_{2s+2-r}$ $(r = 0, 1, \ldots, s)$ for $0 \leqslant h \leqslant 1$. (We recall that $s(0) = \lim_{h\to 0} s(h)$ if $s(h) = h^{-1}\int_{-h/2}^{h/2} f(t)\,dt$; in particular, $s(0) = f(0)$ if $f(t)$ is continuous at $t = 0$.)

Thus the polynomial $P_{2s+2}(x)$ satisfies (2.7). We also note that (2.31) and the equality $b_0 = b_{2s+2}$ imply the existence of a point $x_{2s+2} < x_{2s+1}$ such that

$$P_{2s+2}(x_{2s+2}) > 0. \tag{2.33}$$

The assertion of the theorem concerning the zeros of the polynomial (2.5) for $h \geqslant 1/2$ follows from (2.30), (2.31) and (2.33).

It remains to consider the case $0 \leqslant h < 1/2$. Here we split the integrals from 0 to $1 - h$ (from h to 1) into integrals from 0 to h and from h to $1 - h$ (from h to $1 - h$ and from $1 - h$ to 1). We then transform $P_{2s+2}(x)$ into the form

$$\begin{aligned}(2s+2)(1-x)P_{2s+2}(x) &= \frac{1}{h}\int_0^h [\psi(h-t)-\psi(h+t)]\,x\Big[P_{2s+1}\Big(\frac{1+t}{2}, x\Big)\\ &- P_{2s+1}\Big(\frac{1-t}{2}, x\Big)\Big]dt + \frac{1}{h}\int_h^{1-h}[\psi(t-h)-\psi(t+h)]\,x\Big[P_{2s+1}\Big(\frac{1+t}{2}, x\Big)\\ &- P_{2s+1}\Big(\frac{1-t}{2}, x\Big)\Big]dt + \frac{1}{h}\int_{1-h}^1 \psi(t-h)\,x\Big[P_{2s+1}\Big(\frac{1+t}{2}, x\Big) - P_{2s+1}\Big(\frac{1-t}{2}, x\Big)\Big]dt\\ &+ \frac{1}{h}\int_{1-h}^1 \psi(2-h-t)\Big[P_{2s+1}\Big(\frac{1-t}{2}, x\Big) - x^2 P_{2s+1}\Big(\frac{1+t}{2}, x\Big)\Big]dt.\end{aligned}$$

The rest of the proof is completed in exactly the same way as for $h \geqslant 1/2$.

REMARK 2.3. When $h = 0$ we have $P_{2s+2}(x) = x\,Q_{2s}(x)$, where $Q_{2s}(x)$ is a polynomial of degree $2s$ such that

$$(2s+2)(1-x)\,Q_{2s}(x) = -\int_0^1 \psi'(t)\Big[P_{2s+1}\Big(\frac{1+t}{2}, x\Big) - P_{2s+1}\Big(\frac{1-t}{2}, x\Big)\Big]dt$$

and $\operatorname{sgn} Q_{2s}(x_i) = (-1)^i$, $i = 1, \ldots, 2s + 1$, if $\psi'(t)$ is different from zero on a set of positive measure, nonnegative and summable.

Let us prove the theorem for the polynomial $P^*_{2s}(x)$. We denote the coefficient of x^r in $P^*_{2s}(x)$ by a_r. Then, by virtue of (2.6),

$$a_r = \int_0^{1/2} \varphi^*(t) \frac{1}{h} \int_{-h/2}^{h/2} \sum_{k=0}^{2s} (-1)^k C_{2s}^k (t + r - k - u)_+^{2s-1}\, du + \int_{1/2}^{1} \varphi^*(t) \frac{1}{h} \int_{-h/2}^{h/2} \sum_{k=0}^{2s} (-1)^k C_{2s}^k (t + r - k - 1 - u)_+^{2s-1}\, du. \tag{2.34}$$

Let us show that $P^*_{2s}(x)$ satisfies (2.7). In (2.34) we replace r by $2s - r$, k by $2s - k$, t by $1 - t$, u by $-u$ and make use of the fact that $\varphi^*(1 - t) = \varphi^*(t)$. We then get

$$a_{2s-r} = \int_{1/2}^{1} \varphi^*(t) \frac{1}{h} \int_{-h/2}^{h/2} \sum_{k=0}^{2s} (-1)^k C_{2s}^k (1 - t - r + k + u)_+^{2s-1}\, du + \int_0^{1/2} \varphi^*(t) \frac{1}{h} \int_{-h/2}^{h/2} \sum_{k=0}^{2s} (-1)^k C_{2s}^k (- t - r + k + u)_+^{2s-1}\, du.$$

Since

$$(y - k - u)_+^{2s-1} - (k + u - y)_+^{2s-1} = (y - k - u)^{2s-1}, \tag{2.34a}$$

it follows in the same way as for $P_{2s+2}(x)$ (see (2.32)) that $a_r = a_{2s-r}$ $(r = 0, 1, \ldots, s)$, and hence that (2.7) holds.

To complete the proof, we transform $P^*_{2s}(x)$ into a more suitable form. From (2.6) and (2.8) we find that

$$\begin{aligned} P^*_{2s}(x) &= \frac{1}{h} \int_0^{1/2} \varphi^*(t)\, dt \int_{-h/2}^{h/2} \bar{P}_{2s}(1 + t - u, x)\, du + \frac{1}{h} \int_{1/2}^{1} \varphi^*(t)\, dt \int_{-h/2}^{h/2} \bar{P}_{2s}(t - u, x)\, du \\ &= \left\{ \int_0^{1/2} \varphi^*(t) \left[\bar{P}_{2s+1}\left(1 + t + \frac{h}{2}, x\right) - \bar{P}_{2s+1}\left(1 + t - \frac{h}{2}, x\right) \right] dt \right. \\ &\quad \left. + \int_{1/2}^{1} \varphi^*(t) \left[\bar{P}_{2s+1}\left(t + \frac{h}{2}, x\right) - \bar{P}_{2s+1}\left(t - \frac{h}{2}, x\right) \right] dt \right\} \frac{1}{(2s+1)(1-x)h}. \end{aligned}$$

We replace t by $(1 - t)/2$ (by $(1 + t)/2$) in the first (second) integral and put

$$2\psi^*(t) = \varphi^*\left(\frac{1-t}{2}\right) = \varphi^*\left(\frac{1+t}{2}\right).$$

Then

$$2(2s+1)(1-x)hP^*_{2s}(x)=\int_0^1\psi^*(t)\bar{P}_{2s+1}\left(\frac{3-t+h}{2},x\right)dt$$
$$-\int_0^1\psi^*(t)\bar{P}_{2s+1}\left(\frac{3-t-h}{2},x\right)dt+\int_0^1\psi^*(t)\bar{P}_{2s+1}\left(\frac{1+t+h}{2},x\right)dt \tag{2.35}$$
$$-\int_0^1\psi^*(t)\bar{P}_{2s+1}\left(\frac{1+t-h}{2},x\right)dt.$$

From (2.5), (2.6) and (2.10)–(2.12) we find that:

a) for $0\leqslant t\leqslant 1$

$$\bar{P}_{2s+1}\left(\frac{3-t+h}{2},x\right)=(1-x)^{2s+1}\sum_{l=0}^{\infty}\left(l+\frac{1}{2}+\frac{h-t}{2}\right)_+^{2s}x^l$$
$$=(1-x)^{2s+1}\sum_{l=0}^{\infty}\left(l+\frac{1}{2}+\frac{h-t}{2}\right)^{2s}x^l=P_{2s}\left(\frac{1+h-t}{2},x\right);$$

b) for $0\leqslant t\leqslant 1-h$

$$\bar{P}_{2s+1}\left(\frac{3-t-h}{2},x\right)=P_{2s}\left(\frac{1-h-t}{2},x\right);$$

c) for $1-h\leqslant t\leqslant 1$

$$\bar{P}_{2s+1}\left(\frac{1-t-h}{2},x\right)=xP_{2s}\left(\frac{3-h-t}{2},x\right);$$

d) for $0\leqslant t\leqslant 1-h$

$$\bar{P}_{2s+1}\left(\frac{1+t+h}{2},x\right)=xP_{2s}\left(\frac{1+t+h}{2},x\right);$$

e) for $1-h\leqslant t\leqslant 1$

$$\bar{P}_{2s+1}\left(\frac{1+t+h}{2},x\right)=P_{2s}\left(\frac{t+h-1}{2},x\right);$$

f) for $0\leqslant t\leqslant 1$

$$\bar{P}_{2s+1}\left(\frac{1+t-h}{2},x\right)=xP_{2s}\left(\frac{1+t-h}{2},x\right).$$

We split the first integral in (2.35) into $\int_0^h+\int_h^1$, the second into $\int_0^{1-h}+\int_{1-h}^1$ the third into $\int_0^{1-h}+\int_{1-h}^1$ and the fourth into $\int_0^h+\int_h^1$. We thus have eight integrals arranged in the indicated order. We now replace t by $h-t$ in the first integral, t by $t+h$ in the second, t by $t-h$ in the third, t by $2-t-h$ in the fourth, t by $t-h$ in the fifth, t by $2-t-h$ in the sixth, t by $h-t$ in the seventh and t by $t+h$ in the eighth. We then get

$$
\begin{aligned}
2(2s+1)(1-x)hP_{2s}^{*}(x) &= \int_0^h \psi^*(h-t)\,P_{2s}\left(\frac{1+t}{2},x\right)dt \\
&+ \int_0^{1-h} \psi^*(h+t)\,P_{2s}\left(\frac{1-t}{2},x\right)dt - \int_h^1 \psi^*(t-h)\,P_{2s}\left(\frac{1-t}{2},x\right)dt \\
&- \int_{1-h}^1 \psi^*(2-t-h)\,xP_{2s}\left(\frac{1+t}{2},x\right)dt + \int_h^1 \psi^*(t-h)\,xP_{2s}\left(\frac{1+t}{2},x\right)dt \\
&+ \int_{1-h}^1 \psi^*(2-t-h)\,P_{2s}\left(\frac{1-t}{2},x\right)dt - \int_0^h \psi^*(h-t)\,xP_{2s}\left(\frac{1-t}{2},x\right)dt \\
&- \int_0^{1-h} \psi^*(h+t)\,xP_{2s}\left(\frac{1+t}{2},x\right)dt.
\end{aligned}
\tag{2.36}
$$

For $0 < h \leqslant 1/2$ we write (2.36) in the form

$$
\begin{aligned}
2(2s+1)(1-x)hP_{2s}^{*}(x) &= \int_0^h \psi^*(h-t)\left[P_{2s}\left(\frac{1+t}{2},x\right) - xP_{2s}\left(\frac{1-t}{2},x\right)\right]dt \\
&+ \int_0^h \psi^*(h+t)\left[P_{2s}\left(\frac{1-t}{2},x\right) - xP_{2s}\left(\frac{1+t}{2},x\right)\right]dt \\
&+ \int_h^{1-h} [\psi^*(t+h) - \psi^*(t-h)]\left[P_{2s}\left(\frac{1-t}{2},x\right) - xP_{2s}\left(\frac{1+t}{2},x\right)\right]dt \\
&+ \int_{1-h}^1 [\psi^*(2-t-h) - \psi^*(t-h)]\left[P_{2s}\left(\frac{1-t}{2},x\right) - xP_{2s}\left(\frac{1+t}{2},x\right)\right]dt.
\end{aligned}
\tag{2.37}
$$

For $1/2 < h \leqslant 1$ we have

$$
\begin{aligned}
2(2s+1)(1-x)hP_{2s}^{*}(x) &= \int_0^h \psi^*(h-t)\left[P_{2s}\left(\frac{1+t}{2},x\right) - xP_{2s}\left(\frac{1-t}{2},x\right)\right]dt \\
&+ \int_0^{1-h} \psi^*(h+t)\left[P_{2s}\left(\frac{1-t}{2},x\right) - xP_{2s}\left(\frac{1+t}{2},x\right)\right]dt \\
&+ \int_h^1 [\psi^*(2-t-h) - \psi^*(t-h)]\left[P_{2s}\left(\frac{1-t}{2},x\right) - xP_{2s}\left(\frac{1+t}{2},x\right)\right]dt \\
&+ \int_{1-h}^h \psi^*(2-t-h)\left[P_{2s}\left(\frac{1-t}{2},x\right) - xP_{2s}\left(\frac{1+t}{2},x\right)\right]dt.
\end{aligned}
\tag{2.38}
$$

By virtue of Remark 2.2, the polynomial $P_{2s}(1, x)$ has $2s - 1$ distinct negative zeros $-\infty < y_{2s-1} < \cdots < y_2 < y_1 < 0$, which we denote by $\{y_i\}$. We denote by $\{z_i\}$ the zeros $-\infty < z_{2s} < \cdots < z_2 < z_1 < 0$ of the polynomial $P_{2s}(1/2, x)$.

Since, according to Lemma 2.5, $P_{2s}(1, x)$ and $P_{2s}(1/2, x)$ have alternating zeros, it follows that

$$
-\infty < z_{2s} < y_{2s-1} < z_{2s-1} < \ldots < y_2 < z_2 < y_1 < z_1 < 0. \tag{2.39}
$$

The identity (2.20) implies that $P_{2s}(0, x)$ vanishes at $x = y_i$ $(i = 1, \ldots, 2s - 1)$ as well as at $x = 0$. By virtue of Lemma 2.3, the polynomial $P_{2s}(t, x)$ has $2s$ distinct negative zeros $\{x_i(t)\}$, $i = 1, \ldots, 2s$, for $0 < t < 1$.

From Lemma 2.4 it follows that these zeros strictly decrease with increasing t. Consequently, for $0 < t < 1$

$$y_i < x_i\left(\frac{1+t}{2}\right) < z_i, \qquad i = 1, 2, \ldots, \quad 2s-1,$$
$$z_i < x_i\left(\frac{1-t}{2}\right) < y_{i-1}, \qquad i = 2, 3, \ldots, \quad 2s. \tag{2.40}$$

In addition, $z_1 < x_1((1-t)/2) < 0$ and $-\infty < x_{2s}((1+t)/2) < z_{2s}$. These inequalities together with (2.39) and (2.40) imply

$$x_{i+1}\left(\frac{1+t}{2}\right) < y_i < x_i\left(\frac{1+t}{2}\right), \quad i = 1, 2, \ldots, \quad 2s-1,$$
$$x_{i+1}\left(\frac{1-t}{2}\right) < y_i < x_i\left(\frac{1-t}{2}\right), \quad i = 1, 2, \ldots, \quad 2s-1. \tag{2.41}$$

It was noted in the proof of Lemma 2.4 that $P_k(t, 0) > 0$ for $0 < t < 1$. As a result, we get from (2.41) that

$$\operatorname{sgn} P_{2s}\left(\frac{1+t}{2}, y_i\right) = \operatorname{sgn} P_{2s}\left(\frac{1-t}{2}, y_i\right) = (-1)^i, \qquad i = 1, 2, \ldots, 2s-1,$$
$$0 < t < 1. \tag{2.42}$$

Since $y_i < 0$ and $\psi^*(t)$ ($\psi^*(t) = \varphi^*((1+t)/2) = \varphi^*((1-t)/2)$) is nondecreasing, it follows from (2.42) that all of the integrands in (2.37) and (2.38) either vanish or have the sign $(-1)^i$ at $x = y_i$ ($i = 1, 2, \ldots, 2s-1$). Thus $P^*_{2s}(y_i) = (-1)^i$ ($i = 1, \ldots, 2s-1$). In addition,

$$P^*_{2s}(0) = \int_0^{1/2} \varphi^*(t) \frac{1}{h} \int_{-h/2}^{h/2} (t-u)_+^{2s-1}\, du + \int_{1/2}^{1} \varphi^*(t)\, dt \frac{1}{h} \int_{-h/2}^{h/2} (t-1-u)_+^{2s-1}\, du > 0.$$

Since the polynomial $P^*_{2s}(x)$ is symmetric, we can again, as in the case of $P_{2s+2}(x)$, establish that $P^*_{2s}(x)$ has $2s$ distinct negative zeros.

Since $P^*_{2s}(x)$ and $P_{2s+2}(x)$ are symmetric and have simple zeros, it follows that the point $x = -1$ cannot be a zero of these polynomials, inasmuch as it would then be a multiple zero. The theorem is completely proved.

§3. Exact solutions of the functional interpolation problem for special functionals

We will assume that $n \geqslant 1$ and that the functions under consideration have a locally absolutely continuous $(n-1)$th derivative. Suppose given a sequence $Y = \{y_m\} \in l_p^{(n)}$, i.e.

$$\|Y\|_{l_p^{(n)}} = \begin{cases} \left\{\sum_{m=-\infty}^{\infty} |\Delta^n y_m|^p\right\}^{1/p}, & 1 \leqslant p < \infty \\ \sup_m |\Delta^n y_m|; & p = \infty \end{cases} < \infty, \tag{3.1}$$

where $\Delta^n y_m = \sum_{s=0}^n (-1)^{n-s} C_n^s y_{m+s}$, and a sequence of functionals

$$F_m(f) = \frac{1}{h} \int_{-h/2}^{h/2} f(m+x)\, dx, \qquad 0 \leqslant h \leqslant 1, \qquad m = 0, \pm 1, \pm 2, \ldots. \tag{3.2}$$

When $h = 0$, we put $F_m(f) = f(m)$. We will regard two sequences in $l_p^{(n)}$ as identical if they differ by the values at integral points of a polynomial of degree at most $n - 1$. We will consider in conjunction with $l_p^{(n)}$ the space $L_p^{(n)}$ of functions defined on the real line whose $(n - 1)$th derivative is locally absolutely continuous and whose nth derivative belongs to $L_p(-\infty, \infty)$ $(1 \leqslant p \leqslant \infty)$. We will regard two functions in $L_p^{(n)}$ as identical if they differ by a polynomial of degree at most $n - 1$, and we define the norm in $L_p^{(n)}$ as follows:

$$\|f\|_{L_p^{(n)}} = \begin{cases} \left\{ \int\limits_{-\infty}^{\infty} |f^{(n)}(t)|^p \, dt \right\}^{1/p}, & 1 \leqslant p < \infty, \\ \operatorname*{ess\,sup}_t |f^{(n)}(t)|, & p = \infty. \end{cases}$$

Does there exist a function $f(x) \in L_p^{(n)}$ such that $F_m(f) = y_m$ $(m = 0, \pm 1, \pm 2, \ldots)$ for any sequence $Y = \{y_m\} \in l_p^{(n)}$? If the answer is positive, what is the value of the quantity

$$A_{n,p}(M) = \sup_{\|Y\|_{l_p^{(n)}} \leqslant M} \; \inf_{f \in \Phi(Y)} \|f\|_{L_p^{(n)}}, \tag{3.3}$$

where $\Phi(Y)$ is the set of functions $\{f\}$ satisfying the condition

$$\frac{1}{h} \int\limits_{-h/2}^{h/2} f(m+t)\, dt = y_m \quad (m = 0, \pm 1, \pm 2, \ldots)? \tag{3.4}$$

It will be proved below that $A_{n,p}$ is finite, and its exact value will be found. We have

THEOREM 3.1. *For any natural number n, any $p \in (1, \infty]$ and any $h \in [0, h]$*

$$A_{n,p}(M) = (n-1)! \left(\|S_n(t)\|_{L_q(0,1)}\right)^{-1} M, \qquad \frac{1}{p} + \frac{1}{q} = 1, \tag{3.5}$$

where

$$S_{2s}(t) = \begin{cases} \dfrac{1}{4s}\left[Q_{2s}^*\left(\dfrac{h}{2}+t\right) + Q_{2s}^*\left(\dfrac{h}{2}-t\right)\right], & 0 \leqslant t \leqslant \dfrac{h}{2}, \\ \dfrac{1}{4s}\left[Q_{2s}^*\left(t+\dfrac{h}{2}\right) - Q_{2s}^*\left(t-\dfrac{h}{2}\right)\right], & \dfrac{h}{2} \leqslant t \leqslant 1-\dfrac{h}{2}, \\ -\dfrac{1}{4s}\left[Q_{2s}^*\left(1+\dfrac{h}{2}-t\right) + Q_{2s}^*\left(-1+\dfrac{h}{2}+t\right)\right], & 1-\dfrac{h}{2} \leqslant t \leqslant 1, \end{cases} \tag{3.6}$$

if $n = 2s$, and

$$S_{2s+1}(t) = \begin{cases} \dfrac{1}{4s+2}\left[Q_{2s+1}^*\left(\dfrac{h}{2}+t\right) - Q_{2s+1}^*\left(\dfrac{h}{2}-t\right)\right], & 0 \leqslant t \leqslant \dfrac{h}{2}, \\ \dfrac{1}{4s+2}\left[Q_{2s+1}^*\left(t+\dfrac{h}{2}\right) - Q_{2s+1}^*\left(t-\dfrac{h}{2}\right)\right], & \dfrac{h}{2} \leqslant t \leqslant 1-\dfrac{h}{2}, \\ \dfrac{1}{4s+2}\left[Q_{2s+1}^*\left(1+\dfrac{h}{2}-t\right) - Q_{2s+1}^*\left(-1+t+\dfrac{h}{2}\right)\right], & 1-\dfrac{h}{2} \leqslant t \leqslant 1, \end{cases} \tag{3.7}$$

if $n = 2s + 1$. The polynomial $Q_n^*(x)$ is defined by (2.1) *and* (2.2), *i.e.*

$$Q_n^*(t) = \sum_{l=0}^{n+1} \sum_{k=0}^{n+1-l} (-1)^{n-k+1} C_{n+1}^{k+l} (k+1-t)^n. \tag{3.8}$$

From the proof of Theorem 3.1 we get

THEOREM 3.2. *There exists a bounded linear operator B acting from $l_p^{(n)}$ into $L_p^{(n)}$ and satisfying the conditions $F_m(BY) = y_m$ with norm equal to $A_{n,p}(1)$.*

PROOF OF THEOREM 3.1. Let $f(x)$ be a function satisfying conditions (3.4). Representing it by Taylor's formula with remainder in integral form and calculating the nth-order differences with step equal to unity of the left and right sides of (3.4), we get

$$\begin{aligned}(n-1)!\,\Delta^n y_m &= \frac{(n-1)!}{h}\int_{-h/2}^{h/2} \Delta^n f(m+x)\,dx \\ &= \frac{1}{h}\int_{-h/2}^{h/2} dx \int_0^n \sum_{k=0}^{n} (-1)^{n-k} C_n^k (k-t)_+^{n-1} f^{(n)}(m+x+t)\,dt \\ &= \frac{1}{h}\int_{-h/2}^{h/2} dx \int_{1+x}^{n+1+x} \sum_{k=0}^{n} (-1)^{n-k} C_n^k (k+1-t+x)_+^{n-1} f^{(n)}(m-1+t)\,dt \\ &= \frac{1}{h}\int_{-h/2}^{h/2} dx \int_0^{n+2} \sum_{k=0}^{n} (-1)^{n-k} C_n^k (k+1+x-t)_+^{n-1} f^{(n)}(m-1+t)\,dt \\ &= \sum_{s=0}^{n+1}\int_0^1 f^{(n)}(m-1+s+t)\,dt\,\frac{1}{h}\int_{-h/2}^{h/2}\sum_{k=0}^{n}(-1)^{n-k} C_n^k (k+1+x-t-s)_+^{n-1}dx.\end{aligned} \tag{3.9}$$

We will find a lower bound for the quantity $A_{n,p}(M)$. To this end we make use of the general method for obtaining lower bounds presented in §1. Let N be a natural number. In (3.9) we put

$$\Delta^n y_m = \Delta^n y_m^* = \begin{cases} 0, & m < 0, \\ (-1)^m MN^{-1/p}, & 0 \leqslant m \leqslant N-1, \\ 0, & m \geqslant N, \end{cases}$$

multiply both sides by $(-1)^m$ and sum over m from 0 to $N-1$. Then

$$\begin{aligned}(n-1)!\,M = N^{-1/q} &\sum_{m=0}^{N-1} (-1)^m \sum_{s=0}^{n+1}\int_0^1 f^{(n)}(m-1+s+t)\,dt \\ &\times \frac{1}{h}\int_{-h/2}^{h/2}\sum_{k=0}^{n}(-1)^{n-k} C_n^k (k+1+x-t-s)_+^{n-1}\,dx.\end{aligned} \tag{3.10}$$

In the sums over m and s we introduce new indices of summation by putting $s = s$ and $l = m + s$. Then

$$M(n-1)! = N^{-1/q}\Big\{\int_0^1 \sum_{l=0}^{n} (-1)^l f^{(n)}(l-1+t) \sum_{s=0}^{l} (-1)^s a_s(t)\,dt$$
$$+\int_0^1 \sum_{l=n+1}^{N-1} (-1)^l f^{(n)}(t+l-1) \sum_{s=0}^{n+1} (-1)^s a_s(t)\,dt + \int_0^1 \sum_{l=N}^{N+n} (-1)^l f^{(n)}(t+l-1)$$
$$\times \sum_{s=l-N+1}^{n+1} (-1)^s a_s(t)\,dt\Big\} = N^{-1/q}\Big\{\int_n^{N-1} f^{(n)}(t) S_n(t)\,dt + \int_{-1}^{n} f^{(n)}(t) b_1(t)\,dt \tag{3.11}$$
$$+\int_{N-1}^{N-n} f^{(n)}(t) b_2(t)\,dt\Big\},$$

$$a_s(t) = \frac{1}{h}\int_{-h/2}^{h/2} \sum_{k=0}^{n} (-1)^{n-k} C_n^k (k+1+x-t-s)_+^{n-1}\,dt, \tag{3.12}$$
$$S_n(t+l) = (-1)^l \sum_{s=0}^{n+1} (-1)^s a_s(t),$$

$0 \leqslant t \leqslant 1$, $n+1 \leqslant l \leqslant N-1$, i.e. $|S(t)|$ is a 1-periodic function,

$$b_1(l-1+t) = (-1)^l \sum_{s=0}^{l} (-1)^s a_s(t),$$

$$b_2(N+l+t-1) = (-1)^{l+N} \sum_{s=l+1}^{n+1} (-1)^s a_s(t),$$

$0 \leqslant t \leqslant 1$, $l = 0, 1, \ldots, n$. We note that $b_1(t)$ and $b_2(t)$ do not depend on N. We estimate the right side of (3.11) with the use of Holder's inequality:

$$M(n-1)! \leqslant N^{-1/q}\Big\{\Big[\int_n^{N-1} |f^{(n)}(t)|^p\,dt\Big]^{1/p}\Big[\int_n^{N-1} |S_n(t)|^q\,dt\Big]^{1/q} + \Big[\int_{-1}^{n} |f^{(n)}(t)|^p\,dt\Big]^{1/p}$$
$$\times \Big[\int_{-1}^{n} |b_1(t)|^q\,dt\Big]^{1/q} + \Big[\int_{N-1}^{N+n} |f^{(n)}(t)|^p\,dt\Big]^{1/p}\Big[\int_{N-1}^{N+n} |b_2(t)|^q\,dt\Big]^{1/q}$$
$$\leqslant N^{-1/q}\|f^{(n)}\|_{L_p(-\infty,\infty)}\{(N-n)^{1/q}\|S_n(t)\|_{L_q(0,1)} + O(1)\},$$

where $O(1)$ is a constant depending only on n. Thus

$$A_{n,p}(M) \geqslant \inf_{f\in\Phi(Y^*)} \|f^{(n)}\|_{L_p} \geqslant \frac{M(n-1)!\,N^{1/q}}{(N+1-n)^{1/q}\|S_n(t)\|_{L_q(0,1)} + O(1)}.$$

Since this inequality is valid for any N, we have

$$A_{n,p}(M) \geqslant \frac{M(n-1)!}{\|S_n(t)\|_{L_q(0,1)}}, \tag{3.13}$$

where $S_n(t)$ is defined by (3.12) for $l = 0$.

We will show that estimate (3.13) is sharp. As a preliminary, we will study some of the properties of $S_n(t)$ for $0 \leqslant t \leqslant 1$.

Suppose $n = 2r$. Then from (3.12) we have

$$S_{2r}(t) = \frac{1}{h}\int_{-h/2}^{h/2}\sum_{s=1}^{2r+1}(-1)^s\sum_{k=0}^{2r}(-1)^k C_{2r}^k (k+1+x-t-s)_+^{2r-1}\,dx.$$

We transform $S_{2r}(t)$, taking (2.2) into account. For $0 \leqslant t \leqslant h/2$ we have

$$\begin{aligned} S_{2r}(t) &= \frac{1}{h}\int_{-h/2}^{t}\sum_{s=1}^{2r}(-1)^s\sum_{k=s}^{2r}(-1)^k C_{2r}^k (k+1+x-t-s)^{2r-1}\,dx \\ &+ \frac{1}{h}\int_{t}^{h/2}\sum_{s=2}^{2r+1}(-1)^s\sum_{k=s-1}^{2r}(-1)^k C_{2r}^k (k+1+x-t-s)^{2r-1}\,dx \\ &= -\frac{1}{h}\int_{-h/2}^{t}\sum_{s=0}^{2r-1}(-1)^s\cdot\sum_{k=s+1}^{2r}(-1)^k C_{2r}^k (k+x-t-s)^{2r-1}\,dx \\ &+ \frac{1}{h}\int_{t}^{h/2}\sum_{s=0}^{2r-1}(-1)^s\sum_{k=s+1}^{2r}(-1)^k C_{2r}^k (k+x-t-s-1)^{2r-1}\,dx \\ &= \frac{1}{h}\int_{-h/2}^{t} Q_{2r-1}^*(t-x)\,dx - \frac{1}{h}\int_{t}^{h/2} Q_{2r-1}^*(1-x+t)\,dx; \end{aligned}$$

for $h/2 \leqslant t \leqslant 1 - h/2$

$$\begin{aligned} S_{2r}(t) &= \frac{1}{h}\int_{-h/2}^{h/2}\sum_{s=1}^{2r}(-1)^s\sum_{k=s}^{2r}(-1)^k C_{2r}^k (k+1+x-t-s)^{2r-1}\,dx \\ &= -\frac{1}{h}\int_{-h/2}^{h/2}\sum_{s=0}^{2r-1}(-1)^s\sum_{k=s+1}^{2r}(-1)^k C_{2r}^k (k+x-t-s)^{2r-1}\,dx \\ &= \frac{1}{h}\int_{-h/2}^{h/2} Q_{2r-1}^*(t-x)\,dx; \end{aligned}$$

and for $1 - h/2 \leqslant t \leqslant 1$

$$\begin{aligned} S_{2r}(t) &= \frac{1}{h}\int_{-h/2}^{t-1}\sum_{s=0}^{2r-1}(-1)^s\sum_{k=s+1}^{2r}(-1)^k C_{2r}^k (k+x+1-t-s)^{2r-1}\,dx \\ &+ \frac{1}{h}\int_{t-1}^{h/2}\sum_{s=1}^{2r}(-1)^s\sum_{k=s}^{2r}(-1)^k C_{2r}^k (k+1+x-t-s)^{2r-1}\,dx \\ &= -\frac{1}{h}\int_{-h/2}^{t-1} Q_{2r-1}^*(t-1-x)\,dx + \frac{1}{h}\int_{t-1}^{h/2} Q_{2r-1}^*(t-x)\,dx. \end{aligned}$$

The polynomial $Q_{2r-1}^*(x)$ is odd relative to the point $x = 1/2$ by virtue of Lemma 2.3. Hence the above equalities imply

$$S_{2r}(t)=\begin{cases}\frac{1}{h}\int\limits_{-h/2}^{h/2} Q^*_{2r-1}(|t-x|)\,dx, & 0\leqslant t\leqslant h/2,\\ \frac{1}{h}\int\limits_{-h/2}^{h/2} Q^*_{2r-1}(t-x)\,dx, & h/2\leqslant t\leqslant 1-h/2,\\ -\frac{1}{h}\int\limits_{-h/2}^{h/2} Q^*_{2r-1}(|t-1-x|)\,dx, & 1-h/2\leqslant t\leqslant 1.\end{cases} \quad (3.14)$$

Suppose $n = 2r + 1$. Then, analogously to the preceding, we transform $S_{2r+1}(t)$, $0 \leqslant t \leqslant 1$, into the form

$$S_{2r+1}(t)=\begin{cases}\frac{1}{h}\Big[\int\limits_{-h/2}^{t} Q^*_{2r}(t-x)\,dx-\int\limits_{t}^{h/2} Q^*_{2r}(x-t)\,dx\Big], & 0\leqslant t\leqslant h/2,\\ \frac{1}{h}\int\limits_{-h/2}^{h/2} Q^*_{2r}(t-x)\,dx, & h/2\leqslant t\leqslant 1-h/2,\\ \frac{1}{h}\Big[-\int\limits_{-h/2}^{-1+t} Q^*_{2r}(-1+t-x)\,dx+\int\limits_{-1+t}^{h/2} Q^*_{2r}(1+x-t)\,dx, & 1-h/2\leqslant t\leqslant 1.\end{cases}$$

(3.15)

Taking into account property 1) of the polynomials $Q_n^*(x)$ in Lemma 2.1 as well as property 4), we arrive at (3.6) and (3.7). The following lemma discusses certain properties of the function $S_n(t)$. From (3.6) and (3.7) it is easily seen that $S_n(t)$ is a polynomial of degree $n - 1$.

LEMMA 3.1. *The function $S_n(t)$ is even (odd) relative to the point $t = 1/2$ for odd (even) n. In addition, on the interval $(0, 1/2)$ it does not change in sign, and $|S_n(t)|$ is strictly decreasing (strictly increasing) for n even (odd), so that $|S_n(t)| < 0$ for $t \in (0, 1/2)$.*

PROOF. Suppose $n = 2r$ and $0 \leqslant t \leqslant h/2$. Then from (3.6) we have

$$S_{2r}(t)=\frac{1}{4r}[Q^*_{2r}(h/2+t)+Q^*_{2r}(h/2-t)],$$

and $S_{2r}(t)$ does not change in sign on $(0, h/2]$ since, according to property 2) of Lemma 2.1, $Q_{2r}^*(t)$ does not change in sign on $(0, 1)$. Further, from properties 3) and 1) of Lemma 2.1 we have

$$|S_{2r}(t)|'=\frac{1}{4r}(-1)^r[Q^{*\prime}_{2r}(h/2+t)+Q^{*\prime}_{2r}(h/2-t)] \qquad (3.16)$$
$$=(-1)^r[Q^*_{2r-1}(h/2+t)-Q^*_{2r-1}(h/2-t)],$$

Properties 1)–3) of Lemma 2.1 imply that $|Q_{2r-1}^*(x)|$ is strictly decreasing on $[0, 1/2]$. Also, $Q_{2r-1}^*(x)$ is an odd function relative to the point $x = 1/2$. Therefore $Q_{2r-1}^*(h/2 + t)$ and $-Q_{2r-1}^*(h/2 - t)$ have the same sign for $h/2 + t > 1/2$ by virtue of the oddness of $Q_{2r-1}^*(x)$ relative to $x = 1/2$, and it accordingly follows from (3.16) by virtue of property 3) that $|S_{2r}(t)|' < 0$. If, on the other hand, $h/2 + t \leqslant 1/2$, the analogous assertion follows from properties 2) and 3) and the fact that $|Q_{2r-1}^*(x)|$ is strictly increasing on $[0, 1/2]$.

Suppose $h/2 \leqslant t \leqslant 1/2$. It then follows from (3.6) that the sign of $S_{2r}(t)$ is determined by the sign of $Q_{2r}^*(t + h/2)$. In fact, the function $Q_{2r}^*(x)$ is even relative to $x = 1/2$, does not change in sign on $(0, 1)$ and, as a result of properties 1) and 2) of Lemma 2.1, is increasing on $(0, 1/2)$; also, $t - h/2 < 1 - h/2 - t$ under our restrictions on t. Therefore

$$\left| Q_{2r}^*\left(t+\frac{h}{2}\right)\right| > \left| Q_{2r}^*\left(t-\frac{h}{2}\right)\right|.$$

From this inequality, properties 2) and 3) of Lemma 2.1 and property 1) we deduce that

$$|S_{2r}(t)|' = \frac{1}{4r}(-1)^r\left[Q_{2r}^{*\prime}\left(t+\frac{h}{2}\right) - Q_{2r}^{*\prime}\left(t-\frac{h}{2}\right)\right]$$
$$= (-1)^r\left[Q_{2r-1}^*\left(t+\frac{h}{2}\right) - Q_{2r-1}^*\left(t-\frac{h}{2}\right)\right].$$

Further, in exactly the same way as in the preceding case, we establish that $|S_{2r}(t)|' < 0$ for $h/2 \leqslant t \leqslant 1/2$. Let us prove that $S_{2r}(t)$ is odd relative to $t = 1/2$. Suppose $t = 1/2 + t_1$, where $0 \leqslant t_1 \leqslant (1 - h)/2$. Then from (3.6) and the evenness of $Q_{2r}^*(x)$ relative to $x = 1/2$ we have

$$S_{2r}(t) = S_{2r}(^1/_2 + t_1) = \frac{1}{4r}\left[Q_{2r}^*\left(^1/_2 - t_1 - \frac{h}{2}\right) - Q_{2r}^*\left(^1/_2 + t_1 - \frac{h}{2}\right)\right]$$
$$= \frac{1}{4r}\left[Q_{2r}^*\left(^1/_2 - t_1 - \frac{h}{2}\right) - Q_{2r}^*\left(^1/_2 - t_1 + \frac{h}{2}\right)\right] = -S_{2r}(^1/_2 - t_1).$$

For $t = 1/2 + t_1$, $(1 - h)/2 \leqslant t_1 \leqslant 1/2$, we have

$$S_{2r}(^1/_2 + t_1) = -\frac{1}{4r}\left[Q_{2r}^*\left(^1/_2 + \frac{h}{2} - t_1\right) + Q_{2r}^*\left(-^1/_2 + \frac{h}{2} + t_1\right)\right]$$
$$= -S_{2r}(^1/_2 - t_1),$$

since $1/2 - t_1 \leqslant h/2$.

Since $S_{2r}(t)$ is a continuous function (more precisely, a polynomial of degree $2r - 1$) and its modulus is strictly decreasing on $(0, 1/2)$, it does not change in sign on $(0, 1/2)$. The lemma is proved for the polynomial $S_{2r}(t)$.

We will show that the function $S_{2r+1}(t)$ has analogous properties. Suppose $0 \leqslant t \leqslant h/2$. Since $|Q_{2r+1}^*(t)|$ decreases on $[0, 1/2]$ and $Q_{2r+1}^*(t) = -Q_{2r+1}^*(1 - t)$ by virtue of properties 1), 3) and 2), we note from (3.7) that $S_{2r+1}(t)$ has the same sign on $[0, h/2]$ as $-Q_{2r+1}^*(h/2 - t)$. But the latter function does not change in sign on $[0, h/2]$. Therefore, by virtue of properties 3) and 2) of Lemma 2.1,

$$|S_{2r+1}(t)| = \frac{(-1)^r}{4r+2}\left[Q_{2r+1}^*\left(\frac{h}{2}+t\right) - Q_{2r+1}^*\left(\frac{h}{2}-t\right)\right].$$

From this result and property 1) of Lemma 2.1 we have

$$|S_{2r+1}(t)|' = (-1)^r\left[Q_{2r}^*\left(\frac{h}{2}+t\right) - Q_{2r}^*\left(\frac{h}{2}-t\right)\right]$$

and, taking into account property 3) of the polynomial $Q_{2r}^*(t)$ and its evenness relative to $t = 1/2$, we conclude from the preceding that $|S_{2r+1}(t)|' > 0$ for $0 \leqslant t \leqslant h/2$.

For $h/2 \leqslant t \leqslant 1/2$ we find, analogously to the preceding, that

$$|S_{2r+1}(t)|' = \left\{\frac{(-1)^r}{4r+2}\left[Q^*_{2r+1}\left(t+\frac{h}{2}\right) - Q^*_{2r+1}\left(t-\frac{h}{2}\right)\right]\right\}'$$
$$= (-1)^r\left[Q^*_{2r}\left(t+\frac{h}{2}\right) - Q^*_{2r}\left(t-\frac{h}{2}\right)\right].$$

Further, it follows from properties 2), 3) and 1) of Lemma 2.1 that $|Q^*_{2r}(t)|$ is strictly increasing on $[0, 1/2]$. Therefore

$$\left|Q^*_{2r}\left(t+\frac{h}{2}\right)\right| > \left|Q^*_{2r}\left(t-\frac{h}{2}\right)\right|,$$

if $t + h/2 \leqslant 1/2$ or (since $0 \leqslant t \leqslant 1/2$ and $Q^*_{2r}(t)$ is even relative to $t = 1/2$) if $t + h/2 > 1/2$ but $1 - t - h/2 > t - h/2$. From this result and property 3) of Lemma 2.1 we have $|S_{2r+1}(t)|' > 0$ for $0 \leqslant t < 1/2$. Let us prove that $S_{2r+1}(t)$ is even relative to $t = 1/2$. We put $t = 1/2 + t_1$, where $0 \leqslant t_1 \leqslant (1-h)/2$. Then from (3.7) and the evenness of $Q^*_{2r+1}(t)$ relative to $t = 1/2$ we get

$$S_{2r+1}\left(\frac{1}{2}+t_1\right) = \frac{1}{4r+2}\left[Q^*_{2r+1}\left(\frac{1}{2}+t_1+\frac{h}{2}\right) - Q^*_{2r+1}\left(\frac{1}{2}+t_1-\frac{h}{2}\right)\right]$$
$$= \frac{1}{4r+2}\left[Q^*_{2r+1}\left(\frac{1}{2}-t_1+\frac{h}{2}\right) - Q^*_{2r+1}\left(\frac{1}{2}-t_1-\frac{h}{2}\right)\right] = S_{2r+1}\left(\frac{1}{2}-t_1\right).$$

Suppose $t = 1/2 + t_1$, where $(1-h)/2 \leqslant t_1 \leqslant 1/2$. Then

$$S_{2r+1}\left(\frac{1}{2}+t_1\right) = \frac{1}{4r+2}\left[Q^*_{2r+1}\left(\frac{1}{2}+\frac{h}{2}-t_1\right) - Q^*_{2r+1}\left(-\frac{1}{2}+t_1+\frac{h}{2}\right)\right]$$
$$= S_{2r+1}\left(\frac{1}{2}-t_1\right),$$

since $0 \leqslant 1/2 - t_1 \leqslant 1/2 - (1-h)/2 = h/2$. And the evenness of $S_{2r+1}(t)$ relative to $t = 1/2$ is proved. Further, we deduce in the same way as for $S_{2r}(t)$ that $S_{2r+1}(t)$ does not change in sign on $(0, 1/2)$ or, consequently (by virtue of its evenness), on $[0, 1]$. Lemma 3.1 is completely proved.

Let us find an upper bound for the quantity $A_{n,p}(M)$. We first consider the case of odd $n = 2r+1$. It follows from Lemma 3.1 that (3.13) can be converted into an equality by seeking $f^{(2r+1)}(t)$ in the form

$$f^{(2r+1)}(m+t) = \begin{cases} Z_m|S_{2r+1}(t)|^{q-1}, & 0 \leqslant t \leqslant 1, \quad 1 \leqslant q < \infty, \\ m = 0, \pm 1, \pm 2, \ldots, & \dfrac{1}{p}+\dfrac{1}{q} = 1, \end{cases} \tag{3.17}$$

where the Z_m are constants. From (3.9) and (3.17) we have

$$(2r)!\,\Delta^{2r+1}y_m = \sum_{s=0}^{2r+2} Z_{m+s-1}\int_0^1 |S_{2r+1}(t)|^{q-1}\,dt\,\frac{1}{h}$$
$$\times \int_{-h/2}^{h/2} \sum_{k=0}^{2r+1} (-1)^k C^k_{2r+1}(k+1+x-t-s)^{2r}_+\,dx. \tag{3.18}$$

The function $\varphi(t) = |S_{2r+1}(t)|^{q-1}$ satisfies all of the conditions of Theorem 2.1 by virtue of Lemma 3.1. Therefore the characteristic polynomial of this difference equation has the same properties as the polynomial (2.5). Applying Theorem 2.1 and Theorem C of §1, we find that

$$Z_m = \sum_{k=-\infty}^{\infty} \sum_{s=0}^{r+1} \frac{x_s^{r+|k|}}{P'_{2r+2}(x_s)} (2r)!\, \Delta^{2r+1} y_{m-r-1+k}, \tag{3.19}$$

$$\| Z_m \|_{l_p} \leqslant (2r)!\, \| \Delta^{2r+1} y_m \|_{l_p} \, | P_{2r+2}(-1) |^{-1}, \tag{3.20}$$

where

$$P_{2r+2}(x) = \sum_{s=0}^{2r+2} x^s \int_0^1 | S_{2r+1}(t) |^{q-1} dt \frac{1}{h} \int_{-h/2}^{h/2} \sum_{k=0}^{2r+1} (-1)^k C_{2r+1}^k (k+1+x-t-s)_+^{2r-1} dx. \tag{3.21}$$

The value of this polynomial at $x = -1$ is

$$P_{2r+2}(-1) = \int_0^1 | S_{2r+1}(t) |^{q-1} dt \frac{1}{h} \int_{-h/2}^{h/2} \sum_{s=0}^{2r+2} (-1)^s \sum_{k=0}^{2s+1} (-1)^k C_{2r+1}^k (k+1+x-t-s)_+^{2r} dx,$$

and hence, taking (2.32) and (3.12) into account,

$$| P_{2r+2}(-1) | = \int_0^1 | S_{2r+1}(t) |^q dt. \tag{3.22}$$

From (3.17), (3.19) and (3.21) we get

$$\begin{aligned} \| f^{(2r+1)} \|_{L_p} &= \left\{ \sum_{m=-\infty}^{\infty} \int_0^1 | S_{2r+1}(t) |^{(q-1)p} dt \, | Z_m |^p \right\}^{1/p} \\ &= \left\{ \int_0^1 | S_{2r+1}(t) |^q dt \right\}^{1/p} \| Z_m \|_{l_p} \leqslant (2r)!\, \| \Delta^{2r+1} y_m \|_{l_p} \| S_{2r+1}(t) \|_{L_q(0,1)}^{-1}. \end{aligned} \tag{3.23}$$

We note that the constructed function satisfies (3.16) but not (3.4). In order to satisfy (3.4), we represent $f(x)$ in the form

$$f(x) = \widetilde{P}_{2r}(x) + \frac{1}{(2r)!} \int_0^x (x-t)^{2r} f^{(2r+1)}(t)\, dt,$$

where $f^{(2r+1)}(t)$ is defined by (3.17), (3.15) and (3.19), and find the polynomial $\widetilde{P}_{2r}(x)$ of degree at most $2r$ from the conditions

$$F_k(f) = F_k(\widetilde{P}_{2r}) + F_k\left[\frac{1}{(2r)!} \int_0^x (x-t)^{2r} f^{(2r+1)}(t)\, dt \right],$$

$k = 0, 1, \ldots, 2r$, which uniquely determine $\widetilde{P}_{2r}(x)$ and, consequently, a function $f(x)$ satisfying (3.4). From (3.13) and (3.23) we conclude that Theorem 3.1 is correct for odd n. Since, in deducing an upper bound, we constructed a linear operator acting from $l_p^{(n)}$ into $L_p^{(n)}$, Theorem 3.2 is also correct in this case.

Let us return to the case $n = 2r$. From Lemma 3.1 we conclude that (3.13) will be an equality if $f^{(2r)}(t) = C|S_{2r}(t)|$. Therefore $f^{(2r)}(t)$ will be sought in the form

$$f^{(2r)}(m+t)=\begin{cases} Z_m\,|S_{2r}(1+t)|^{q-1}, & -1/2\leqslant t\leqslant 0,\\ Z_m\,|S_{2r}(t)|^{q-1}, & 0\leqslant t\leqslant 1/2,\\ 1\leqslant q<\infty,\ \dfrac{1}{p}+\dfrac{1}{q}=1, & m=0,\pm 1,\pm 2,\ldots, \end{cases} \tag{3.24}$$

where the z_m are constants.

From (3.9) and (3.24) we get

$$(2r-1)!\,\Delta^{2r}y_m=\sum_{s=1}^{2r+1}Z_{m-1+s}\int_0^{1/2}\varphi^*(t)\frac{1}{h}\int_{-h/2}^{h/2}\sum_{k=0}^{2r}(-1)^k C_{2r}^k(k+1+x-s-t)_+^{2r-1}dx\,dt$$

$$+\sum_{s=0}^{2r}Z_{m+s}\int_{1/2}^{1}\varphi^*(t)\,dt\frac{1}{h}\int_{-h/2}^{h/2}\sum_{k=0}^{2r}(-1)^k C_{2r}^k(k+1+x-t-s)_+^{2r-1}dx$$

$$=\sum_{s=0}^{2r}Z_{m+s}\Big\{\int_0^{1/2}\varphi^*(t)\,dt\frac{1}{h}\int_{-h/2}^{h/2}\sum_{k=0}^{2r}(-1)^k C_{2r}^k(k+x-s-t)_+^{2r-1}dx$$

$$+\int_{1/2}^{1}\varphi^*(t)\,dt\frac{1}{h}\int_{-h/2}^{h/2}\sum_{k=0}^{2r}(-1)^k C_{2r}^k(k+1+x-t-s)_+^{2r-1}dx\Big\},\tag{3.25}$$

where $\varphi^*(t)=|S_{2r}(t)|^{q-1}$

The characteristic polynomial $P_{2r}^*(x)$ of the difference equation (3.25) satisfies all of the conditions of Theorem 2.1 by virtue of Lemma 3.1 and (2.6). Consequently, Theorem C of §1 can be applied to the difference equation (3.25), and we get

$$Z_m=\sum_{k=-\infty}^{\infty}\sum_{s=0}^{r}\frac{x_s^{r-1+|k|}}{P_{2r}^{*\prime}(x_s)}(2r-1)!\,\Delta^{2r}y_{m-r+k},\tag{3.26}$$

$$\|Z_m\|_{l_p}\leqslant(2r-1)!\,\|\Delta^{2r}y_m\|_{l_p}\,|P_{2r}^*(-1)|^{-1}.\tag{3.27}$$

Further, we have

$$P_{2r}^*(-1)=\int_0^{1/2}\varphi^*(t)\,dt\frac{1}{h}\int_{-h/2}^{h/2}\sum_{s=0}^{2r}(-1)^s\sum_{k=0}^{2r}(-1)^k C_{2r}^k(k+x-s-t)_+^{2r-1}dx$$

$$+\int_{1/2}^{1}\varphi^*(t)\,dt\frac{1}{h}\int_{-h/2}^{h/2}\sum_{s=0}^{2r}(-1)^s\sum_{k=0}^{2r}(-1)^k C_{2r}^k(k+x+1-s-t)_+^{2r-1}dx.\tag{3.28}$$

Since $(k-s-y)_+^{2r-1}-(s+y-k)_+^{2r-1}=(k-s-y)^{2r-1}$, it follows from (3.28), the identity

$$\sum_{k=0}^{2r}(-1)^k C_{2r}^k(k-s-y)^{2r-1}\equiv 0$$

and (3.12) that

$$|P_{2r}^*(-1)|=\int_0^1|S_{2r}(t)|^q\,dt.\tag{3.29}$$

From (3.24), (3.27) and (3.29) we get, analogously to (3.23), that

$$\|f^{(2r)}\|_{L_p}\leqslant(2r-1)!\,\|\Delta^{2r}y_m\|_{l_p}\,\|S_{2r}(t)\|_{L_q(0,1)}^{-1}.\tag{3.30}$$

The rest is the same as in the preceding case. Theorems 3.1 and 3.2 are completely proved.

In (1.8) we put

$$g(x)=\begin{cases}0, & x\leqslant -h/2,\\ 1, & x\geqslant h/2,\\ \text{a linear function between } -h/2 \text{ and } h/2.\end{cases}$$

In deducing a lower bound, we made special use of the sequence $\{\bar{y}_k^{(N)}\}$. Inequalities (3.30), (3.23) and (3.13) imply that this sequence tends to an extremal for the right side of (1.14) as $N\to\infty$, i.e.

$$\inf_{\|\zeta\|_{l_q}=1}\|\Phi(\zeta,t)\|_{L_q}=\frac{1}{(n-1)!}\int_0^1|S_n(t)|^q\,dt,$$

where $\Phi(\zeta, t)$ is defined by (1.13), in which $g(x)$ has the form indicated above.

It is not difficult to see that Theorem 3.1 is correct in the more general case when

$$F_k(f)=\frac{1}{h}\int_{-h/2}^{h/2}f(kh_1+t)\,dt,$$

where $0<h_1<\infty$ and $0\leqslant h\leqslant h_1$. But this case can be reduced to the case $h_1=1$ by putting $\psi(x)=f(xh_1)$.

Suppose in (1.8) that $h=1$ and $g(x)$ takes constant values for $x\leqslant -1/2$ and $x\geqslant 1/2$. Then

$$\Delta^n y_k=\int_{-\infty}^{\infty}f^{(n)}(k+t)\,dt\int_{-1/2}^{1/2}\varphi_n(x-t)\,dg(x). \tag{3.31}$$

Further, since $\varphi_n(x-t)=0$ for $t\leqslant x$ and $t\geqslant x+n$ by virtue of (1.9), we can write

$$\begin{aligned}\Delta^n y_k&=\int_{-1}^{n+1}f^{(n)}(k+t)\,dt\int_{-1/2}^{1/2}\varphi_n(x-t)\,dg(x)\\ &=\frac{1}{(n-1)!}\sum_{s=0}^{n+1}\int_0^1 f^{(n)}(k-1+s+t)\,dt\int_{-1/2}^{1/2}\sum_{m=0}^{n}(-1)^m C_n^m(x+m+1-s-t)_+^{n-1}\,dg(x).\end{aligned} \tag{3.32}$$

Proceeding now as in the derivation of (3.10)–(3.13), we get

$$\begin{aligned}A_{n,p}(M,g)&=\sup_{\|Y\|_{l_p^{(n)}}\leqslant M}\ \inf_{f\in F(Y,g)}\|f^{(n)}_{(x)}\|_{L_p}\\ &\geqslant M(n-1)!\,|P_{\Upsilon_n}(-1,g)|^{-1/q}\left(\frac{1}{p}+\frac{1}{q}=1,\ 1<p\leqslant\infty\right),\end{aligned} \tag{3.33}$$

where $F(Y,g)$ is the class of functions with a locally absolutely continuous $(n-1)$th derivative that satisfy the conditions

$$y_k=\int_{-1/2}^{1/2}f(k+x)\,dg(x),\qquad k=0,\pm1,\pm2,\dots, \tag{3.34}$$

$$P_{\Upsilon_{2r-1}}(u,g)=\sum_{l=0}^{2r}u^l\int_0^1|\theta_{2r-1}(t)|^{q-1}\operatorname{sgn}\theta_{2r-1}(t)\,\psi_{l,2r-1}(t)\,dt, \tag{3.35}$$

$$P_{\gamma_{2r}}(u,g)=\sum_{l=0}^{2r}u^l\left\{\int_0^{1/2}|\theta_{2r}(t)|^{q-1}\operatorname{sgn}\theta_{2r}(t)\psi_{l,2r}(t)\,dt\right. \tag{3.36}$$

$$\left.+\int_{1/2}^{1}|\theta_{2r}(t)|^{q-1}\operatorname{sgn}\theta_{2r}(t)\psi_{l-1,2r}(t)\,dt\right\},\qquad \frac{1}{p}+\frac{1}{q}=1,\quad 1\leqslant q<\infty$$

$$\psi_{l,n}(t)=\int_{-1/2}^{1/2}\sum_{s=0}^{n}(-1)^{n-s}C_n^s(s-l+1-t+x)_+^{n-1}\,dg(x), \tag{3.37}$$

$$\theta_n(t)=\sum_{l=0}^{n}(-1)^l\psi_{l,n}(t),\qquad 0\leqslant t\leqslant 1. \tag{3.38}$$

We seek the function needed for obtaining an upper bound of $A_{n,p}(M,g)$ in the form

$$f^{(n)}(k+t)=Z_k^{(n)}|\theta_n(t)|^{q-1}\operatorname{sgn}\theta_n(t),\qquad 0\leqslant t<1,\qquad n=2r-1, \tag{3.39}$$

$$f^{(n)}(k+t)=\begin{cases}Z_k^{(n)}|\theta_n(1+t)|^{q-1}\operatorname{sgn}\theta_n(1+t), & -1/2<t<0,\\ Z_{k+1}^{(n)}|\theta_n(t)|^{q-1}\operatorname{sgn}\theta_n(t), & 0\leqslant t<1/2,\qquad n=2r,\end{cases} \tag{3.40}$$

where $p^{-1}+q^{-1}=1$ and $1<p\leqslant\infty$. Substituting these values into (3.32), we obtain the following difference equations for the $Z_k^{(n)}$:

$$(2r-2)!\,\Delta^{2r-1}y_k=\sum_{s=0}^{2r}Z_k^{(2r-1)}\int_0^1|\theta_{2r-1}(t)|^{q-1}\operatorname{sgn}\theta_{2r-1}(t)\psi_{s,2r-1}(t)\,dt, \tag{3.41}$$

$$(2r-1)!\,\Delta^{2r}y_k=\sum_{s=0}^{2r}Z_k^{(2r)}\left\{\int_0^{1/2}|\theta_{2r}(t)|^{q-1}\operatorname{sgn}\theta_{2r}(t)\psi_{s,2r}(t)\,dt\right. \tag{3.42}$$

$$\left.+\int_{1/2}^{1}|\theta_{2r}(t)|^{q-1}\operatorname{sgn}\theta_{2r}(t)\psi_{s-1,2r}(t)\,dt\right\}.$$

If the zeros of the polynomial $P_{\gamma_n}(u,g)$ are mutually distinct, negative and do not include $u=-1$, we can solve the difference equations (3.41), (3.42) with the use of Theorem B and (1.30). We then find, analogously to (3.23) and (3.30), an upper bound of $A_{n,p}(M,g)$ that coincides with a lower bound. Thus the following result holds.

THEOREM 3.3. *If the zeros of the polynomial $P_{\gamma_n}(u,g)$ are mutually distinct, negative and do not include $u=-1$, then*

$$A_{n,p}(M,g)=(n-1)!\,M\,|P_{\gamma_n}(-1,g)|^{-1/q}\left(\frac{1}{p}+\frac{1}{q}=1,\ 1<p\leqslant\infty\right). \tag{3.43}$$

§4. Interpolation-in-the-mean splines

It is unnatural to speak of approximating a function $f(x)\in L_p$ by interpolation splines, since it is not essential that one know the values of such a function at individual points. We will therefore study here questions concerning the approximation of functions from L_p by interpolation-in-the-mean splines.

Suppose given two real numbers h_1 and h such that $0 < h_1 \leqslant h < \infty$.

DEFINITION 4.1. We will say that a function

$$S_n(x) = S_n(x, f) = S_n(x, h_1, h; f)$$

is an *interpolation-in-the-mean spline* for $f(x)$ if:

a) $S_n(x)$ together with its derivatives up to order $n-1$ is locally absolutely continuous on $(-\infty, \infty)$;

$$\text{b) } S_n^{(n)}(x) = Z_m^{(n)}, \quad mh + \frac{1+(-1)^n}{4} h \leqslant x \leqslant mh + h + \frac{1+(-1)^n}{4} h \tag{4.1}$$

for $m = 0, \pm 1, \pm 2, \ldots$, where the $Z_m^{(n)}$ are constants;

$$\text{c) } \frac{1}{h_1} \int_{-h_1/2}^{h_1/2} [S_n(mh+t) - f(mh+t)]\, dt = 0, \quad m = 0, \pm 1, \pm 2, \ldots \tag{4.2}$$

We will prove the existence and uniqueness of an interpolation-in-the-mean spline.

THEOREM 4.1. *If the sequence*

$$\alpha_m = \frac{1}{h_1} \int_{-h_1/2}^{h_1/2} \Delta_h^n f(mh+t)\, dt$$

is bounded, then the interpolation-in-the-mean spline $S_n(x, h_1, h; f)$ $(0 < h_1 \leqslant h)$ exists and, if its nth derivative is bounded, is unique.

PROOF. Since the difference equations (3.18) and (3.25) have only one bounded solution for $q = 1$ and a bounded right side, the uniqueness of an interpolation-in-the-mean spline is proved.

Further, two splines of degree n with the same nth derivative differ by a polynomial $P_{n-1}(x)$ of degree at most $n-1$. Therefore, for the proof of existence, it remains to show that the polynomial

$$P_{n-1}(x) = S_n(x, h_1, h; f) - \frac{1}{(n-1)!} \int_0^x (x-t)^{n-1} S_n^{(n)}(t, h_1, h; f)\, dt \tag{4.3}$$

is uniquely determined by conditions (4.2). We put

$$V_i(f) = \frac{1}{h_1} \int_{-h_1/2}^{h_1/2} f(ih+t)\, dt \tag{4.4}$$

and require that

$$V_i(f) = V_i(S_n), \quad i = 0, 1, \ldots, \quad n-1,$$

i.e.

$$V_i(P_{n-1}) - V_i(f) = V_i(\varphi) - V_j(\psi), \quad i = 0, 1, \ldots, \quad n-1, \tag{4.5}$$

where

$$\varphi(x) = \frac{1}{(n-1)!} \int_0^x (x-t)^{n-1} S_n^{(n)}(t, h_1, h; f)\, dt$$

and $\psi(x) = f(x) - \varphi(x)$.

The numbers $V_i(\psi)$ are known to us. Since $0 < h_1 \leqslant h$, there exists in each of the intervals $(ih - h_1/2, ih + h_1/2)$, $i = 0, 1, \ldots, n-1$, a point x_i at which $V_i(P_{n-1}) = P_{n-1}(x_i)$ with, obviously, $x_i \neq x_j$ for $i \neq j$. Thus a polynomial satisfying conditions (4.5) exists; and, since $V_i(S_n) = V_i(f)$ for $i = 0, \pm 1, \pm 2, \ldots$, by virtue of the difference equations (3.18) and (3.25), the existence of an interpolation-in-the-mean spline is proved, and Theorem 4.1 is thereby completely proved.

The existence and uniqueness actually hold if $|\Delta^n \alpha_k| \leqslant M(|k| + 1)^\beta$, where β is a fixed nonnegative number. Questions concerning the existence of interpolating splines for sequences not growing more rapidly than $(|k| + 1)^\beta$, $\beta > 0$, were considered by Schoenberg [16].

We will first study the convergence of such splines to continuous functions in the uniform metric.

THEOREM 4.2. *If a function $f(x)$ is continuous on $(-\infty, \infty)$ together with its lth derivative and $\omega_{n-l+1}(f^{(l)}, \delta) \longrightarrow 0$ as $\delta \longrightarrow 0$, then for $0 \leqslant l \leqslant n$*

$$|f^{(i)}(x) - S_n^{(i)}(x, h_1, h; f)| \leqslant C_{n,i}\omega_{n-i+1}(f^{(i)}, h),$$

$i = 0, 1, \ldots, l$, and the constant $C_{n,i}$ depends only on n and i.

PROOF. Suppose first $0 \leqslant i \leqslant n-1$. By virtue of (4.2) and the continuity of $f(x)$ and $S_n(x; f)$, there exists in each of the intervals $(kh - h_1/2, kh + h_1/2)$ at least one point x_k^0 at which $f(x_k^0) = S_n(x_k^0; f)$. Further, if $f'(x)$ is continuous, then, by Rolle's theorem, there exists in each of the intervals $(kh - h_1/2, kh + h + h_1/2)$ at least one point x_k^1 at which $f'(x_k^1) = S_n'(x_k^1; f)$; more generally, if $f^{(i)}(x)$ is continuous and $i \leqslant n-1$, then $S_n^{(i)}(x; f)$ is also continuous and, by virtue of a repeated application of Rolle's theorem, there exists in each of the intervals $(kh - h_1/2, kh + ih + h_1/2)$ at least one point x_k^i at which

$$f^{(i)}(x_k^i) = S_n^{(i)}(x_k^i; f), \qquad k = 0, \pm 1, \pm 2, \ldots.$$

We now consider the nonoverlapping intervals

$$\alpha_s = 3s(i+1)(n-i+1)h < x < 3(s+1)(i+1)(n-i+1)h = \alpha_{s+1}.$$

By virtue of the preceding, there exist in each of them at least $n - i + 1$ points $x_{s,j}$ $(j = 0, 1, \ldots, n-i)$ at which $S_n^{(i)}(x_{s,j}; f) = f^{(i)}(x_{s,j})$ and

$$(i+1)h \leqslant |x_{s,j+1} - x_{s,j}| \leqslant 3(i+1)h, \qquad j = 0, 1, \ldots, n-i-1,$$
$$s = 0, \pm 1, \pm 2, \ldots.$$

Thus, analogously to (3.10) in [17] (see also [18] and [19]), we have

$$f^{(i)}(x) - S_n(x, h) = -\frac{1}{(n-i-1)!}\int_{\alpha_s}^{\alpha_{s+1}} \varphi_i(x, t) S_n^{(n)}(t, h_1, h; f)\, dt$$
$$+ \sum_{j=1}^{n-i+1} \{f^{(i)}(x) - f^{(i)}(x_{s,j})\} \frac{\omega_{n-i+1}(x)}{(x - x_{s,j})\,\omega'_{n-i+1}(x_{s,j})},$$

$$\varphi_i(x, t) = (x-t)_+^{n-i-1} - \sum_{j=1}^{n-i+1} \frac{\omega_{n-i+1}(x)(x_{s,j} - t)_+^{n-i-1}}{(x - x_{s,j})\,[\omega'_{n-i+1}(x_{s,j})]},$$

where $\omega_{n-i+1}(x) = (x - x_{s,1}) \cdots (x - x_{s,n-i+1})$. If $x, t \in [\alpha_s, \alpha_{s+1}]$, then $|\varphi_i(x, t)| \leqslant A(n, i)h^{n-i-1}$ under the fulfilment of conditions (4.5), where the constant $A(n, i)$ depends only on n and i. Moreover,

$$\int_{\alpha_s}^{\alpha_{s+1}} \varphi_i(x, t)\, dt = 0.$$

Therefore, using a result of Whitney in the same way as in [17], §3, we get

$$|f^{(i)}(x) - S_n^{(i)}(x, h_1, h; f)| \leqslant C_1(i, n)\, \omega_{n-i+1}(f^{(i)}, h) + C_2(i, n) \max_{\alpha_s \leqslant t \leqslant \alpha_{s+1}} |S_n^{(n)}(t, h_1, h; f) - S_n^{(n)}(\alpha_s, h_1, h; f)|. \tag{4.6}$$

We put

$$y_m = \frac{1}{h_1} \int_{-h_1/2}^{h_1/2} f(mh + t)\, dt$$

and

$$\widetilde{S}_n(x, u, 1; f) = S_n(xh; h_1, h; f), \tag{4.7}$$

where $u = h_1/h$, $0 < u \leqslant 1$. Since $0 < h_1 \leqslant h$, it follows by virtue of (2.7) and (2.1) that the spline $\widetilde{S}_n(x, u, 1; f)$ satisfies the conditions

$$\begin{gathered} \widetilde{S}_n^{(n)}(x) = \widetilde{S}_n^{(n)}(x, u, 1; f) = h^n S_n^{(n)}(xh, h_1, h; f) = h^n Z_m^{(n)}, \\ m + \frac{1 + (-1)^n}{4} \leqslant x < m + 1 + \frac{1 + (-1)^n}{4}, \qquad m = 0, \pm 1, \pm 2, \ldots \end{gathered} \tag{4.8}$$

Thus the spline $\widetilde{S}_n(x)$ is representable in the form (3.17) or (3.24) with $q = 1$ and Z_m replaced by $h^n Z_m^{(n)}$. Consequently, for the determination of the $Z_m^{(n)}$ we have the difference equations obtained from (3.18) and (3.25) by putting $q = 1$, $h = u$ and multiplying the left sides by h^{-2r-1} and h^{-2r} respectively. Since the function identically equal to unity satisfies all of the conditions of Theorem 2.1, we see from (3.19) and (3.26) that

$$Z_m^{(n)} = \sum_{k=-\infty}^{\infty} \sum_{s=0}^{\left[\frac{n}{2}\right]} \frac{x_s^{\left[\frac{n}{2}\right]-1+|k|}}{P'_{\Upsilon_n}(x_s)} (n-1)!\, \frac{\Delta^n y_{m+k-\left[\frac{n}{2}\right]}}{h^n}, \tag{4.9}$$

where

$$P_{\Upsilon_n}(x) = \sum_{s=0}^{2r} x^s \int_0^1 dt \frac{1}{h} \int_{-h/2}^{h/2} \sum_{k=0}^{2r-1} (-1)^k C_{2r-1}^k (k + 1 + x - t - s)_+^{2r-2}\, dx \tag{4.10}$$

for $n = 2r - 1$, and

$$\begin{aligned} P_{\Upsilon_n}(x) = \sum_{s=0}^{2r} x^s \Bigg\{ & \int_0^{1/2} dt \frac{1}{h} \int_{-h/2}^{h/2} \sum_{k=0}^{2r} (-1)^k C_{2r}^k (k + x - t - s)_+^{2r-1}\, dx \\ & + \int_{1/2}^{1} dt \frac{1}{h} \int_{-h/2}^{h/2} \sum_{k=0}^{2r} (-1)^k C_{2r}^k (k + 1 + x - t - s)_+^{2r-1}\, dx \Bigg\} \end{aligned} \tag{4.11}$$

for $n = 2r$. Suppose

$$\max_{\alpha_s \leqslant t \leqslant \alpha_{s+1}} |S_n^{(n)}(t, h_1, h; f) - S_n^{(u)}(\alpha_s, h_1, h; f)|$$

is achieved at some point $t_s \in [\alpha_s, \alpha_{s+1}]$. We can assume that the point t_s has the form $t_s = kh$, where k is an integer in the interval $[\alpha_s/h, \alpha_{s+1}/h]$. In this connection,

$$\left|k - \frac{\alpha_s}{h}\right| \leqslant \left|\frac{\alpha_{s+1}}{h} - \frac{\alpha_s}{h}\right| = 3(i+1)(n-i+1). \tag{4.12}$$

Thus

$$\begin{aligned} &\max_{\alpha_s \leqslant t \leqslant \alpha_{s+1}} |S_n^{(n)}(t, h_1, h; f) - S_n^{(n)}(\alpha_s, h_1, h; f)| \\ &= |S_n^{(n)}(kh, h_1, h; f) - S_n^{(n)}(\alpha_s, h_1, h; f)| = \left|\sum_{r=\frac{\alpha_s}{h}}^{k-1} (Z_{r+1}^{(n)} - Z_r^{(n)})\right| \\ &\leqslant \left(k - \frac{\alpha_s}{h}\right) \max_r |Z_{r+1}^{(n)} - Z_r^{(n)}| \leqslant 3(i+1)(n-i+1) \max_r |Z_{r+1}^{(n)} - Z_r^{(n)}|. \end{aligned} \tag{4.13}$$

From (4.9) we have

$$\begin{aligned} \max_m |Z_{m+1}^{(n)} - Z_m^{(n)}| &= \max_m \left| \sum_{k=-\infty}^{\infty} \sum_{s=0}^{\left[\frac{n}{2}\right]} \frac{(n-1)!\, x_s^{\left[\frac{n}{2}\right]-1+|k|}}{P'_{\gamma_n}(x_s)} \frac{\Delta^{n+1} y_{m+k-\left[\frac{n}{2}\right]}}{h^n} \right| \\ &\leqslant h^{-n} \sup_m |\Delta^{n+1} y_m| \sum_{k=-\infty}^{\infty} \left| \sum_{s=0}^{\left[\frac{n}{2}\right]} \frac{(n-1)!\, x_s^{\left[\frac{n}{2}\right]-1+|k|}}{P'_{\gamma_n}(x_s)} \right| \\ &= h^{-n} \sup_m |\Delta^{n+1} y_m| (n-1)! \, |P_{\gamma_n}(-1)|^{-1}. \end{aligned} \tag{4.14}$$

Further,

$$|\Delta^{n+1} y_m| = \frac{1}{h_1} \left| \int_{-h_1/2}^{h_1/2} \Delta^{n+1} f(mh + t)\, dt \right| \leqslant \omega_{n+1}(f, h). \tag{4.15}$$

We recall that $\Delta^1 f(mh) = \Delta f(mh) = f(mh + h) - f(mh)$ and $\Delta^{n+1} f(mh) = \Delta\Delta^n f(mh)$ for $n \geqslant 1$.

In our case the characteristic polynomials of the difference equations (3.18) and (3.25) coincide respectively with the polynomials (2.5) and (2.6) for $\varphi(t) \equiv \varphi^*(t) \equiv 1$ and $h = u$, where $u = h_1/h$.

We will show that $|P_{\gamma_n}(-1)|$ is bounded from below by a positive constant not depending on u.

From (3.22) and (3.28), taking into account that $q = 1$ in our case, we have

$$|P_{\gamma_n}(-1)| = \int_0^1 |S_{n-1}(t)|\, dt, \tag{4.16}$$

where $S_{n-1}(t)$ is defined by (3.14) and (3.15) for $h = u$. Further, according to Lemma 3.1, the function $|S_{n-1}(t)|$ is even relative to $t = 1/2$ and is strictly monotonic on $[0, 1/2]$. Therefore

$$|P_{\gamma_n}(-1)| = 2\int_0^{1/2} |S_{n-1}(t)|\, dt = 2\left|\int_0^{1/2} S_{n-1}(t)\, dt\right|. \tag{4.17}$$

In the subsequent proof we will use the following properties of the polynomial $Q_n^*(t)$:

a) $Q_n^*(t)$ is even or odd relative to $t = 1/2$, depending on the evenness or oddness of n;

b) on the interval $[0, 1/2]$ the function $|Q_{2r}^*(t)|$ is strictly increasing while $|Q_{2r-1}^*(t)|$ is strictly decreasing; also, $Q_{2r}^*(0) = Q_{2r-1}(1/2) = 0$.

Property a) was formulated in Lemma 2.1. Property b) also readily follows from this lemma; also, its validity was established in the proof of Lemma 3.1.

Using obvious inequalities and Lemma 3.1, we get

$$\int_0^{1/2} |S_{2r}(t)|\,dt > \int_0^{1/4} |S_{2r}(t)|\,dt > \frac{1}{4}\left|S_{2r}\left(\frac{1}{4}\right)\right|, \tag{4.18}$$

$$\int_0^{1/2} |S_{2r+1}(t)|\,dt > \int_{1/4}^{1/2} |S_{2r+1}(t)|\,dt > \frac{1}{4}\left|S_{2r+1}\left(\frac{1}{4}\right)\right|. \tag{4.19}$$

Suppose $u < 1/2$. Then from (3.14) and property b) we have

$$\begin{aligned}\left|S_{2r}\left(\frac{1}{4}\right)\right| &= \frac{1}{u}\left|\int_{-u/2}^{u/2} Q_{2r-1}^*\left(\frac{1}{4}-x\right)dx\right| \\ &> \frac{1}{u}\left|\int_0^{u/2} Q_{2r-1}^*\left(\frac{1}{4}-x\right)dx\right| > \frac{1}{2}\left|Q_{2r-1}^*\left(\frac{1}{4}\right)\right| > 0,\end{aligned} \tag{4.20}$$

and, analogously, from (3.15)

$$\begin{aligned}\left|S_{2r+1}\left(\frac{1}{4}\right)\right| &= \frac{1}{u}\left|\int_{-u/2}^{u/2} Q_{2r}^*\left(\frac{1}{4}-x\right)dx\right| = \frac{1}{u}\left|\int_{-u/2}^{u/2} Q_{2r}^*\left(\frac{1}{4}+x\right)dx\right| \\ &> \frac{1}{u}\left|\int_0^{u/2} Q_{2r}^*\left(\frac{1}{4}+x\right)dx\right| > \frac{1}{2}\left|Q_{2r}^*\left(\frac{1}{4}\right)\right| > 0.\end{aligned} \tag{4.21}$$

When $1/2 \leqslant u \leqslant 1$, we find from (3.14) and property b) that

$$\left|S_{2r}\left(\frac{1}{4}\right)\right| = \frac{1}{u}\left|\int_{-u/2}^{u/2} Q^*_{2r-1}\left(\left|\frac{1}{4}-x\right|\right)dx\right| = \frac{1}{u}\left|\int_0^u Q^*_{2r-1}\left(\left|\frac{1}{4}+x-\frac{u}{2}\right|\right)dx\right|$$

$$= \frac{1}{u}\left|\int_0^{\frac{u}{2}-\frac{1}{4}} Q^*_{2r-1}\left(\frac{u}{2}-\frac{1}{4}-x\right)dx + \int_{\frac{u}{2}-\frac{1}{4}}^{u} Q^*_{2r-1}\left(\frac{1}{4}+x-\frac{u}{2}\right)dx\right|$$

$$= \frac{1}{u}\left|\int_0^{\frac{u}{2}-\frac{1}{4}} Q^*_{2r-1}\left(\frac{u}{2}-\frac{1}{4}-x\right)dx + \int_{\frac{u}{2}-\frac{1}{4}}^{\frac{u}{2}+\frac{1}{4}} Q^*_{2r-1}\left(\frac{1}{4}-\frac{u}{2}+x\right)dx\right. \tag{4.22}$$

$$\left.+\int_{\frac{u}{2}+\frac{1}{4}}^{u} Q^*_{2r-1}\left(\frac{1}{4}-\frac{u}{2}+x\right)dx\right| = \frac{1}{u}\left|\int_0^{\frac{u}{2}-\frac{1}{4}}\left[Q^*_{2r-1}(y)-Q^*_{2r-1}\left(\frac{1}{2}-y\right)\right]dy\right.$$

$$+\int_{\frac{u}{2}-\frac{1}{4}}^{\frac{u}{2}+\frac{1}{4}} Q^*_{2r-1}\left(\frac{1}{4}-\frac{u}{2}+x\right)dx\right| = \frac{1}{u}\left|\int_0^{\frac{u}{2}-\frac{1}{4}}\left[Q^*_{2r-1}(y)-Q^*_{2r-1}\left(\frac{1}{2}-y\right)\right]dx\right|$$

$$+\frac{1}{u}\left|\int_0^{1/2} Q^*_{2r-1}(y)\,dy\right| > \frac{1}{u}\left|\int_0^{1/2} Q^*_{2r-1}(y)\,dy\right| \geqslant \left|\int_0^{1/2} Q^*_{2r-1}(y)\,dy\right| > 0,$$

and from (3.15) that

$$\left|S_{2r+1}\left(\frac{1}{4}\right)\right| = \frac{1}{u}\left|\int_{-u/2}^{1/4} Q^*_{2r}\left(\frac{1}{4}-x\right)dx - \int_{1/4}^{u/2} Q^*_{2r}\left(x-\frac{1}{4}\right)dx\right|$$

$$\geqslant\left|\int_0^{\frac{1}{4}+\frac{u}{2}} Q^*_{2r}(y)\,dy - \int_0^{\frac{u}{2}-\frac{1}{4}} Q^*_{2r}(y)\,dy\right| = \left|\int_{\frac{u}{2}-\frac{1}{4}}^{\frac{u}{2}+\frac{1}{4}} Q^*_{2r}(y)\,dy\right| = \left|\int_{\frac{u}{2}-\frac{1}{4}}^{1/2} Q^*_{2r}(y)\,dy\right. \tag{4.23}$$

$$\left.+\int_{1/2}^{\frac{u}{2}+\frac{1}{4}} Q^*_{2r}(y)\,dy\right| = \left|\int_{\frac{u}{2}-\frac{1}{4}}^{1/2} Q^*_{2r}(y)\,dy + \int_0^{\frac{u}{2}-\frac{1}{4}} Q^*_{2r}\left(\frac{1}{2}+y\right)dy\right|$$

$$>\left|\int_0^{1/2} Q^*_{2r}(y)\,dy\right| > 0.$$

Since $Q^*_n(y)$ does not depend on y, it follows from (4.20)–(4.23), (4.18), (4.19), (4.16) and (4.17) that

$$|P_{\gamma_n}(-1)| \geqslant C > 0, \tag{4.24}$$

where C does not depend on u, i.e. on h or h_1.

Inequalities (4.6), (4.13)–(4.15), (4.24) and the inequality

$$\omega_{n+1}(f,\ h) \leqslant h^i\omega_{n-i+1}(f^{(i)},\ h)$$

imply the assertion of the theorem for $0 \leqslant i \leqslant l < n$. The theorem is also valid for

$l = n$ but for i satisfying $0 \leqslant i < l$. Thus it remains to consider the case $i = l = n$.

With respect to a given x, we find a k such that $kh - h \leqslant x \leqslant kh + nh + h$. In the same way as before, we establish that there exist at least two points x_1 and x_2 in this interval at which

$$f^{(n-1)}(x_i) = S_n^{(n-1)}(x_i; f), \qquad i = 1, 2.$$

Since, by assumption, $f^{(n)}(x)$ is continuous while $S_n^{(n-1)}(x; f)$ is piecewise linear, it follows that at an extremal point $x_0 \in (x_1, x_2)$ of the function $\varphi(x) = f^{(n-1)}(x) - S_n^{(n-1)}(x; f)$ either $\varphi'(x_0) = 0$ or $\varphi'(x_0 + 0)$ and $\varphi'(x_0 - 0)$ have opposite signs. In other words, there exists a point $y \in (x_1, x_2)$ at which either $\varphi'(y) = 0$ or $\varphi'(y)$ and $f^{(n)}(x) - S_n^{(n)}(x; f)$ have opposite signs, so that

$$| f^{(n)}(x) - S_n^{(n)}(x; f) | \leqslant | f^{(n)}(x) - f^{(n)}(y) + S_n^{(n)}(y; f) - S_n^{(n)}(x; f) | \leqslant \omega(f^{(n)}, \\ | x - y |) + | S_n^{(n)}(y; f) - S_n^{(n)}(x; f) |. \tag{4.25}$$

Taking into account the fact that $|x - y| \leqslant 2(n + 1)h$ and estimating in the same way as before the second summand in the right side of (4.25), we deduce the assertion of the theorem for $i = l = n$. The theorem is completely proved. An analogous result holds for periodic interpolation-in-the-mean splines.

We will now prove an analog of Theorem 3.3 in [20] for interpolation-in-the-mean splines.

THEOREM 4.3. *If $f(x)$ is a function having a locally absolutely continuous $(l - 1)$th derivative, $0 \leqslant l \leqslant n$, and*

$$\omega(f^{(l)}, h)_p = \sup_{|t| \leqslant h} \left\{ \int_{-\infty}^{\infty} | f^{(l)}(x + t) - f^{(l)}(x) |^p \, dx \right\}^{1/p} \to 0$$

as $h \to 0$, then

$$\| f^{(i)}(x) - S_n^{(i)}(x, h_1, h; f) \|_{L_q(-\infty, \infty)} \leqslant C h^{l - i + \frac{1}{q} - \frac{1}{p}} \omega(f^{(l)}, h)_p \tag{4.26}$$

for $i = 0, 1, \ldots, l - 1$ and $1 \leqslant p \leqslant q \leqslant \infty$, where $f^{(0)}(x) = f(x)$ and C is a constant depending only on i, l and n.

PROOF. Let V_i have the same meaning as in (4.4). We represent the difference $f - S_n$ by Taylor's formula with remainder in integral form. Also, we put $S_n(x) = S_n(x, h_1, h; f)$. Then

$$f(x) - S_n(x) = P_{n-1}(x) + \int_0^x \left\{ \frac{(x - u)^{l-1}}{(l - 1)!} f^{(l)}(u) - \frac{(x - u)^{n-1}}{(n - 1)!} S_n^{(n)}(u) \right\} du$$

for $0 \leqslant x \leqslant nh$. Since $V_j(f - S_n) = 0$ for $j = 0, 1, \ldots, n$ (there are too many conditions here, but these equalities are valid for any integral j by virtue of the definition of S_n if they are valid for any n successive values of j) and the functionals V_j are linear and independent, we can determine the polynomial $P_{n-1}(x)$ from the relation [21]

$$\begin{vmatrix} P_{n-1}(x) & 1 & x & \dots & x^n \\ V_0(-\psi) & V_0(1) & V_0(x) & \dots & V_0(x^n) \\ \cdot & \cdot & \cdot & \cdot & \cdot \\ V_n(-\psi) & V_n(1) & V_n(x) & \dots & V_n(x^n) \end{vmatrix} = 0. \tag{4.27}$$

Using the rule for integrating a determinant, we get

$$\int_{-h_1/2}^{h_1/2} dt_0 \dots \int_{-h_1/2}^{h_1/2} \Big[P_{n-1}(x) W(t_0, h+t_1, \dots, nh+t_n)$$
$$+ \sum_{s=0}^{n} (-1)^s \psi(sh+t_s) W(x, t_0, \dots, sh-h+t_{s-1}, sh$$
$$+h+t_{s+1}, \dots, nh+t_n)\Big] dt_n = 0,$$

where $W(x_1, \dots, x_n)$ is the Vandermonde determinant. Thus

$$f(x) - S_n(x) = \int_0^{nh} [\varphi_l(x, u) f^{(l)}(u) - \varphi_n(x, u) S_n^{(n)}(u)]\, du,$$

where

$$\varphi_l(x, u) = \frac{(x-u)_+^{l-1}}{(l-1)!} - \sum_{s-0}^{n} (-1)^s \Big[\int_R W(t)\, dt\Big]^{-1} \int_R W_s(x, t)(sh+t_s-u)_+^{l-1}\, dt$$

and

$$R = \Big\{t_i : i = 0, 1, \dots, n, |t_i| \leqslant \frac{h_1}{2}\Big\},\ dt = dt_0\, dt_1 \dots dt_n, \tag{4.28}$$
$$W(t) = W(t_0, h+t_1, \dots, nh+t_n),$$
$$W_s(x, t) = W(x, t_0, \dots, sh-h+t_{s-1}, sh+h+t_{s+1}, \dots, nh+t_n). \tag{4.29}$$

Suppose $mnh \leqslant x' \leqslant (m+1)nh$. We put $x = x' - mnh$. Then $0 \leqslant x \leqslant nh$. Further, $f(x) - S_n(x)$ can be represented by Taylor's formula in a neighborhood of the point mnh for $m = 0, \pm 1, \pm 2, \dots$. Then, repeating verbatim the above arguments, we get

$$f(mnh+x) - S_n(mnh+x)$$
$$= \int_0^{nh} [\varphi_l(x, u) f^{(l)}(mnh+u) - \varphi_n(x, u) S_n^{(n)}(mnh+u)]\, du,$$
$$0 \leqslant x \leqslant nh,\ m = 0, \pm 1, \pm 2, \dots. \tag{4.30}$$

For what follows, we require certain estimates, i.e. we must note that

$$
\begin{aligned}
&\left|\int_{-h_1/2}^{h_1/2} dt_0 \ldots \int_{-h_1/2}^{h_1/2} W(t_0, h+t_1, \ldots, nh+t_n)\, dt_n\right| \\
&= \left|\int_{-h_1/2}^{h_1/2} dt_0 \ldots \int_{-h_1/2}^{h_1/2} \prod_{h \geqslant i > j \geqslant 0} (ih + t_i - jh - t_j)\, dt_n\right| \\
&\geqslant \int_{-h_1/2}^{h_1/2} dt_0 \ldots \int_{-h_1/2}^{h_1/2} \prod_{h \geqslant i > j \geqslant 0} (h + t_i - t_j)\, dt_n \qquad (4.31) \\
&\geqslant \int_{-h_1/2}^{h_1/2} dt_0 \ldots \int_{-h_1/2}^{h_1/2} \left(h - \frac{h_1}{2} - t_0\right)^n \left(h - \frac{h_1}{2} - t_1\right)^{n-1} \ldots \left(h - \frac{h_1}{2} - t_{n-1}\right) dt_n \\
&= h_1 \frac{1}{(n+1)!} \{[h^{n+1} - (h-h_1)^{n+1}] \ldots [h^2 - (h-h_1)^2]\} \\
&= \frac{h_1}{(n+1)!} h^{\frac{(n+1)(n+2)-2}{2}} \left\{\left[1 - \left(1 - \frac{h_1}{h}\right)^{n+1}\right] \ldots \left[1 - \left(1 - \frac{h_1}{h}\right)^2\right]\right\} \\
&\geqslant \frac{h_1}{(n+1)!} h^{\frac{(n+1)(n+2)-2}{2}} \left(\frac{h_1}{h}\right)^n = \frac{h_1^{n+1}}{(n+1)!} h^{\frac{n(n+1)}{2}},
\end{aligned}
$$

since $1-(1-u)^k \geqslant u$ for $k \geqslant 1$ and $0 \leqslant u \leqslant 1$.

Further, for $0 \leqslant x \leqslant nh$

$$
\begin{aligned}
I_s &= \left|\int_{-h_1/2}^{h_1/2} dt_0 \ldots \int_{-h_1/2}^{h_1/2} W(x, t_0, \ldots, sh - h + t_{s-1}, sh + h + t_{s+1}, \ldots, nh + t_n)\, dt_n\right| \\
&= \left| h_1 \int_{-h_1/2}^{h_1/2} dt_0 \ldots \int_{-h_1/2}^{h_1/2} dt_{s-1} \int_{-h_1/2}^{h_1/2} dt_{s+1} \ldots \int_{-h_1/2}^{h_1/2} \prod_{n \geqslant i > j > 0}{}' (ih + t_i - jh - t_j)\, dt_n \right|,
\end{aligned}
\qquad (4.32)
$$

where Π' means that $ih + t_i$ $(jh + t_j)$ must be replaced by x when $i = s$ $(j = s)$. Further, $|ih + t_i - jh - t_j| \leqslant (n+1)h$ for $i, j \neq s$. If, on the other hand, i or j is equal to s, then the corresponding factor does not exceed $(n+2)h$. Therefore $|ih + t_i - jh - t_j| \leqslant (n+2)h$ for all i, j. Thus

$$
I_s \leqslant C_n h_1^{n+1} h^{\frac{(n+1)n}{2}}, \qquad (4.33)
$$

where C_n depends only on n.

Since $I_s = I_s(x)$ is a polynomial of degree n satisfying (4.33) for $0 \leqslant x \leqslant nh$, it follows by virtue of Markov's inequality (see, for example, [21]) that

$$
|I_s^{(i)}(x)| \leqslant C_n h_1^{n+1} h^{\frac{(n+1)n}{2} - i} \quad (i = 0, 1, \ldots, n). \qquad (4.34)
$$

Suppose $1 \leqslant l < n$ and $f(x) = x^l$ in (4.30). Then $S_n(x) \equiv x^l$ by virtue of the uniqueness of an interpolation-in-the-mean spline with bounded nth derivative, and hence

$$
\int_0^{nh} \varphi_l(x, u)\, du = 0, \ 0 \leqslant x \leqslant nh, \ l = 1, 2, \ldots . \qquad (4.35)
$$

From (4.30) and (4.35) we have

$$R(x) = f(mnh + x) - S_n(mnh + x)$$
$$= \int_0^{nh} \varphi_l(x, u)\Big[f^{(l)}(mnh + u) - \frac{1}{nh}\int_0^{nh} f^{(l)}(mnh + v)\,dv\Big]du$$
$$- \int_0^{nh} \varphi_n(x, u)\Big[S_n^{(n)}(mnh + u) - \frac{1}{nh}\int_0^{nh} S_n^{(n)}(mnh + v)\,dv\Big]du, \tag{4.36}$$

and

$$R^{(i)}(x) = \frac{1}{nh}\int_0^{nh} \varphi_l^{(i)}(x, u)\int_0^{nh} [f^{(l)}(mnh + u) - f^{(l)}(mnh + v)]\,dv\,du \tag{4.37}$$
$$- \frac{1}{nh}\int_0^{nh} \varphi_n^{(i)}(x, u)\int_0^{nh} [S_n^{(n)}(mnh + u) - S_n^{(n)}(mnh + v)]\,dv\,du, \quad i = 0, 1, \ldots, l-1.$$

From (4.37), (4.27)–(4.29) and (4.31)–(4.33) we have

$$\|R^{(i)}(x)\|_{L_q} \leqslant C(l, n)\Big\{h^{l-2-i}\Big[\sum_{m=-\infty}^{\infty}\int_0^{nh}\Big|\int_0^{nh} du\int_0^{nh}\Big|f^{(l)}(mnh + u)$$
$$- f^{(l)}(mnh + v)|\,dv\,|^q dx\Big]^{1/q} + h^{n-2-i+1/q}\Big[\sum_{m=-\infty}^{\infty}\Big|\int_0^{nh} du\int_0^{nh}\Big|S_n^{(n)}(mnh + u)$$
$$- S_n^{(n)}(mnh + v)|\,dv\,|^q\Big]^{1/q}\Big\}.$$

Adding and subtracting $f^{(l)}(mnh + u + v)$ $(S_n^{(n)}(mnh + u + v))$ within the inner absolute value signs of the first (second) summand and applying Minkowski's inequality, we get

$$\|R^{(i)}(x)\|_{L_q} \leqslant C(l, n)\Big\{h^{l-2-i+1/q}\Big[\sum_{m=-\infty}^{\infty}\Big|\int_0^{nh} du\int_0^{nh}\Big|f^{(l)}(mnh + u)$$
$$- f^{(l)}(mnh + u + v)\Big|dv\,|^q\Big]^{1/q} + h^{n-2-i+1/q}\Big[\sum_{m=-\infty}^{\infty}\Big|\int_0^{nh} du\int_0^{nh}\Big|S_n^{(n)}(mnh + u)$$
$$- S_n^{(n)}(mnh + u + v)\Big|dv\,|^q\Big]^{1/q}\Big\} = C(l, n)\,[h^{l-2-i+1/q}R_1 + h^{n-2-i+1/q}R_2]. \tag{4.38}$$

We estimate R_1 for $1 \leqslant p \leqslant q \leqslant \infty$ by applying the generalized Minkowski inequality and Holder's inequality:

$$R_1 \leqslant \int_0^{nh} dv\Big\{\sum_{m=-\infty}^{\infty}\Big[\int_0^{nh} |f^{(l)}(mnh + u + v) - f^{(l)}(mnh + u)|\,du\Big]^q\Big\}^{1/q}$$
$$\leqslant \int_0^{nh} dv\,(nh)^{1-1/p}\Big\{\sum_{m=-\infty}^{\infty}\Big[\int_0^{nh} |f^{(l)}(mnh + u + v) - f^{(l)}(mnh + u)|^p du\Big]^{q/p}\Big\}^{1/q}$$
$$\leqslant (nh)^{1-1/p}\max_{|v|\leqslant nh}\sup_m\Big[\int_0^{nh} |f^{(l)}(mnh + u + v) - f^{(l)}(mnh + u)|^p du\Big]^{1/p-1/q} \tag{4.39}$$
$$\int_0^{nh}\Big[\int_{-\infty}^{\infty} |f^{(l)}(u + v) - f^{(l)}(u)|^p du\Big]^{1/q} dv \leqslant (nh)^{2-1/p}\,\omega(f^{(l)}, nh)_p.$$

Analogously,

$$R_2 \leqslant (nh)^{2-1/p}\,\omega(S_n^{(n)}, nh)_p \leqslant 2\,(nh)^{2-1/p}\,\|S_n^{(n)}(t)\|_{L_q}. \tag{4.40}$$

Further, by virtue of (4.9), Theorem C of §1, (4.24) and Lemma 1.1 we have for $l < n$

$$\|S_n^{(n)}(t)\|_{L_p}=\Big\{\sum_{k=-\infty}^{\infty} h\,|Z_k^{(n)}|^p\Big\}^{1/p}\leqslant h^{1/p-n}\,|P_{\gamma_n}(-1)|\,\Big\|\frac{1}{h_1}\int_{-h_1/2}^{h_1/2}\Delta_h^n f(kh+t)\,dt\Big\|_{l_p}. \tag{4.41}$$

Here

$$\Big\|\frac{1}{h_1}\int_{-h_1/2}^{h_1/2}\Delta_h^n f(kh+t)\,dt\Big\|_{l_p}\leqslant\frac{1}{h_1}\int_{-h_1/2}^{h_1/2}\|\Delta_h^n f(kh+t)\|_{l_p}\,dt$$

$$=\frac{1}{h_1}\int_{-h_1/2}^{h_1/2}\Big\|\int_0^{lh}\varphi_l(x-u)\,\Delta_h^{n-l}f^{(l)}(kh+t+u)\,du\Big\|_{l_p}dt,$$

where $\varphi_l(x-u)$ is defined by (1.9), and $|\varphi_l(x-u)|\leqslant C(l)h^{l-1}$ for $0\leqslant x\leqslant lh$, $|u|\leqslant h_1/2$, $h_1\leqslant h$. Thus

$$\Big\|\frac{1}{h_1}\int_{-h_1/2}^{h_1/2}\Delta_h^n f(kh+t)\,dt\Big\|_{l_p}\leqslant C(l)\frac{h^{l-1}}{h_1}\int_{-h_1/2}^{h_1/2}dt\,\Big\|\int_0^{lh}|\Delta_h^{n-l}f^{(l)}(kh+t+u)|\,du\Big\|_{l_p}$$

$$\leqslant C(l)\frac{h^{l-1/p}}{h_1}\,l\int_{-h_1/2}^{h_1/2}\|\Delta_h f^{(l)}(t+u)\|_{L_p}\,dt\;2^{n-l-1}\leqslant C_1(l)\,h^{l-1/p}\,\omega(f^{(l)},h)_p\quad(h_1\leqslant h). \tag{4.42}$$

Inequalities (4.38)–(4.42) imply the assertion of Theorem 4.3. Theorem 5 of [20] can be obtained as a special case of Theorem 4.3 by letting $h_1\to 0$.

We note that Varga's article [22] is devoted to questions concerning functional spline interpolation in the case of a finite interval.

THEOREM 4.4. *If a function $f(x)\in L_p(-\infty,\infty)$ has a locally absolutely continuous $(n-1)$th derivative, and an nth derivative $f^{(n)}(x)\in L_p(-\infty,\infty)$, and $h_1/h\geqslant\alpha<0$, where α is an absolute constant, then there exists a unique interpolation-in-the-mean spline $S_n(x,h_1,h;f)$, with bounded nth derivative in $L_p(-\infty,\infty)$, for which*

$$\|f^{(i)}(x)-S_n^{(i)}(x,h_1,h;f)\|_{L_p(-\infty,\infty)}\leqslant C(i,n)\,\omega(f^{(i)},h)_p\qquad(i=0,n). \tag{4.43}$$

Theorem 4.4 complements Theorem 4.3.

PROOF. The representation (4.9), in which $P_{\gamma_n}(x)$ is defined by (4.10) and (4.11), is valid for $S_n^{(n)}(x,h_1,h;f)=Z_m^{(n)}$, $m=0,\pm1,\pm2,\dots$, $mh+\alpha\leqslant x\leqslant mh+h+\alpha$, where $\alpha=0$ for $n=2s+1$ and $\alpha=h/2$ for $n=2s$. It follows from the uniqueness of a bounded solution of the difference equations (3.18) and (3.25) when $q=1$ that $S_n(x,h_1,h;f)\equiv x^n/n!$ for $f(x)=x^n/n!$, i.e. $S_n^{(n)}(x,h_1,h;f)\equiv 1$, and hence, by virtue of (4.9),

$$\sum_{k=-\infty}^{\infty}\sum_{s=1}^{\left[\frac{n}{2}\right]}\frac{x_s^{\left[\frac{n}{2}\right]-1+|k|}}{P'_{\gamma_n}(x_s)}(n-1)!=1. \tag{4.44}$$

We will first consider the case $i=n$. Here

$$\frac{1}{h^n}\Delta_h^n f\Big(mh+kh-\Big[\frac{n}{2}\Big]h+t\Big)$$

$$=\frac{1}{h^n}\int_0^{nh}\varphi_n(u)\,f^{(n)}\Big(mh+kh-\Big[\frac{n}{2}\Big]h+t+u\Big)du, \tag{4.45}$$

where $\varphi_n(u)$ is defined by (1.9). The function $\varphi_n(u)$ is often called a *B*-spline (see, for example, [23] or [24]).

It is well known [24] that

$$\frac{1}{h^n}\int_0^{nh}\varphi_n(u)\,du = 1. \tag{4.46}$$

The latter fact follows trivially from (4.45) if one puts $f(x) = x^n$. From (1.9) it trivially follows that

$$|\varphi_n(u)| \leqslant Ch^{(n-1)} \tag{4.47}$$

for $0 \leqslant u \leqslant nh$, where C depends only on n.

From (4.9), (4.44) and (4.45) it follows that

$$S_n^{(n)}(x, h_1, h; f) = \frac{1}{h_1 h^n}\int_{-h_1/2}^{h_1/2} dt \int_0^{nh}\varphi_n(n) f^{(n)}\left(mh - \left[\frac{n}{2}\right]h + t + u\right) du \tag{4.48}$$
$$- \sum_{k=-\infty}^{\infty}\gamma_k(\beta_{k+1} - \beta_k),$$

where

$$\gamma_k = \gamma_{-k-1} = \frac{1-\alpha_0}{2} - \sum_{r=1}^{k}\alpha_r = \frac{1-\alpha_0}{2} - \sum_{r=1}^{k}\alpha_{-r}, \qquad k = 0, 1, 2, \ldots, \tag{4.49}$$

$$\alpha_r = \alpha_{-r} = \sum_{s=1}^{[h/2]}\frac{x_s^{[h/2]-1+|r|}}{P'_{\gamma_n}(x_s)}(n-1)! \tag{4.50}$$

and

$$\beta_k = \beta_{k,m} = \frac{1}{h_1 h^n}\int_{-h_1/2}^{h_1/2} dt\int_0^{nh}\varphi_n(u) f^{(n)}\left(mh - \left[\frac{n}{2}\right]h + kh + t + u\right) du. \tag{4.51}$$

Taking into account (4.49), Minkowski's inequality and the generalized Minkowski inequality, we get

$$I = \| f^{(n)}(x) - S_n^{(n)}(x, h_1, h; f)\|_{L_p} \leqslant \frac{1}{h_1}\int_{-h_1/2}^{h_1/2} dt\Big\{\sum_{m=-\infty}^{\infty}\int_{mh+\alpha}^{mh+h+\alpha}\Big| f^{(n)}(x)$$
$$- \frac{1}{h^n}\int_0^{nh}\varphi_n(u) f^{(n)}\left(mh - \left[\frac{n}{2}\right]h + t + u\right) du\Big|^p dx\Big\}^{1/p} \tag{4.52}$$
$$+ \frac{1}{h_1}\int_{-h_1/2}^{h_1/2}\Big\{\sum_{m=-\infty}^{\infty}\int_{mh+\alpha}^{mh+h+\alpha}\Big|\sum_{k=-\infty}^{\infty}\gamma_k\frac{1}{h^n}\int_0^{nh}\varphi_n(u)\Big[f^{(n)}\left(mh - \left[\frac{n}{2}\right]h + kh + t + u + h\right)$$
$$- f^{(n)}\left(mh - \left[\frac{n}{2}\right]h + kh + t + u\right)\Big] du\Big|^p dx\Big\}^{1/p} dt = I_1 + I_2.$$

We estimate I_2:

$$I_2 \leqslant \frac{h^{1/p-1}}{h_1} C(n) \int_{-h_1/2}^{h_1/2} dt \Big\{ \sum_{m=-\infty}^{\infty} \Big[\sum_{k=-\infty}^{\infty} |\gamma_k| \int_0^{nh} \Big| f^{(n)}\Big(mh - \Big[\frac{n}{2}\Big]h + kh + h + t + u\Big)$$

$$- f^{(n)}\Big(mh - \Big[\frac{n}{2}\Big]h + kh + t + u\Big)\Big| du \Big]^p \Big\}^{1/p} \leqslant \frac{h^{1/p-1}}{h_1} C(n) \int_{-h_1/2}^{h_1/2} dt \sum_{k=-\infty}^{\infty} |\gamma_k|$$

$$\times \Big\{ \sum_{m=-\infty}^{\infty} \Big[\int_0^{nh} \Big| f^{(n)}\Big(mh - \Big[\frac{n}{2}\Big]h + kh + h + t + u\Big) - f^{(n)}\Big(mh - \Big[\frac{n}{2}\Big]h + kh$$

$$+ t + u\Big) \Big| du \Big]^p \Big\}^{1/p} \leqslant \frac{C(n)}{h_1} \int_{-h_1/2}^{h_1/2} dt \sum_{k=-\infty}^{\infty} |\gamma_k| \Big\{ \sum_{m=-\infty}^{\infty} \int_0^{nh} | f^{(n)}(mh + h + t + u)$$

$$- f^{(n)}(mh + t + u)|^p \, du \Big\}^{1/p} \tag{4.53}$$

and

$$I_2 \leqslant C(n) \sum_{k=-\infty}^{k=\infty} |\gamma_k| \, \omega(f^{(n)}, h)_p \leqslant C_1(n)\, \omega(f^{(n)}, h)_p. \tag{4.54}$$

We will show that the constant $C_1(n)$ depends only on n. By virtue of (4.49), (4.50) and (4.44) we have

$$\sum_{k=-\infty}^{\infty} |\gamma_k| = 2 \sum_{k=0}^{\infty} |\gamma_k| = 2 \sum_{k=1}^{\infty} \Big| \sum_{s=k}^{\infty} \alpha_s \Big|. \tag{4.55}$$

We have already noted that the polynomial $P_{\gamma_n}(x)$ can be represented in the form (2.5) or (2.6). Therefore it is symmetric and all of its zeros are negative and mutually distinct. It follows by virtue of Lemma 1.1 and Theorem C that the numbers α_s have alternating signs. We will prove that $|\alpha_s| > |\alpha_{s+1}|$. To this end we make use of the identity

$$\sum_{s=1}^{m} \frac{x_s^{m-1}\Big(x_s - \dfrac{1}{x_s}\Big)}{P'_{\gamma_n}(x_s)(x + x_s)\Big(x + \dfrac{1}{x_s}\Big)} = \frac{x^{m-1}(-1)^{m-1}}{P_{\gamma_n}(-x)}, \qquad m = \Big[\frac{n}{2}\Big],$$

which was obtained in the proof of Lemma 2 in [5]. Multiplying this identity by $x - 1$, we get

$$\sum_{s=1}^{m} \frac{x_s^{m-1}(1 + x_s)}{P'_{\gamma_n}(x_s)} \left[\frac{1}{x\Big(1 + \dfrac{x_s}{x}\Big)} - \frac{1}{1 + x_s x} \right] = \frac{(-1)^m x^{m-1}(1 - x)}{P_{\gamma_n}(-x)}$$

$$= \frac{|x_1| \dots |x_m|}{x} \prod_{s=1}^{m} \sum_{k=0}^{\infty} \Big(\frac{|x_s|}{x}\Big)^k \sum_{n=0}^{\infty} (|x_s| x)^u (1 - x), \ |x_s| < 1. \tag{4.56}$$

We prove in addition some trivial facts:

$$\sum_{k=0}^{\infty} \Big(\frac{|x_s|}{x}\Big)^k \sum_{u=0}^{\infty} (|x_s| x)^u = \sum_{k=-\infty}^{\infty} x^k \sum_{r=0}^{\infty} |x_s|^{2r+|k|} = \sum_{k=-\infty}^{\infty} \gamma_{s,k} x^k. \tag{4.57}$$

Clearly, here $\gamma_{s,k} = \gamma_{s,-k}$ and $\gamma_{s,k} > \gamma_{s,k+1}$ for $k \geqslant 0$.

Suppose given the following pair of Laurent series converging in an annulus containing the unit circle:

$$f_i(z) = \sum_{k=-\infty}^{\infty} \alpha_k^{(i)} z^k \quad (i = 1, 2),$$

in which $\alpha_k^{(i)} = \alpha_{-k}^{(i)}$ and $\alpha_k^{(i)} > \alpha_{k+1}^{(i)} > 0$ for $i = 1, 2$ and $k = 0, 1, \ldots$. Then the function $\varphi(z) = f_1(z) f_2(z)$ has the Laurent series $\varphi(z) = \Sigma_{-\infty}^{\infty} \gamma_k z^k$, where $\gamma_k = \gamma_{-k}$ and $\gamma_k > \gamma_{k+1} > 0$ for $k = 0, 1, \ldots$. In fact,

$$\gamma_k = \sum_{s=-\infty}^{\infty} \alpha_s^{(1)} \alpha_{s+k}^{(2)}$$

and it follows from the conditions on $\alpha_k^{(i)}$ that $\gamma_k > 0$ and

$$\gamma_{-k} = \sum_{s=-\infty}^{\infty} \alpha_s^{(1)} \alpha_{s-k}^{(2)} = \sum_{s=-\infty}^{\infty} \alpha_{-s}^{(1)} \alpha_{-s-k}^{(2)} = \sum_{s=-\infty}^{\infty} \alpha_s^{(1)} \alpha_{s+k}^{(2)} = \gamma_k.$$

Further, for $k \geqslant 0$ we have

$$\gamma_k - \gamma_{k+1} = \sum_{s=0}^{\infty} \alpha_s^{(1)} (\alpha_{s+k}^{(2)} - \alpha_{s+k+1}^{(2)}) + \sum_{s=1}^{k} \alpha_s^{(1)} (\alpha_{k-s}^{(2)} - \alpha_{k+1-s}^{(2)}) - \sum_{s=k+1}^{\infty} \alpha_s^{(1)} (\alpha_{s-k-1}^{(2)} - \alpha_{s-k}^{(2)})$$

$$= \sum_{s=0}^{\infty} (\alpha_s^{(1)} - \alpha_{s+2k+1}^{(1)}) (\alpha_{k+s}^{(2)} - \alpha_{k+s+1}^{(2)}) + \sum_{s=0}^{k-1} (\alpha_{k-s}^{(1)} - \alpha_{k+s+1}^{(1)}) (\alpha_s^{(2)} - \alpha_{s+1}^{(2)}) > 0.$$

From this result and (4.56) we have

$$\begin{aligned} \frac{1}{x} \sum_{k=0}^{\infty} x^{-k} (-1)^k \sum_{s=1}^{m} \frac{x_s^{m-1+|k|} (1 + x_s)}{P'_{\gamma_n}(x_s)} - \sum_{k=0}^{\infty} (-1)^k x^k \sum_{s=1}^{m} \frac{x_s^{m-1+k} (1 + x_s)}{P'_{\gamma_n}(x_s)} \\ = |x_1| \ldots |x_m| (1 - x) \sum_{k=-\infty}^{\infty} c_k x^{k-1} = |x_1| \ldots |x_m| \sum_{k=-\infty}^{\infty} (c_{k+1} - c_k) x^k, \end{aligned} \tag{4.58}$$

where $c_k = c_{-k}$ and $c_k > c_{k+1} > 0$ for $\gamma = 0, 1, \ldots$. Equating the coefficients of negative powers of x in (4.58), we find that

$$(-1)^k \sum_{s=1}^{m} \frac{x_s^{m-1+k} (1 + x_s)}{P'_{\gamma_n}(x_s)} = |\alpha_s| - |\alpha_{s+1}| = c_k - c_{k+1} > 0.$$

Thus the terms of the series $\Sigma_0^{\infty} \alpha_s$ alternate in sign and monotonically tend to zero, so that

$$\left| \sum_{s=k}^{\infty} \alpha_s \right| < |\alpha_k|.$$

This result together with (4.55) and Lemma 1.1 implies

$$\sum_{k=-\infty}^{\infty} |\gamma_k| < 2 \sum_{k=1}^{\infty} |\alpha_k| < \sum_{k=-\infty}^{\infty} |\alpha_k| = |P_{\gamma_n}(-1)|^{-1}.$$

Hence, taking (4.24) into account,

$$\sum_{k=-\infty}^{\infty} |\gamma_k| < |P_{\gamma_n}(-1)|^{-1} \leqslant C,$$

where C depends only on n.

We estimate I_1:

$$I_1=\frac{1}{h_1}\int_{-h_1/2}^{h_1/2}dt\Big\{\sum_{m=-\infty}^{\infty}\int_0^h\Big|h^{-n}\int_0^{nh}\Big[f^{(n)}(mh+\alpha+x)-f^{(n)}\Big(mh-\Big[\frac{n}{2}\Big]h+t+u\Big)\Big]$$

$$\times\varphi_n(u)du\Big|^p dx\Big\}^{1/p}\leqslant\frac{C(n)}{hh_1}\int_{-h_1/2}^{h_1/2}dt\Big\{\sum_{m=-\infty}^{\infty}\int_0^h\Big[\frac{1}{nh}\int_0^{nh}du\int_0^{nh}\Big|f^{(n)}(mh+\alpha+x)$$

$$-f^{(n)}(mh+\alpha+x+u+v)+$$

$$+f^{(n)}(mh+\alpha+x+u+v)-f^{(n)}\Big(mh-\Big[\frac{n}{2}\Big]h+t+u\Big)\Big|dv\Big]^p dx\Big\}^{1/p}.$$

Applying Hölder's inequality with exponent p twice and then Minkowski's inequality to the right side of the latter inequality, we get

$$I_1\leqslant C_1(n)\omega(f^{(n)},h)_p.\tag{4.59}$$

Thus the assertion of the theorem for $i=n$ follows from (4.52), (4.54) and (4.59).

To complete the proof of the theorem, it remains to consider the case $i=0$. We put

$$S_n(x,h_1,h;f)=P_{n-1}(x)+\frac{1}{(n-1)!}\int_0^x(x-t)^{n-1}S_n(t,h_1,h;f)\,dt \tag{4.60}$$
$$=P_{n-1}(x)+\varphi(x),\ 0\leqslant x\leqslant nh.$$

Hence

$$P_{n-1}(x)=S_n(x,h_1,h;f)-\varphi(x)=\psi(x).$$

Since $V_i(P_{n-1})=V_i(\psi)$ for $i=0,\pm1,\pm2,\ldots$, it follows analogously to (4.27) and (4.30) that

$$P_{n-1}(x)=\sum_{s=0}^{n-1}(-1)^s\Big\{\int_{-h_1/2}^{h_1/2}dt_0\ldots\int_{-h_1/2}^{h_1/2}W(t_0,h+t_1,\ldots,nh-h+t_{n-1})\,dt_{n-1}\Big\}^{-1}$$
$$\times\int_{-h_1/2}^{h_1/2}dt_0\ldots\int_{-h_1/2}^{h_1/2}\psi(sh+t_s)\,W\,dt_{n-1}. \tag{4.61}$$

Here and below

$$W=W(s,t_0,\ldots,sh-h+t_{s-1},sh+h+t_{s+1},\ldots,nh-h+t_{n-1}).$$

In particular, if $f(x)\equiv1$, it follows by virtue of the uniqueness of an interpolation-in-the-mean spline that $S_n(x,h_1,h;f)\equiv1$ and $S_n^{(n)}(x)\equiv0$, i.e.

$$1=\sum_{s=0}^{n-1}(-1)^s\frac{\int_{-h_1/2}^{h_1/2}dt_0\ldots\int_{-h_1/2}^{h_1/2}W\,dt_{n-1}}{\int_{-h_1/2}^{h_1/2}dt_0\ldots\int_{-h_1/2}^{h_1/2}W(t_0,h+t_1,\ldots,nh-h+t_{n-1})\,dt_{n-1}}$$
$$=\sum_{s=0}^{n-1}(-1)^s\frac{h_1\int_{-h_1/2}^{h_1/2}dt_0\ldots\int_{-h_1/2}^{h_1/2}dt_{s-1}\int_{-h_1/2}^{h_1/2}dt_{s+1}\ldots\int_{-h_1/2}^{h_1/2}W\,dt_{n-1}}{\int_{-h_1/2}^{h_1/2}dt_0\ldots\int_{-h_1/2}^{h_1/2}W(t_0,h+t_1,\ldots,nh-h+t_{n-1})\,dt_{n-1}} \tag{4.62}$$

Suppose $mnh\leqslant x'\leqslant(m+1)nh$. We put $x=x'-mnh$. Then $0\leqslant x\leqslant nh$.

Further, the spline $S_n(x, h_1, h; f)$ can be represented by Taylor's formula with remainder in integral form in a neighborhood of the point mnh for $m = 0, \pm 1, \pm 2, \ldots$. Then, by repeating verbatim the above arguments, we can write

$$P_{n-1}(x') = P_{n-1}(mnh + x) = $$
$$= \sum_{s=0}^{n-1} (-1)^s \left\{ \int_{-h_1/2}^{h_1/2} dt_0 \ldots \int_{-h_1/2}^{h_1/2} W(t_0, h + t_1, \ldots, nh - h + t_{n-1})\, dt_{n-1} \right\}^{-1}$$
$$\times \int_{-h_1/2}^{h_1/2} dt_0 \ldots \int_{-h_1/2}^{h_1/2} \psi(mnh + sh + t_s)\, W(x, t_0, \ldots, sh - h + t_{s-1},$$
$$sh + h + t_{s+1}, \ldots, nh - h + t_{n-1})\, dt_{n-1}. \tag{4.63}$$

Let us estimate the deviation $\|f(x) - S_n(x, h_1, h; f)\|_{L_p}$ We represent $P_{n-1}(x)$ in the form of a difference of two polynomials:

$$P_{n-1}(x) = Q_{n-1}(x) - R_{n-1}(x), \tag{4.64}$$

where $Q_{n-1}(x)$ $(R_{n-1}(x))$ is defined by (4.63) with $P_{n-1}(x)$ replaced by $Q_{n-1}(x)$ $(R_{n-1}(x))$ and $\psi(x)$ replaced by $f(x)$ $(\varphi(x))$, where

$$\varphi(x') = \frac{1}{(n-1)!} \int_{mnh}^{x+mnh} (x + mnh - t)^{n-1} S_n^{(n)}(t, h_1, h; f)\, dt \tag{4.65}$$

for $x' = mnh + x$ and $0 \leqslant x < nh$. We have

$$\|f(x) - S_n(x, h_1, h; f)\|_{L_p} = \|f(x) - Q_{n-1}(x) + R_{n-1}(x)$$
$$- \varphi(x)\|_{L_p} \leqslant \|f(x) - Q_{n-1}(x)\|_{L_p} + \|R_{n-1}(x) - \varphi(x)\|_{L_p} = \widetilde{I}_1 + \widetilde{I}_2. \tag{4.66}$$

Taking into account (2.62) and the definition of $Q_{n-1}(x)$, we estimate $\widetilde{I}_1$:

$$\widetilde{I}_1 = \left\{ \sum_{m=-\infty}^{\infty} \int_{mnh}^{(m+1)nh} |f(x) - Q_{n-1}(x)|^p\, dx \right\}^{1/p}$$
$$= \left\{ \sum_{m=-\infty}^{\infty} \int_0^{nh} |f(mnh + x) - Q_{n-1}(mnh + x)|^p\, dx \right\}^{1/p} \tag{4.67}$$
$$= \left\{ \sum_{m=-\infty}^{\infty} \int_0^{nh} \left| \sum_{s=0}^{n-1} (-1)^s l_s(x) \frac{1}{h_1} \int_{-h_1/2}^{h_1/2} [f(x + mnh) - f(mnh) + sh + t_s)]\, dt_s \right|^p dx \right\}^{1/p},$$

where

$$l_s(x) = h_1 \left\{ \int_{-h_1/2}^{h_1/2} dt_0 \ldots \int_{-h_1/2}^{h_1/2} W(t_0, h + t_1, \ldots, nh - h + t_{n-1})\, dt_{n-1} \right\}^{-1} \int_{-h_1/2}^{h_1/2} dt_0 \ldots$$
$$\ldots \int_{-h_1/2}^{h_1/2} dt_{s-1} \int_{-h_1/2}^{h_1/2} dt_{s+1} \ldots \int_{-h_1/2}^{h_1/2} W(x, t_0, \ldots, sh - h + t_{s-1}, sh + h + t_{s+1}, \ldots,$$
$$\ldots, hn - h + t_{n-1})\, dt_{n-1}. \tag{4.68}$$

Taking into consideration inequalities (4.31) and (4.33) with n replaced by $n - 1$ and equalities (4.67) and (4.68), and applying the generalized Minkowski inequality, we find that

$$\widetilde{I}_1 \leqslant C(n) \Big\{ \sum_{m=-\infty}^{\infty} \Big(\int_0^{nh} \Big| \sum_{s=0}^{n-1} \frac{1}{h_1} \int_{-h_1/2}^{h_1/2} [f(x+mnh) - f(mnh+sh+t_s)]\, dt_s \Big|^p dx \Big\}^{1/p}$$

$$\leqslant C(n) \sum_{s=0}^{n-1} \Big\{ \sum_{m=-\infty}^{\infty} \int_0^{nh} \Big| \frac{1}{h_1} \int_{-h_1/2}^{h_1/2} [f(x+mnh) - \frac{1}{h_1} \int_{-h_1/2}^{h_1/2} f(x+t_s+v+mnh)\, dv$$

$$+ \frac{1}{h_1} \int_{-h_1/2}^{h_1/2} f(x+t_s+v+mnh)\, dv - f(mnh+sh+t_s)\Big] dt_s \Big|^p dx \Big\}^{1/p}. \tag{4.69}$$

Taking into account the fact that $0 < h_1 \leqslant h$ and $h_1/h \geqslant \alpha > 0$, where α is an absolute constant, we get in the same way as in the estimation of I_1 in the case $i = n$ that

$$\widetilde{I}_1 \leqslant C\omega(f, h)_p. \tag{4.70}$$

We estimate $\widetilde{I}_2$ analogously to $\widetilde{I}_1$:

$$\widetilde{I}_2 = \Big\{ \sum_{m=-\infty}^{\infty} \int_0^{nh} \Big| \sum_{s=0}^{n-1} (-1)^s l_s(x) \frac{1}{h_1} \int_{-h_1/2}^{h_1/2} [\varphi(x+mnh) - \varphi(mnh$$

$$+ sh + t_s)]\, dt_s \Big|^p dx \Big\}^{1/p} \leqslant C(n) \sum_{s=0}^{n-1} \Big\{ \sum_{m=-\infty}^{\infty} \int_0^{nh} \Big| \frac{1}{h_1} \int_{-h_1/2}^{h_1/2} [\varphi(x$$

$$+ mnh) - \varphi(mnh+sh+t_s)]\, dt_s \Big|^p dx \Big\}^{1/p} \leqslant h^{n-1} C_1(n) \sum_{s=0}^{n-1}$$

$$\times \Big\{ \Big[\sum_{m=-\infty}^{\infty} \int_0^{nh} \Big(\frac{1}{h_1} \int_{-h_1/2}^{h_1/2} \int_0^{nh} | S_n^{(n)}(mnh+t, h_1, h; f) |\, dt\, dt_s \Big)^p dx \Big]^{1/p} \tag{4.71}$$

$$+ \Big[\sum_{m=-\infty}^{\infty} \int_0^{nh} \Big(\frac{1}{h_1} \int_{-h_1/2}^{h_1/2} \int_{-h}^{nh+h} | S_n^{(n)}(mnh+t, h_1, h; f) |\, dt\, dt_s \Big)^p$$

$$\times dx \Big]^{1/p} \Big\} \leqslant h^n C_2(n) \Big\{ \Big[\sum_{m=-\infty}^{\infty} \int_0^{nh} | S_n^{(n)}(mnh+t, h_1, h; f) |^p dt \Big]^{1/p}$$

$$+ \Big[\sum_{m=-\infty}^{\infty} \int_{-h}^{nh+h} | S_n^{(n)}(mnh+t, h_1, h; f) |^p dt \Big]^{1/p} \Big\}$$

$$\leqslant 3C_2(n) h^n \Big\{ \sum_{m=-\infty}^{\infty} \int_0^{nh} | S_n^{(n)}(mnh+t, h_1, h; f) |^p dt \Big\}^{1/p}$$

$$= 3C_3(n) h^{n+\frac{1}{p}} \| Z_k^{(n)} \|_{l_p},$$

where

$$Z_k^{(n)} = S_n^{(n)}(kh+t, h_1, h; f), \ \alpha \leqslant t \leqslant \alpha + h,$$

and α, as before, is equal to 0 (to $h/2$) for n odd (even).

By virtue of (4.9) and Lemma 1.1,

$$\begin{aligned}\|Z_k^{(n)}\|_{l_p} &\leqslant |P_{\gamma_n}(-1)|^{-1}\left\|\frac{1}{h_1}\int_{-h_1/2}^{h_1/2}\Delta_h^n f(kh+t)\,dt\right\|_{l_p}\\ &\leqslant h_1^{\frac{1}{p}}h^{-n}|P_{\gamma_n}(-1)|^{-1}\left\|\left(\int_{-h_1/2}^{h_1/2}|\Delta_h^n f(kh+t)|^p\,dt\right)^{1/p}\right\|_{l_p}\\ &\leqslant 2^{n-1}h_1^{\frac{1}{p}}h^{-n}|P_{\gamma_n}(-1)|^{-1}\,\omega(f,h)_p,\end{aligned} \tag{4.72}$$

where $P_{\gamma_n}(x)$ is defined by (2.10) or (2.11), depending on the evenness or oddness of n.

The assertion of Theorem 4.4 for $i = 0$ follows from (4.66) and (4.70)–(4.72) for $1 \geqslant h_1/h \geqslant \alpha > 0$, where α is an absolute constant, and the theorem is completely proved.

Let $\widetilde{W}_p^l$ $(1 \leqslant p \leqslant \infty)$ denote the class of 2π-periodic functions with an absolutely continuous $(l-1)$th derivative $(l \geqslant 1)$ and an lth derivative that is p-power integrable on $[0, 2\pi]$. The following result is valid.

THEOREM 4.5. *If* $1 \leqslant l \leqslant n, f(x) \in \widetilde{W}_p^l$ *and* $S_n(x, h_1, 2\pi/r; f)$ $(0 < h_1 \leqslant 2\pi/r,$ $r \geqslant 1)$ *is a periodic interpolation-in-the-mean spline, then*

$$\begin{aligned}&\left\|f^{(i)}(x) - S_n^{(i)}\left(x, h_1, \frac{2\pi}{r}; f\right)\right\|_{L_q(0,2\pi)}\\ &\qquad\leqslant C(i,l,n)\,h^{\gamma-i}\,\omega\left(f^{(l)}, \frac{2\pi}{r}\right)_{L_p(0,2\pi)}, \quad i = 0, 1, \ldots, l-1,\end{aligned} \tag{4.73}$$

where $\gamma = l$ *for* $1 \leqslant q \leqslant p \leqslant \infty$ *and* $\gamma = l + q^{-1} - p^{-1}$ *for* $1 \leqslant p \leqslant q \leqslant \infty$, r *is an integer and*

$$\omega(f,h)_{L_p(0,2\pi)} = \sup_{|t|\leqslant h}\left\{\int_0^{2\pi}|f(x+t)-f(x)|^p\,dx\right\}^{1/p}.$$

PROOF. The case $p \leqslant q$ is proved in the same way as Theorem 4.3, except that in the right side of (4.38) and the subsequent inequalities one should sum over s from 0 to $r-1$, replace h by $2\pi/r$ and use the inequality

$$\left\{\int_0^{2\pi n}|f|^p\,dx\right\}^{1/p} \leqslant n^{1/p}\|f\|_{L_p(0,2\pi)} \leqslant n\|f\|_{L_p(0,2\pi)}.$$

The case $1 \leqslant q \leqslant p \leqslant \infty$ follows from Hölder's inequality, the inequality

$$\|f^{(i)} - S_n^{(i)}\|_{L_q(0,2\pi)} \leqslant (2\pi)^{\frac{1}{q}-\frac{1}{p}}\|f^{(i)} - S_n^{(i)}\|_{L_p(0,2\pi)}$$

and the inequality just proved for $q = p$.

Theorem 4.5 is new only for the case of even n and for the case of odd n when $q > p \in [1, 2]$. The remaining cases were proved by Varga [22] and Varga and Swartz [25] for more general knots, for generalized splines (the L-splines introduced in the general case by Ahlberg, Nilson and Walsh [26]; a special case of such splines was considered earlier by Schoenberg [27]) and for nonperiodic functions under certain additional restrictions on the spline. In this work, the approximation is considered only on a finite interval of the real line. The restriction $q \geqslant \max(p, 2)$ in the work of Varga and Swartz is connected with the fact that their inequalities are proved with the use of the second integral relation (see, for example, Ahlberg, Nilson and Walsh [4]). Our proof of Theorem 4.5 does not make use of integral relations. Of importance here is the knowledge of an explicit expression for the highest nonzero derivative of the spline.

§5. Approximation of higher derivatives

Suppose given a continuous 2π-periodic function having a continuous 2π-periodic fourth derivative. If a 2π-periodic cubic spline $S_3(x)$ interpolates a function $f(x)$ on a mesh Δ: $0 = x_0 < x_1 < \cdots < x_n = 2\pi$, if the knots of the mesh are nearly equidistant, i.e.

$$\lim_{n\to\infty} \max_{i,j} \frac{x_{i+1}-x_i}{x_{j+1}-x_j} = 1,$$

and if by $S_3^{(4)}(x)$ is understood the expression

$$S_3^{(4)}(x) = \frac{S_3'''\left(\frac{x_i+x_{i+1}}{2}\right) - S_3'''\left(\frac{x_{i-1}+x_i}{2}\right)}{(x_{i+1}-x_{i-1})/2}, \quad \frac{x_{i-1}+x_i}{2} \leqslant x \leqslant \frac{x_i+x_{i+1}}{2},$$

$i = 0, 1, \ldots, n$, where the nodes are assumed to be periodically positive on the whole real line, then, as was shown by Ahlberg, Nilson and Walsh [4],

$$\| f^{(4)}(x) - S_3^{(4)}(x) \|_{C(0,2\pi)} \to 0 \tag{5.1}$$

for $n \to \infty$.

We will show here that the analogous result holds in the case of a uniform mesh for splines of arbitrary degree interpolating a function defined on $(-\infty, \infty)$.

Suppose $f(x)$ is a continuous function and $S_n(x, h; f)$ is a spline of degree n interpolating $f(x)$ at the nodes $\{kh\}$, i.e.

a) $S_n(kh, h; f) = f(kh)$, $k = 0, \pm 1, \pm 2, \ldots$;

b) $S_n(x, h; f)$ coincides on each of the intervals

$$[mh - \alpha_n h, mh + h - \alpha_n h], \quad \alpha_n = \frac{1+(-1)^n}{4}, \quad m = 0, \pm 1, \pm 2, \ldots,$$

with a polynomial of degree at most n;

c) $S_n(x, h; f)$ has a continuous $(n-1)$th derivative.

We put

$$Z_m^{(n)} = S_n^{(n)}(x, h; f), \; x \in [mh - \alpha_n h, mh + h - \alpha_n h],$$

$m = 0, \pm 1, \pm 2, \ldots$. Then a representation of form (4.9) is valid for $Z_m^{(n)}$ [28], i.e.

$$Z_m^{(n)} = \sum_{k=-\infty}^{\infty} \sum_{s=1}^{[n/2]} \frac{x_s^{[n/2]-1+|k|}}{P'_{\gamma_n}(x_s)} (n-1)! \frac{\Delta_h^n f\left(mh+kh-\left[\frac{n}{2}\right]h\right)}{h^n}, \tag{5.2}$$

where $-1 < x_s < 0$,

$$\sum_{k=-\infty}^{\infty} \left| \sum_{s=0}^{[n/2]} \frac{x_s^{[n/2]-1+|k|}}{P'_{\gamma_n}(x_s)} \right| = | P_{\gamma_n}(-1) |^{-1} > 0$$

and $P_{\gamma_n}(-1)$ depends only on n and not on f or h.

As the $(n+1)$th derivative of $S_n(x, h; f)$, we will understand the expression

$$S_n^{(n+1)}(x, h; f) = \frac{Z_{m+1}^{(n)} - Z_m^{(n)}}{h}, \quad mh \leqslant x < mh + h,$$

$m = 0, \pm 1, \pm 2, \ldots$.

The following assertion holds.

THEOREM 5.1. *If $f(x)$ has a uniformly continuous $(n+1)$th derivative on $(-\infty, \infty)$, then*

$$\| f^{(n+1)}(x) - S_n^{(n+1)}(x, h) \|_{C(-\infty,\infty)} \leqslant C(n)\, \omega(f^{(n+1)}, h)$$

for $h \to 0$, where $C(n)$ depends only on n.

PROOF. Since the numbers $\{Z_m^{(n)}\}$ are defined by (5.2), for $mh \leqslant x < mh + h$ we have

$$\begin{aligned} | f^{(n+1)}(x) - S_n^{(n+1)}(x, h) | &= \left| f^{(n+1)}(x) - \frac{Z_{m+1}^{(n)} - Z_m^{(n)}}{h} \right| \\ &= \left| f^{(n+1)}(x) - \sum_{k=-\infty}^{\infty} \sum_{s=1}^{[n/2]} \frac{x_s^{[n/2]-1+|k|}}{P'_{\gamma_n}(x_s)} (n-1)! \frac{\Delta_h^{n+1} f(mh + kh - [n/2]h)}{h^{n+1}} \right|. \end{aligned} \tag{5.3}$$

Since the analog of (4.44) is satisfied in this case and, in addition, (4.46), (4.49) and (4.50) hold, it follows analogously to (4.48) that

$$\begin{aligned} &\sum_{k=-\infty}^{\infty} \sum_{s=1}^{[n/2]} \frac{x_s^{[n/2]-1+|k|}}{P'_{\gamma_n}(x_s)} (n-1)! \frac{\Delta_h^{n+1} f(mh + kh - [n/2]h)}{h^{n+1}} \\ &= \frac{1}{h^{n+1}} \int_0^{nh+h} \varphi_{n+1}(u) f^{(n+1)}\left(mh - \left[\frac{n}{2}\right]h + u\right) du - \sum_{k=-\infty}^{\infty} \gamma_k (\beta_{k+1} - \beta_k), \end{aligned} \tag{5.4}$$

where

$$\gamma_k = \gamma_{-k-1} = \frac{1-\alpha_0}{2} - \sum_{r=1}^{k} \alpha_r = \frac{1-\alpha_0}{2} - \sum_{r=1}^{k} \alpha_{-r},$$

$$k = 0, 1, 2, 3, \ldots,$$

$$\alpha_r = \alpha_{-r} = \sum_{s=1}^{[n/2]} \frac{x_s^{[n/2]-1+|r|}}{P'_{\gamma_n}(x_s)} (n-1)!$$

and

$$\beta_k = \beta_{k,m} = \frac{1}{h^{n+1}} \int_0^{nh+h} \psi_{n+1}(u) f^{(n+1)}\left(mh - \left[\frac{n}{2}\right]h + kh + u\right) du. \tag{5.5}$$

Thus, from (5.4) and (5.5) we have (for $mh \leqslant x < mh + h$)

$$| f^{(n+1)}(x) - S_n^{(n+1)}(x, h) |$$

$$\leqslant \frac{1}{h^{n+1}} \left| \int_0^{nh+h} \varphi_{n+1}(u) \left[f^{(n+1)}(x) - f^{(n+1)}\left(mh - \left[\frac{n}{2}\right]h + u\right)\right] du \right| \tag{5.6}$$
$$+ \omega(f^{(n+1)}, h) \,|\gamma_k|\, \max_k |\beta_{k+1} - \beta_k| \leqslant \omega\left(f^{(n+1)}, \frac{n+2}{2} h\right)$$
$$+ \sum_{k=-\infty}^{\infty} \sum_{k=-\infty}^{\infty} |\gamma_k| \leqslant C(n)\, \omega(f^{(n+1)}, h).$$

We note in addition that (4.46) was used in the derivation of (5.6). The theorem is proved.

An analysis of the proof readily reveals that the following more general theorem actually holds.

THEOREM 5.2. *If the function $f(x)$ has a uniformly continuous $(n + l)$th derivative $(l \geqslant 1)$ and the spline $S_n(x, h; f)$ interpolates $f(x)$ at the knots $\{kh\}$ for $k = 0, \pm 1, \pm 2, \ldots$, then*

$$\| f^{(n+l)}(x) - S_n^{(n+l)}(x, h; f) \|_{C(-\infty,\infty)} \leqslant C(n)\, \omega(f^{(n+l)}, h), \tag{5.7}$$

where by $S_n^{(n+l)}(x, h; f)$ is meant the expression

$$S_n^{(n+l)}(x) = \frac{\Delta^l Z_m^{(n)}}{h^l}, \qquad mh \leqslant x < mh + h.$$

We note that, as was shown by Zav'jalov [29], an analog of (5.1) also holds for multicubic splines.

We also note that in Theorems 5.1 and 5.2 the sequence of differences $\Delta_h^n f(kh)$ $(k = 0, \pm 1, \pm 2, \ldots)$ can be unbounded, whereas the sequences $\Delta_h^{n+l+1} f(kh)$ are bounded; and hence $|\Delta_h^n f(kh)|$ for any fixed h does not grow more rapidly than $(|k| + 1)^{l+1}$. But then, as was noted in §3, there exists only one spline $S_n(x, h; f)$ whose nth derivative does not grow more rapidly than $(|k| + 1)^{l+1}$. This result follows trivially from [5] (see also [16]).

6. Concluding remarks

Here we mention certain results which to some extent adjoin the investigations of the author. In connection with the results of §§1 and 3, we note the investigations begun by Atteia [30]–[32] and continued by Carasso [33] and Laurent [34]–[36]. They consider two Hilbert spaces X and Y and a bounded linear operator T acting from X into Y with kernel of finite dimension q. In addition, there is given a sequence of linear functionals $\{k_i\}_1^N$, $N \geqslant q$, on X and a sequence of real numbers $\{\alpha_i\}_1^N$. It is proved that if Φ_α denotes the totality of elements $\{x\}$, $x \in X$, for which $k_i(x) = \alpha_i$ for $i = 1, 2, \ldots, N$, then there exists a unique element $a \in X$ such that

$$\| Ta \|_Y = \inf_{x \in \Phi_\alpha} \| Tx \|_Y. \tag{6.1}$$

This element a is called a generalized spline.

The results of §§1 and 3 can be interpreted in similar terms. However, in our case X and Y are not Hilbert spaces and the number of functionals $\{k_i\}$ is not finite. Also, there is introduced, in addition, a certain (in our case, convex) set Δ of sequences $\alpha = \{\alpha_i\}$, and one seeks the quantity

$$\sup_{\alpha\in\Delta}\inf_{x\in\Phi_\alpha}\|Tx\|_Y, \tag{6.2}$$

where, again, T is a linear operator acting from X into Y.

Thus, in §§1 and 3 one can put $Y = L_p(-\infty, \infty)$, $1 \leqslant p \leqslant \infty$, and let $X = X_p^{(n)}$ be the space of functions $f(x)$ with locally absolutely continuous $(n-1)$th derivative and with nth derivative in $L_p(-\infty, \infty)$. The norm in X can be defined, for example, as follows:

$$\|f\|_X = \|f\|_{L_p} + \|\Delta^n f\|_{L_p},$$

where

$$\Delta^n f(x) = \sum_{i=0}^{n}(-1)^{n-i}C_n^i f(x+i),$$

$$T = \frac{d^n}{dx^n}, \; k_i(f) = \int_{-\infty}^{\infty} f(i+t)\,dg(t), \qquad i = 0, \pm 1, \pm 2, \ldots,$$

and, for $n \geqslant 1$, $g(t)$ is a given function of bounded variation.

We note that investigations of the author connected with special cases of problem (6.2) were published [5] at the same time as the first investigations of Atteia [30], [31].

In regard to §4, we note that the results obtained here contain, as a special case for $h_1 \longrightarrow 0$, the results of Ahlberg, Nilson and Walsh (see [4]) on the approximation of periodic functions on a uniform mesh, as well as earlier investigations of the author (see [18]–[20], [28] and [37]) on the interpolation of functions on the real line (see also [38]). Here one is concerned with orders of estimates. Of course, the results of §4 do not generalize results of an asymptotically exact, or exact, nature [39]–[43].

The results of §5 remain valid for interpolation-in-the-mean splines. The proof is not essentially different.

BIBLIOGRAPHY

1. V. S. Rjaben'kiĭ, *On the stability of finite difference schemes, and on the application of the method of finite differences to the solution of the Cauchy problem for a system of partial differential equations*, Candidate's Dissertation, Moscow State Univ., Moscow, 1952. (Russian)

2. V. S. Rjaben'kiĭ and A. F. Filippov, *On the stability of difference equations*, GITTL, Moscow, 1956; German transl., VEB Deutscher Verlag, Berlin, 1960. MR **19**, 865; **23** # A437.

3. S. L. Sobolev, *Lectures on the theory of cubature formulas*. Part II, Novosibirsk. Gos. Univ., Novosibirsk, 1965. MR **35** # 3884.

4. J. H. Ahlberg, E. N. Nilson and J. L. Walsh, *The theory of splines and their application*, Academic Press, New York, 1967. MR **39** # 684.

5. Ju. N. Subbotin, *On the connection between finite differences and the corresponding derivatives*, Trudy Mat. Inst. Steklov. 78 (1965), 24–42 = Proc. Steklov Inst. Math. 78 (1965), 23–42 (1967). MR **32** # 7978.

6. ———, *Functional interpolation in the mean with smallest nth derivative*, Trudy Mat. Inst. Steklov. **88** (1967), 30–60 = Proc. Steklov Inst. Math. **88** (1967), 31–63 (1969). MR **36** # 3002.

7. I. J. Schoenberg, *Cardinal interpolation and spline functions*, J. Approximation Theory 2 (1969), 167–206. MR **41** # 2266.

8. R. G. Cooke, *Infinite matrices and sequence spaces*, Macmillan, London, 1950. MR **12**, 694.

9. M. G. Kreĭn, *Integral equations on a half-line with kernel depending upon the difference of the arguments*, Uspehi Mat. Nauk **13** (1958), no. 5 (83), 3–120; English transl., Amer. Math. Soc. Transl. (2) **22** (1962), 163–288. MR **21** # 1507.

10. A. Zygmund, *Trigonometric series*, 2nd rev. ed., Vol. I, Cambridge Univ. Press, New York, 1959. MR **21** # 6498.

11. A. O. Gel'fond, *The calculus of finite differences*, 3rd rev. ed., "Nauka", Moscow, 1967; French transl. of 2nd ed., Dunod, Paris, 1963. MR **28** # 376; **35** # 7021.

12. N. E. Nörlund, *Vorlesungen über Differenzrechnung*, Springer-Verlag, Berlin, 1924.

13. Alexander Peyerimhoff, *On convergence fields of Nörlund means*, Proc. Amer. Math. Soc. **7** (1956), 335–347. MR **17**, 1199.

14. Einar Hille, *Analytic function theory*. Vol. II, Ginn, Boston, 1962. MR **34** # 1490.

15. N. G. Čebotarev and N. N. Meĭman, *The Routh-Hurwitz problem for polynomials and entire functions*, Trudy Mat. Inst. Steklov. **26** (1949). (Russian) MR **11**, 509.

16. I. J. Schoenberg, *Cardinal interpolation and spline functions*. II, J. Approximation Theory **6** (1972), 404–420. MR **49** # 5649.

17. S. B. Stečkin and Ju. N. Subbotin, Addenda to the Russian translation of [4], "Mir," Moscow, 1972, pp. 270–309. (Russian) MR **50** # 5280.

18. Ju. N. Subbotin, *Spline approximation and smooth bases in* $C(0, 2\pi)$, Mat. Zametki 12 (1972), 43–51 = Math. Notes **12** (1972), 459–463. MR **47** # 5493.

19. ———, *The application of splines in approximation theory*, Linear Operators and Approximation (Proc. Conf., Oberwolfach, 1971), Birkhäuser, Basel, 1972, pp. 405–418. (Russian) MR **50** # 5498.

20. ———, *Approximation of the functions of class* $W^k H^p_\omega$ *by splines of order m*, Dokl. Akad. Nauk SSSR **195** (1970), 1039–1041 = Soviet Math. Dokl. **11** (1970), 1626–1628. MR **42** # 8139.

21. V. L. Gončarov, *Theory of interpolation and approximation of functions*, 2nd ed., GITTL, Moscow, 1954. (Russian) MR **16**, 803.

22. R. S. Varga, *Error bounds for spline interpolation*, Approximations with Special Emphasis on Spline Functions (Proc. Sympos., Madison, Wisc., 1969), Academic Press, New York, 1969, pp. 367–388. MR **40** # 6130.

23. I. J. Schoenberg, *Contributions to the problem of approximation of equidistant data by analytic functions*. Parts A, B, Quart. Appl. Math. **4** (1946), 45–99, 112–141. MR **7**, 487; **8**, 55.

24. John R. Rice, *The approximation of functions*, Vol. 2, Addison-Wesley, Reading, Mass., 1969. MR **39** # 5989.

25. B. K. Swartz and R. S. Varga, *Error bounds for spline and L-spline interpolation*, J. Approximation Theory **6** (1972), 6–49.

26. J. H. Ahlberg, E. N. Nilson and J. L. Walsh, *Convergence properties of generalized splines*, Proc. Nat. Acad. Sci. USA 54 (1965), 344–350. MR **36** # 6847.

27. I. J. Schoenberg, *On trigonometric spline interpolation*, J. Math. Mech. **13** (1964), 795–825. MR **29** # 2589.

28. Ju. N. Subbotin, *Piecewise-polynomial interpolation*, Mat. Zametki **1** (1967), 63–70 = Math. Notes **1** (1967), 41–45. MR **35** # 4645.

29. Ju. S. Zav'jalov, *Interpolation by bicubic splines*, Vyčisl. Sistemy Vyp. 38 (1970), 74–101. (Russian) MR **44** # 5663.

30. Marc Atteia, *Généralisation de la définition et des propriétés des "spline fonctions"*, C. R. Acad. Sci. Paris **260** (1965), 3550–3553. MR **35** # 3340.

31. ———, *"Spline-fonctions" généralisées*, C. R. Acad. Sci. Paris **261** (1965), 2149–2152. MR **35** # 3341.

32. ———, *Fonctions "spline" définies sur un ensemble convexe*, Numer. Math. **12** (1968), 192–210. MR **41** # 2265.

33. Claude Carasso, *Méthode générale de construction des fonctions spline*, Rev. Française Informat. Recherche Opérationelle 1 (1967), no. 5, 119–127. MR **37** # 662.

34. Pierre-Jean Laurent, *Théorèmes de caractérisation en approximation convexe*, Mathematica (Cluj) **10 (33)** (1968), 95–111. MR **41** # 701.

35. Pierre-Jean Laurent, *Construction of spline functions in a convex set*, Approximation with Special Emphasis on Spline Functions (Proc. Sympos., Madison, Wisc., 1969), Academic Press, New York, 1969, pp. 415–446. MR **40** # 6147.

36. ———, *Approximation et optimisation*, Hermann, Paris, 1972.

37. Ju. N. Subbotin, *Piecewise-polynomial interpolation*, Search for Extrema (Proc. Third All-Union Sympos. Extremal Problems), Tomsk. Gos. Univ., Tomsk, 1969, p. 248. (Russian) RŽMat. **1970** # 5Б 137.

38. J. H. Ahlberg and E. N. Nilson, *Polynomial splines on the real line*, J. Approximation Theory **3** (1970), 398–409. MR **45** # 782.

39. V. M. Tihomirov, *Best methods of approximation and interpolation of differentiable functions in the space* $C[-1, 1]$, Mat. Sb. **80 (122)** (1969), 290–304 = Math. USSR Sb. **9** (1969), 275–289. MR **41** # 703.

40. E. W. Cheney and F. Schurer, *Convergence of cubic spline interpolants*, J. Approximation Theory **10** (1974), 103–111. MR **50** # 2760.

41. Blair K. Swartz, $O(h^{2n+2-l})$ *bounds on some spline interpolation errors*, Bull. Amer. Math. Soc. **74** (1968), 1072–1078. MR **38** # 4869.

42. ———, $O(h^{k-j}\omega(D^k f, h))$ *bounds on some spline interpolation errors*, Doctoral Dissertation, New York Univ., New York, 1970.

43. V. L. Velikin, *Approximation by spline functions on some classes of continuous and differentiable functions*, Candidate's Dissertation, Dnepropetrovsk, 1971. (Russian)

Translated by S. SMITH

Trudy Mat. Inst. Steklov.
138 (1975)

Proc. Steklov Inst. Math.
138 (1975)

APPROXIMATION BY SPLINES WITH GIVEN DISTRIBUTION DENSITY OF KNOTS

UDC 517.5

N. I. ČERNYH

ABSTRACT. It is determined how the orders of spline approximations of functions defined on the real axis (semiaxis) depend on the density of the distribution of the knots. Various classes of functions are considered.

Bibliography: 3 titles.

1. Introduction

Let $I = [0, \infty)$ or $(-\infty, \infty)$ and let $1 \leqslant q, p \leqslant \infty$. The set of all functions $g(x)$ which are locally measurable on I and have finite norm

$$\| g(x) \|_{q(I)} = \left\{ \int_I | g(x) |^q \, dx \right\}^{1/q}$$
$$\| g(x) \|_{\infty(I)} = \operatorname*{ess\,sup}_{x \in I} | g(x) |)$$

will be denoted, as usual, by $L_q(I)$. Let $W_p^k(I)$ be the class of functions $f(x)$ which have a locally absolutely continuous derivative of order $k-1$ on I and for which $f^{(k)}(x) \in L_p(I)$, and let $\widetilde{W}_p^k(I) = \{f \in W_p^k(I): \|f^{(k)}\|_{p(I)} \leqslant 1\}$, $k = 1, 2, \ldots$. Norms and classes are defined similarly for a finite interval Δ.

A function $s(x) = s_l(x)$ $(l = 0, 1, \ldots)$ defined on I is called a spline of order l if it is $l-1$ times continuously differentiable $(l > 0)$, and if $s^{(l)}(x)$ is a piecewise constant function such that the set of points at which it is discontinuous is either finite or infinite but has no finite limit points. The points at which $s^{(l)}(x)$ is discontinuous will be called the knots of the spline $s(x)$, which, for convenience, will also include the point $x = 0$.

This definition of the knots of a spline differs somewhat from the standard one (cf. [1], Chapter 1) and is more suitable for the study of the problem of approximation by splines with nonfixed knots.

By virtue of this definition, the knots of a spline $s(x)$ can be ordered in a finite or infinite sequence

$$0 = t_0 < t_1 < t_2 < \ldots \qquad \text{for} \quad I = [0, \infty),$$
$$\ldots < t_{-1} < t_0 = 0 < t_1 < \ldots \quad \text{for} \quad I = (-\infty, \infty),$$

where $t_i \to +\infty$ $(t_{-i} \to -\infty)$ if there are infinitely many knots with positive (negative) indices. A spline $s_l(x)$ is a polynomial of degree l on each interval $[t_{i-1}, t_i]$ (or $[t_n, \infty)$, or $(-\infty, t_{-m}]$) which is free of knots.

AMS (MOS) subject classifications (1970). Primary 41A15.

In this paper, we take as a measure of the complexity of a spline $s(x)$ (for fixed l) various characteristics of the distribution density of its knots, we study the question of the finiteness of the approximations in the metric of $L_q(I)$ of functions $f(x) \in W_p^k(I)$ by splines with knots of given distribution density, and we determine precise orders and give bounds for the corresponding best approximations of functions by splines when the density of knots is increased. From our results there follows, in particular, a serial solution to the problem of approximating a function by means of splines with the sparsest distribution of knots which will guarantee a prescribed accuracy of approximation.

This work represents a development of the methods in [2], which was written in collaboration with Ju. N. Subbotin. Freud and Popov [3] employed a method of spline approximation on an interval which is close to ours.

We give precise definitions of the characteristics of the density of knots of splines which constitute the best approximation, and we note some of their simplest properties.

Let $R = \{t_i\} \subset I$. We denote the number of points t_i on the interval $[x, x') \subset I$ by $m_R(x, x')$. In the case $I = [0, \infty)$, we will call the function $m_R(u) = m_R(0, u)$ the distribution function of the set R. In case $I = (-\infty, \infty)$, we define the distribution function by the relation $m_R(u) = m_R(-u/2, u/2)$ $(u > 0)$. If R is the set of knots of the spline $s(x)$, then $m_R(x, x')$ and $m_R(u)$ will also be denoted by $m_s(x, x')$ and $m_s(u)$. The distribution function $m_s(u)$ is a piecewise constant nondecreasing function (continuous on the left for $I = [0, \infty)$) which completely defines the position of the knots $\{t_i\}$, $m_s(u) \geqslant 1$.

Let $n(u) \geqslant 1$ $(u > 0)$ be a nondecreasing function. We denote the set of splines $s(x) = s_l(x)$ such that $m_s(u) \leqslant n(u)$ by $s_{n(u)}(I)$. For any locally measurable function $f(x)$ defined on I, we put

$$e_{n(u)}(f)_{q(I)} = \inf_{s \in s_{n(u)}(I)} \| f(x) - s(x) \|_{q(I)}. \tag{1}$$

We do not assume here that $f(x) \in L_q(I)$. If $f(x)$ is such that, for any spline $s(x) \in s_{n(u)}(I)$, we have $f(x) - s(x) \notin L_q(I)$, then we will assume that $e_{n(u)}(f)_{q(I)} = +\infty$.

It is clear that $n(u) \leqslant n_1(u)$ (for all $u > 0$) implies both the inclusion

$$s_{n(u)}(I) \subset s_{n_1(u)}(I) \tag{2}$$

and the inequality

$$e_{n_1(u)}(f)_{q(I)} \leqslant e_{n(u)}(f)_{q(I)}. \tag{3}$$

As a measure of the maximum number of knots of the spline $s(x)$ on an arbitrary interval of length u, we introduce the function

$$M_s(u) = \sup_{x \in I} m_s(x, x + u)$$

($M_R(u)$ is defined similarly for an arbitrary set $R = \{t_i\} \subset I$). The inequality $M_s(u) \leqslant N(u)$ $(u > 0)$ characterizes the "sparseness" of the distribution of the knots of $s(x)$

uniformly on I. Therefore $M_s(u)$ will be called the spacing function of the knots of $s(x)$.

We note that, as a measure of compactness, $M_s(u)$ is less precise than $m_s(u)$. In particular, it is impossible to use it to decompose into subclasses the set of splines whose knots are concentrated near $x = +\infty$ ($x = -\infty$) in such a way that, for example, $t_i - t_{i-1} \longrightarrow 0$ as $|i| \longrightarrow \infty$. Obviously for all such splines $s(x)$ we have $M_s(u) \equiv +\infty$. This fact can be stated more precisely in the form of the following alternative: for any spline $s(x)$, either $M_s(u) \equiv +\infty$ or $M_s(u) = O(u)$ for large u. In fact, since $M_s(u)$ is monotone and $M_s(u_1 + u_2) = M_s(u_1) + M_s(u_2)$, it follows that if $M_s(u_0) < \infty$, then for any $u \geqslant u_0$

$$M_s(u) \leqslant M_s\left(\left[\frac{u}{u_0}\right]u_0 + u_0\right) \leqslant \left(\left[\frac{u}{u_0}\right] + 1\right) M_s(u_0) \leqslant \left(\frac{u}{u_0} + 1\right) M_s(u_0).$$

Hence if, for a nondecreasing function $N(u) \geqslant 1$, the ratio $N(u)/u$ is unbounded as $u \longrightarrow +\infty$, then the condition $M_s(u) \leqslant N(u)$ is equivalent to

$$M_s(u) \leqslant \begin{cases} \left(\frac{u}{u_0} + 1\right) N(u_0), & u > u_1, \\ N(u), & u \leqslant u_1, \end{cases}$$

where u_0 is the abscissa of the leftmost point of the intersection of the closure Γ of the graph of $N(u)$ with its supporting line $y = ku$, and u_1 is the abscissa of the rightmost point of the intersection of Γ with the line $y = (1 + u/u_0)N(u_0)$. By virtue of this, in studying best approximations by splines $s(x)$ for which $M_s(u) \leqslant N(u)$, we may restrict ourselves from the outset to those monotone majorants $N(u) \geqslant 1$ of the spacing function $M_s(u)$ which grow no faster than $u \cdot$ const as $u \longrightarrow \infty$. For such $N(u)$ we define the classes of splines

$$S_{N(u)}(I) = \{s(x) = s_l(x) : M_s(u) \leqslant N(u) \quad (u > 0)\}$$

and put

$$E_{N(u)}(f)_{q(I)} = \inf_{s \in S_{N(u)}(I)} \| f(x) - s(x) \|_{q(I)} \tag{4}$$

($(E_{N(u)}(f)_{q(I)} = +\infty$ if for every $s \in S_{N(u)}(I)$, $f(x) - s(x) \notin L_q(I)$).

The condition $N(u) \geqslant 1$ is necessary because $M_s(u) \geqslant 1$. If $n(u) \geqslant N(u)$ (or even $n(u) \equiv N(u)$), then

$$s_{n(u)}(I) \supset S_{N(u)}(I) \tag{5}$$

$$e_{n(u)}(f)_{q(I)} \leqslant E_{N(u)}(f)_{q(I)}. \tag{6}$$

In addition, for $N_1(u) \geqslant N(u)$ we have

$$S_{N_1(u)}(I) \supset S_{N(u)}(I) \tag{7}$$

and

$$E_{N_1(u)}(f)_{q(I)} \leqslant E_{N(u)}(f)_{q(I)}. \tag{8}$$

Since there exist functions $f(x)$ such that only splines with knots heavily concentrated near $x = +\infty$ ($x = \pm\infty$) can insure a finite approximation $\|f(x) - s(x)\|_{q(I)} < \infty$, it is desirable to introduce, in addition to the spacing function $M_s(u)$, a "measure of the concentration" of the knots of a spline. For any $c > 0$ we put

$$\mu_s^{(c)}(u) = \inf_{0 \leqslant x \leqslant cu} m_s(x, x+u) \ (|x| < cu \quad \text{for} \quad I = (-\infty, \infty)),$$
$$\sigma_{\nu(u)}^c(I) = \{s(x) = s_l(x) : \mu_s^{(c)}(u) \leqslant \nu(u) \ (u > 0)\}$$

and we define

$$\sigma_{\nu(u)}(I) = \bigcup_{c>0} \sigma_{\nu(u)}^c(I),$$
$$\varepsilon_{\nu(u)}(f)_{q(I)} = \inf_{s \in \sigma_{\nu(u)}(I)} \|f(x) - s(x)\|_{q(I)}. \tag{9}$$

The condition $s(x) \in \sigma_{\nu(u)}(I)$ means that for some $c = c(s) > 0$, for any $u > 0$, there exists an interval $[x_u, x_u + u)$ of length u and lying inside the interval $(0, (c+1)u) \subset I$ $(x_u \leqslant cu)$, such that not more than $\nu(u)$ knots of $s(x)$ lie in $[x_u, x_u + u)$. Since for $u > u_0$ we have $\inf_c \mu_s^{(c)}(u) \geqslant Au$ (or $\equiv 0$), we will consider the approximation $\epsilon_{\nu(u)}(f)_{q(I)}$ only for those nondecreasing functions $\nu(u) > 0$ which do not grow more slowly than $u \cdot$ const as $u \longrightarrow \infty$.

For $\nu_1(u) \geqslant \nu(u)$ we have

$$\sigma_{\nu_1(u)}(I) \supset \sigma_{\nu(u)}(I) \tag{10}$$

and

$$\varepsilon_{\nu_1(u)}(f)_{q(I)} \leqslant \varepsilon_{\nu(u)}(f)_{q(I)}. \tag{11}$$

If, furthermore, $\nu(u) \leqslant n(u)$, then

$$S_{n(u)}(I) \subset \sigma_{\nu(u)}(I) \tag{12}$$

and

$$\varepsilon_{\nu(u)}(f)_{q(I)} \leqslant e_{n(u)}(f)_{q(I)}. \tag{13}$$

In particular, if $\nu(u) \geqslant n(u) \geqslant N(u)$, then

$$\varepsilon_{\nu(u)}(f)_{q(I)} \leqslant e_{n(u)}(f)_{q(I)} \leqslant E_{N(u)}(f)_{q(I)}, \tag{14}$$

where, in the case $\nu(u) \equiv n(u) \equiv N(u)$, we assume that, for large u, $0 < c_1 < \nu(u)/u < c_2 < \infty$.

The classes of splines introduced and the best approximation both depend on the order of the splines, which henceforth will be assumed to be l unless an explicit statement to the contrary is made.

We will investigate the finiteness and the rate of the approximations $\epsilon_{\nu(u)}(f)_{q(I)}$, $e_{n(u)}(f)_{q(I)}$ and $E_{N(u)}(f)_{q(I)}$ as they relate to the smoothness of $f(x)$. We will decompose the set of all splines $s(x)$ of order l into classes (corresponding to the above definitions) which recognize only the power of the order of growth of the knot distribution densities; that is, in studying the best approximations ϵ, e and E, we confine ourselves to majorants of the form

$$\nu(u) = Au^{\alpha}\ (1 \leqslant \alpha),\ n(u) = Au^{\alpha} + 1\ (0 < \alpha),\ N(u) = Au^{\alpha} + 1\ (0 < \alpha \leqslant 1), \tag{15}$$

where $A > 0$.

Every upper bound of such best approximations will be constructive; that is, for each function $f(x)$ under consideration, we will describe a method of constructing splines which belong to the appropriate class and which guarantee the appropriate smallness of the best approximations. The algorithm for constructing the approximating splines allows us (if we have appropriate information concerning the function) to construct, for a given $\delta > 0$, a spline on any finite interval, which is extendable (constructibly) to every larger interval in such a way that, for unbounded extension, we get a spline on I with optimally spaced knots (with respect to the class of functions) such that $\|f(x) - s(x)\|_{q(I)} \leqslant \delta$.

Let the derivative $f^{(k-1)}(x)$ of $f(x)$ be locally absolutely continuous on I, let $l \geqslant k - 1$ $(k = 1, 2, \ldots)$, and let $\{\xi_i\}$ be a finite or infinite system of points on I such that $\xi_{i-1} < \xi_i$. Fundamental to the construction of approximating splines will be those splines $s(x) = s_l(x, f, k)$ which satisfy the following conditions:

1) Discontinuities in the highest derivative $s^{(l)}(x)$ are permitted only at the points $\{\xi_i\}$ and (for $l > 1$) at given intermediate points $t_j = t_j^{(i)} \in (\xi_{i-1}, \xi_i)$ $(j = 1, \ldots, l - 1)$;

2) At the points $\xi_i = t_{l_1}^{(i)} = t_0^{(i+1)}$ $(l_1 = \max(1, l))$, $s(x)$ satisfies the interpolation conditions:

a) for $k = l$ (> 0) and $k = l + 1$,

$$s^{(r)}(\xi_i) = f^{(r)}(\xi_i)\ (r = 0, 1, \ldots, l_1 - 1;\ i = 0, \pm 1, \ldots),$$
$$(s(\xi_i - 0) = f(\xi_i) \text{ for } l = 0);$$

b) for $0 < k < l$,

$$s^{(r)}(\xi_i) = f^{(r)}(\xi_i)\ (r = 0, 1, \ldots, k - 1;\ i = 0, \pm 1, \ldots),$$

$$s^{(r)}(\xi_i) = 0\ (r = k, \ldots, l - 1;\ i = 0, \pm 1, \ldots).$$

We will always choose ξ_i and $t_j^{(i)}$ so that $s(x)$ belongs to the required class. The spline $s(x)$ is determined uniquely on the union of the intervals $[\xi_{i-1}, \xi_i]$ by the conditions 1) and 2). In fact, we can write $s(x)$, for $x \in [\xi_{i-1}, \xi_i]$, in the form

$$s(x) = \sum_{r=0}^{k-1} f^{(r)}(\xi_{i-1}) \frac{(x - \xi_{i-1})^r}{r!} + \sum_{j=0}^{l_1-1} d_j \frac{(x - t_j)_+^l}{l!}, \tag{16}$$

where $t_j = t_j^{(i)}$ and

$$(x - t)_+^{\beta} = \begin{cases} (x - t)^{\beta}, & x > t, \\ 0, & x \leqslant t,\ \beta \geqslant 0. \end{cases} \tag{17}$$

Then 2) is automatically fulfilled at the ξ_{i-1} (we note that, for $k = l + 1$, the derivative $s^{(k-1)}(\xi_{i-1}) = s^{(k-1)}(\xi_{i-1} + 0)$ is equal to $f^{(k-1)}(\xi_{i-1}) + d_0$, and not to $f^{(k-1)}(\xi_{i-1})$, as in the other cases). The condition at the point ξ_i can be rewritten

$$\sum_{j=0}^{l_1-1} d_j \frac{(\xi_i - t_j)^{l-r}}{(l-r)!} = \int_{\xi_{i-1}}^{\xi_i} (-1)^{l-k+1} \left\{ \frac{d^{l+1-k}}{dt^{l+1-k}} \frac{(\xi_i - t)^{l-r}}{(l-r)!} \right\} f^{(k)}(t)\, dt.$$

$$(r = 0, 1, \ldots, l_1 - 1).$$

Defining the parameters $d_0, \ldots, d_{l_1-1}$ by this linear system and substituting them into (16), we get a representation for $s(x)$ for $\xi_{i-1} \leqslant x \leqslant \xi_i$ in the form

$$s(x) = \sum_{r=0}^{k-1} f^{(r)}(\xi_{i-1}) \frac{(x - \xi_{i-1})^r}{r!} + \frac{(-1)^{l+1-k}}{l!} \int_{\xi_{i-1}}^{\xi_i} \frac{d^{l+1-k}}{dt^{l+1-k}} \left\{ \sum_{j=0}^{l_1-1} \frac{(x - t_j)_+^l \omega_i(t)}{(t - t_j)\, \omega_i'(t_j)} \right\} f^{(k)}(t)\, dt, \tag{18}$$

where

$$\omega_i(t) = (t - t_0)(t - t_1) \ldots (t - t_l) \qquad (t_j = t_j^{(i)}). \tag{19}$$

From Taylor's theorem and (18) and (19) it follows that on each interval $[\xi_{i-1}, \xi_i]$ we have, for a spline $s(x)$ which satisfies 1) and 2),

$$f(x) - s(x) = \frac{(-1)^{l+1-k}}{l!} \int_{\xi_{i-1}}^{\xi_i} \frac{d^{l+1-k}}{dt^{l+1-k}} \left\{ (x - t)_+^l - \sum_{j=0}^{l_1-1} \frac{(x - t_j)_+^l \omega_i(t)}{(t - t_j)\, \omega_i'(t_j)} \right\} f^{(k)}(t)\, dt, \tag{20}$$

$$f^{(r)}(x) - s^{(r)}(x) = \frac{(-1)^{l+1-k}}{(l-r)!} \int_{\xi_{i-1}}^{\xi_i} \frac{d^{l+1-k}}{dt^{l+1-k}} \left\{ (x - t)_+^{l-r} - \sum_{j=0}^{l_1-1} \frac{(x - t_j)_+^{l-r} \omega_i(t)}{(t - t_j)\, \omega_i'(t_j)} \right\} f^{(k)}(t)\, dt \tag{21}$$

$$(r = 1, 2, \ldots, k - 1)$$

§2. Finiteness of approximations in $L_q(I)$ of the classes $W_p^k(I)$ for $1 \leqslant p \leqslant q \leqslant \infty$

In this section, we consider the question of the finiteness of the best approximations $\epsilon_\nu(f)_{q(I)}$, $e_n(f)_{q(I)}$ and $E_N(f)_{q(I)}$ for the $\nu(u)$, $n(u)$ and $N(u)$ in (15), for $A = 1$ and $p \leqslant q$, and also a similar question in connection with the problem of approximating both $f(x)$ and its derivatives simultaneously, given that $f(x) \in W_p^k(I)$. We put

$$\begin{aligned} &\omega(t) = \tau\left(\tau - \frac{1}{l}\right) \ldots (\tau - 1)\ (l > 0), \\ &\omega(t) = \tau\ (l = 0), \\ &l_1 = \max(1, l), \end{aligned} \tag{22}$$

$$\rho_r(z,\tau)=\rho_{r,l,k}(z,\tau)$$

$$=\frac{1}{(l-r)!}\frac{d^{l+1-k}}{d\tau^{l+1-k}}\left\{(z-\tau)_+^{l-r}-\sum_{j=0}^{l}\frac{(z-j/l_1)_+^{l-r}\omega(\tau)}{(\tau-j/l_1)\,\omega'(j/l_1)}\right\}; \tag{23}$$

$$C_r^{p,q}=C_r^{p,q}(k,l)=\begin{cases}\|(\|\rho_r(z,\tau)\|_{q_z[0,1]})\|_{p'[0,1]} & \text{for} \quad p'\geqslant q,\\ \|(\|\rho_r(z,\tau)\|_{p_\tau[0,1]})\|_{q[0,1]} & \text{for} \quad p'\leqslant q,\end{cases} \tag{24}$$

where $p'=p/(p-1)$, and $\|\rho_r(t,\tau)\|_{q_z[0,1]}$ is the norm in $L_q[0,1]$ of $\rho_r(z,\tau)$, regarded as a function of z.

THEOREM 1. *Let $I=[0,\infty)$ or $(-\infty,\infty)$, let $1\leqslant p\leqslant q\leqslant\infty$, let $l\geqslant 0$ and $1\leqslant k\leqslant l+1$ be integers; let*

$$\alpha_k=\alpha_k(p,q)=1-\frac{1}{k}\left(\frac{1}{p}-\frac{1}{q}\right);$$

let $N_\alpha(u)=u^\alpha+1$; and let $\widetilde{W}_p^k(I)$ be the class of functions $f\in W_p^k(I)$, $\|f^{(k)}\|_{p(I)}\leqslant 1$. If $\alpha_k(p,q)\leqslant\alpha\leqslant 1(0<\alpha\leqslant 1$ for $k=1$, $p=1$ and $q=\infty)$, then

$$\sup_{f\in\widetilde{W}_p^k(I)} E_{N_\alpha(u)}(f)_{q(I)}\leqslant(\alpha^{-\alpha}l_1)^{\frac{k\alpha_k}{\alpha}}C_0^{p,q}(k,l), \tag{25}$$

where $l_1=l$ for $l>0$, and $l_1=1$ for $l=0$. Moreover, under the same assumptions there exist constants $0<D_r=D_r(\alpha,p,q,k,l)<\infty$ $(r=0,1,\ldots,k-1)$ such that

$$\sup_{f\in\widetilde{W}_p^k(I)}\ \inf_{s\in S_{N_\alpha(u)}(I)}\ \max_{0\leqslant r\leqslant k-1}\{D_r^{-1}\|f^{(r)}(x)-s^{(r)}(x)\|_{q(I)}\}\leqslant 1. \tag{26}$$

As $D_r(\alpha,p,q,k,l)$ we may take

$$C_r^{p,q}(k,l)\,(\alpha^{-\alpha}l_1)^{\frac{1}{\alpha}\left(k-r+\frac{1}{q}-\frac{1}{p}\right)}.$$

PROOF. For any function $f(x)\in W_p^k(I)$, $\|f^{(k)}\|_{p(I)}=1$, and for any α, $\alpha_k(p,q)\leqslant\alpha\leqslant 1$, $\alpha>0$, we construct a spline $s(x)\in S_{N_\alpha(u)}(I)$ such that $f(x)-s(x)\in L_q(I)$, and we estimate the error $\|f(x)-s(x)\|_{q(I)}$. We carry this out for $I=[0,\infty)$.

We begin with the construction of the knots of $s(x)$. Let $0<\alpha<1$ and $\alpha_k(p,q)\leqslant\alpha$. We put

$$F(x)=\int_0^x|f^{(k)}(t)|^p\,dt\ (p<\infty), \tag{27}$$

$$\gamma=-\frac{\alpha}{1-\alpha},\quad a=l_1^{\frac{1}{1-\alpha}}\alpha^{-\frac{\alpha}{1-\alpha}} \tag{28}$$

and $\xi_0=0$. If the points $\xi_0<\xi_1<\cdots<\xi_{i-1}$ $(i=1,2,\ldots)$ are already constructed and $F(\xi_{i-1})\neq 1$, then we define ξ_i to be the root of the equation $F(x)=F(\xi_{i-1})+a(x-\xi_{i-1})^\gamma$ $(x>\xi_{i-1})$. Since $\gamma<0$, $F(x)$ is a continuous nondecreasing function, and $F(x)\to 1$ $(x\to+\infty)$, $F(\xi_{i-1})<1$, it follows that this equation has a unique solution $x = \xi_i$, and $\xi_i > \xi_{i-1}$ and $F(\xi_i) > F(\xi_{i-1})$. If $F(\xi_{i-1}) = 1$, then $F(x) \equiv 1$ for $x \geqslant \xi_{i-1}$, and we will assume that the process of constructing the $\{\xi_i\}$ is complete at the $(i-1)$th step $(i \geqslant 2)$. Continuing this process as long as possible, we get a finite or infinite sequence of points $0 = \xi_0 < \xi_1 < \xi_2 < \cdots$. Since $\xi_i - \xi_{i-1} = \{a/[F(\xi_i) - F(\xi_{i-1})]\}^{-1/\gamma} \geqslant a^{-1/\gamma}$, it follows in the latter case that $\xi_i \longrightarrow \infty$ as $i \longrightarrow \infty$.

If $\alpha = 1$ (in particular, if $p = \infty$), we put $\xi_0 = 0, \xi_1 = l_1, \xi_2 = 2l_1, \ldots$.

In what follows, we construct a spline $s(x)$ whose only admissible knots are the points $\{\xi_i\}$ just constructed and the points obtained by partitioning the intervals $[\xi_{i-1}, \xi_i]$ into l_1 equal parts. We also estimate the spacing function of these knots $\{t_j\}$.

For $\alpha = 1$, the set of admissible knots is the set of natural numbers, and therefore

$$M_s(u) \leqslant 1 + [u] \leqslant 1 + u. \tag{29}$$

Let $0 < \alpha < 1$. We denote the number of intervals (ξ_{i-1}, ξ_i) with length $\Delta\xi_i = \xi_i - \xi_{i-1} \leqslant l_1 h$ $(h > 0)$ by $N_{l_1 h}$. By virtue of the choice of the ξ_i we have

$$a\,(\Delta\xi_i)^\gamma = \int_{\xi_{i-1}}^{\xi_i} |f^{(k)}(t)|^p\,dt. \tag{30}$$

Therefore

$$N_{l_1 h}\,a\,(l_1 h)^\gamma \leqslant a \sum_{\Delta\xi_i \leqslant l_1 h} \Delta\xi_i^\gamma \leqslant \int_0^\infty |f^{(k)}(t)|^p\,dt = 1, \qquad N_{l_1 h} \leqslant a^{-1}(l_1 h)^{-\gamma}. \tag{31}$$

Since each interval (ξ_{i-1}, ξ_i) is decomposed into l_1 intervals (t_{j-1}, t_j), it follows that the number m_h of intervals (t_{j-1}, t_j) of length $\Delta t_j = t_j - t_{j-1} \leqslant h$ does not exceed $l_1 N_{l_1 h}$, i. e. $m_h \leqslant a^{-1} l_1^{1-\gamma} h^{-\gamma}$, $m_h \leqslant [a^{-1} l_1^{1-\gamma} h^{-\gamma}]$, where $[\beta]$ is the integral part of β.

We order the sequence of intervals (t_{j-1}, t_j) in order of increasing length. Let $u > 0$. It is possible that some of the first intervals in this ordered sequence $\{(t_{j_\nu - 1}, t_{j_\nu})\}$ overlap the interval $(x, x+u)$ $(x \geqslant 0)$ in such a way that the number $m(x, x+u)$ of points t_j in $[x, x+u)$ is the largest possible. Hence $m(x, x+u) - 1$ surely does not exceed the number of intervals (Δh) whose minimal lengths h satisfy $m_h \leqslant [a^{-1} l_1^{1-\gamma} h^{-\gamma}]$ and which lie entirely inside $[x, x+u)$. But the shortest of these intervals can have length only $h_1 = a^{-1/\gamma} l_1^{(1-\gamma)/\gamma}$, since $\gamma < 0$, and therefore $m_h = 0$ for $h < h_1$. Putting aside the interval $(x, x+h_1)$ on $[x, x+u)$, we must take an interval of length $h_2 = (2a)^{-1/\gamma} l_1^{(1-\gamma)/\gamma}$ as the next one, since $m_h \leqslant 1$ for $h < h_2$, and we have already put aside an interval of such a length $(h_1 < h_2)$. Moreover, we can put aside only one interval of length h_2 (if $h_1 + h_2 \leqslant u$) on $(x, x+u)$, and the interval of length h_1 already chosen is included in this number. Then we can put aside only one interval of length $h_3 = (3a)^{-1/\gamma} l_1^{(1-\gamma)/\gamma}$ (if $u \geqslant h_1 + h_2 + h_3$), and so on.

We choose $n \geqslant 0$ so that

$$\sum_{j=0}^{n} (ja)^{-1/\gamma} l_1^{(1-\gamma)/\gamma} \leqslant u < \sum_{j=1}^{n+1} (ja)^{-1/\gamma} l_1^{(1-\gamma)/\gamma}.$$

Then

$$u \geqslant \int_0^n (at)^{-1/\gamma} l_1^{(1-\gamma)/\gamma} dt = \frac{\gamma}{\gamma-1} a^{-1/\gamma} \left(\frac{n}{l_1}\right)^{(\gamma-1)/\gamma}$$

and, according to the arguments above, $m(x, x+u) - 1 \leqslant n$, so that

$$m(x, x+u) \leqslant \alpha^{-\alpha} a^{\alpha-1} l_1 u^\alpha + 1 = u^\alpha + 1,$$

and

$$M_s(u) \leqslant u^\alpha + 1. \tag{32}$$

We construct a spline $s(x) = s_l(x, f, k, p, \alpha)$ which satisfies the conditions 1) and 2) of the preceding section for $t_j = t_j^{(i)} = \xi_{i-1} + (j/l_1)(\xi_i - \xi_{i-1})$ $(j = 0, 1, \ldots, l_1)$ and for the ξ_i $(i = 0, 1, \ldots)$ chosen. When $i = 0$ in these conditions, we understand right-hand derivatives at $\xi_0 = 0$. These derivatives exist in all cases which we consider.

As has been proved, $s(x)$ is uniquely determined on the union of the intervals $[\xi_{i-1}, \xi_i]$ by these conditions, and (20) and (21) are satisfied for $x \in [\xi_{i-1}, \xi_i]$. If there are a finite number of these intervals and if $\max \xi_i = \xi_n$, then we put $s(x) \equiv f(x)$ for $x > \xi_n$. We can do this because, in this case, $F(x) \equiv 1$ and $f^{(k)}(x) \equiv 0$ for $x \geqslant \xi_n$, and consequently $f(x)$ is a polynomial of degree $k - 1 \leqslant l$. It is easy to verify that condition 2) on the derivatives $s^{(r)}(x)$ at $\xi_i = \xi_n$ will be satisfied automatically.

We estimate the error $\|f^{(r)}(x) - s^{(r)}(x)\|_{q(I)}$ for $1 \leqslant p \leqslant q \leqslant \infty$. Since in the case $\xi_n = \max \xi_i < \infty$ we have $f(x) - s(x) \equiv 0$ for $x \geqslant \xi_n$, it follows that the problem of estimating $\|f^{(r)} - s^{(r)}\|_{q(I)}$ in the general case reduces to that of estimating $\|f^{(r)} - s^{(r)}\|_{q[\Delta\xi_i]}$ $([\Delta\xi_i] = [\xi_{i-1}, \xi_i])$.

In order to reduce the calculations, we let

$$\frac{1}{(l-r)!}\left\{(x-t)_+^{l-r} - \sum_{j=0}^{l} \frac{(x-t_j)_+^{l-r}}{(t-t_j)\,\omega_i'(t_j)}\right\} = R_{l-r,i}(x, t)\,(x, t \in [\Delta\xi_i]). \tag{33}$$

By virtue of (21) and the generalized Minkowski inequality for $1 \leqslant q \leqslant \infty$, we have

$$\|f^{(r)} - s^{(r)}\|_{q[\Delta\xi_i]} \leqslant \int_{\xi_{i-1}}^{\xi_i} |f^{(k)}(t)| \left\| \frac{d^{l+1-k}}{dt^{l+1-k}} R_{l-r,i}(x, t) \right\|_{q_x[\Delta\xi_i]} dt,$$

whence, applying Hölder's inequality with exponents p and p' $(1/p + 1/p' = 1,\ 1 \leqslant p \leqslant \infty)$, we get

$$\|f^{(r)}(x) - s^{(r)}(x)\|_{q[\Delta\xi_i]}$$

$$\leqslant \|f^{(k)}(t)\|_{p[\Delta\xi_i]} \left\| \left\{ \left\| \frac{d^{l+1-k}}{dt^{l+1-k}} R_{l-r,i}(x, t) \right\|_{q_x[\Delta\xi_i]} \right\} \right\|_{p'[\Delta\xi_i]},$$

where the x in $q_x(\Delta\xi_i)$ denotes the fact that the norm is taken with respect to x. We determine the explicit dependence of

$$C^{(i)}_{k,\,l,\,r,\,p,\,q} = \left\| \left\{ \left\| \frac{d^{l+1-k}}{dt^{l+1-k}} R_{l-r,\,i}(x,\,t) \right\|_{q_x[\Delta\xi_i]} \right\} \right\|_{p'[\Delta\xi_i]}$$

on i. To do this, we introduce the change of variables $t = \Delta\xi_i\tau + \xi_{i-1}$, $x = \Delta\xi_i z + \xi_{i-1}$. Then $R_{l-r,i}(x,\,t)$ becomes $\Delta\xi_i^{l-r}R_{l-r}(z,\,\tau)$, where

$$R_{l-r}(z,\,\tau) = \frac{1}{(l-r)!}\left\{(z-\tau)_+^{l-r} - \sum_{j=0}^{l} \frac{(z-\tau_j)_+^{l-r}\,\omega(\tau)}{(\tau-\tau_j)\,\omega'(\tau_j)}\right\},$$

$\tau_j = j/l_1$ and $\omega(\tau) = \Pi_0^l(\tau - \tau_j)$. The functions $\omega(\tau)$ and $R_{l-r}(z,\,\tau)$ do not depend on i. Since

$$\frac{d^{l+1-k}}{dt^{l+1-k}} R_{l-r,\,i}(x,\,t) = \frac{d^{l+1-k}}{d\tau^{l+1-k}} \{\Delta\xi_i^{l-r} R_{l-r}(z(x),\,\tau(t))\}\ \frac{d^{l+1-k}\tau}{dt^{l+1-k}}$$
$$= \Delta\xi_i^{(l-r)-(l+1-k)} \frac{d^{l+1-k}}{d\tau^{l+1-k}} R_{l-r}(z,\,\tau),$$

we get, as a result of the change of variables,

$$\|f^{(r)}(x) - s^{(r)}(x)\|_{q[\Delta\xi_i]} \leqslant \|f^{(k)}(t)\|_{p[\Delta\xi_i]}\ \Delta\xi_i^{k-r-1+1/p'+1/q}$$
$$\times \left\| \left\{ \left\| \frac{d^{l+1-k}}{d\tau^{l+1-k}} R_{l-r}(z,\,\tau) \right\|_{q_z[0,\,1]} \right\} \right\|_{p'[0,\,1]}$$

For finite p' and q, the exponents $1/p'$ and $1/q$ arise from the change of variables of integration in the integrals in $C^{(i)}_{k,l,r,p,q}$. For $p' = \infty$ ($q = \infty$, respectively), we assume that $\Delta\xi_i^{1/p'} = 1$ ($\Delta\xi_i^{1/q} = 1$).

To estimate $\|f^{(r)} - s^{(r)}\|_{q[\Delta\xi_i]}$ in the case $p' < q$, it is better to apply Hölder's inequality with exponents p and p' to the inside integral introduced into $\|f^{(r)} - s^{(r)}\|^q_{q[\Delta\xi_i]}$ by means of (21). As a result, we get a somewhat better estimate:

$$\|f^{(r)}(x) - s^{(r)}(x)\|_{q[\Delta\xi_i]} \leqslant \|f^{(k)}(t)\|_{p[\Delta\xi_i]}\,\Delta\xi_i^{k-r-1+1/p'+1/q}$$
$$\times \left\| \left\{ \left\| \frac{d^{l+1-k}}{dt^{l+1-k}} R_{l-r}(z,\,\tau\} \right\|_{p'_\tau[0,\,1]} \right\} \right\|_{q[0,\,1]}$$

Putting

$$C_r = C_r^{p,\,q}(k,\,l) = \begin{cases} \left\| \left\{ \left\| \dfrac{d^{l+1-k}}{dt^{l+1-k}} R_{l-r}(z,\,\tau) \right\|_{q_z[0,\,1]} \right\} \right\|_{p'[0,\,1]}, & p' \geqslant q, \\ \left\| \left\{ \left\| \dfrac{d^{l+1-k}}{dt^{l+1-k}} R_{l-r}(z,\,\tau) \right\|_{p'[0,\,1]} \right\} \right\|_{q[0,\,1]}, & p' \leqslant q, \end{cases}$$

we get, finally,

$$\|f^{(r)}(x) - s^{(r)}(x)\|_{q[\Delta\xi_i]} \leqslant C_r\Delta\xi_i^{k-r+1/q-1/p}\ \|f^{(k)}(t)\|_{p[\Delta\xi_i]}\,. \tag{34}$$

Now we estimate $\|f^{(r)} - s^{(r)}\|_{q(I)}$. We consider the cases $q = \infty$ and $q < \infty$, and also $\alpha = 1$ and $\alpha < 1$, separately.

Because of the choice of the knots, we have $\Delta\xi_i = l_1$ for $\alpha = 1$, and therefore

$$\| f^{(r)}(x) - s^{(r)}(x) \|_{q[\Delta\xi_i]} \leqslant C_r l_1^{k-r+1/q-1/p} \| f^{(k)}(t) \|_{p[\Delta\xi_i]},$$

whence, for $q = \infty$,

$$\begin{aligned} \| f^{(r)}(x) - s^{(r)}(x) \|_{\infty(I)} &\leqslant C_r l_1^{k-r-1/p} \max_i \| f^{(k)}(t) \|_{p[\Delta\xi_i]} \\ &\leqslant C_r l_1^{k-r-1/p} \| f^{(k)}(t) \|_{p(I)} \leqslant C_r l_1^{k-r-1/p}, \end{aligned} \tag{35}$$

and, for $q < \infty$ ($p \leqslant q$), taking into account that $\|f^{(k)}\|_{p[\Delta\xi_i]} \leqslant 1$,

$$\begin{aligned} \| f^{(r)}(x) - s^{(r)}(x) \|_{q(I)} &= \Big\{ \sum_i \| f^{(r)}(x) - s^{(r)}(x) \|_{q[\Delta\xi_i]}^q \Big\}^{1/q} \\ &\leqslant C_r l_1^{k-r-1/p+1/q} \Big\{ \sum_i \| f^{(k)}(t) \|_{p[\Delta\xi_i]}^q \Big\}^{1/q} \leqslant C_r l^{k-r-1/p+1/q} \Big\{ \sum_i \| f^{(k)}(t) \|_{p[\Delta\xi_i]}^p \Big\}^{1/q} \\ &\leqslant C_r l_1^{k-r-1/p+1/q} \| f^{(k)}(t) \|_{p(I)}^{p/q} \leqslant C_r l_1^{k-r-1/p+1/q} \end{aligned} \tag{36}$$

If $\alpha_k \leqslant \alpha < 1$, then (cf. (30))

$$\Delta\xi_i = a^{-1/\gamma} \| f^{(k)}(t) \|_{p[\Delta\xi_i]}^{p/\gamma},$$

where $\gamma = -\alpha/(1-\alpha)$ and $p < \infty$ (since it follows from the hypotheses of the theorem that $q = \infty$, $\alpha_k = 1$, and $\alpha = 1$ for $p = \infty$). Therefore for $\alpha < 1$ (34) is equivalent to

$$\| f^{(r)}(x) - s^{(r)}(x) \|_{q[\Delta\xi_i]} \leqslant C_r a^{-(k-r+1/q-1/p)/\gamma} \| f^{(k)}(t) \|_{p[\Delta\xi_i]}^{\mu},$$

where

$$\mu = 1 + \left(k - r + \frac{1}{q} - \frac{1}{p}\right)\frac{p}{\gamma} = p\left\{\frac{1}{q} + r\left(\frac{1}{\alpha} - 1\right) + k\frac{\alpha - \alpha_k}{\alpha}\right\} \geqslant \frac{p}{q}.$$

Hence, as before, we get for $q = \infty$

$$\| f^{(r)}(x) - s^{(r)}(x) \|_{\infty(I)} \leqslant C_r a^{-(k-r+1/q-1/p)/\gamma} \max_i \| f^{(k)}(t) \|_{p[\Delta\xi_i]}^{\mu} \leqslant C_r a^{-(k-r+1/q-1/p)/\gamma}, \tag{37}$$

and for $q < \infty$

$$\begin{aligned} \| f^{(r)}(x) - s^{(r)}(x) \|_{q(I)} &\leqslant \Big\{ \sum_i \| f^{(r)}(x) - s^{(r)}(x) \|_{q[\Delta\xi_i]}^q \Big\}^{1/q} \\ &\leqslant C_r a^{-(k-r+1/q-1/p)/\gamma} \Big\{ \sum_i \| f^{(k)}(t) \|_{p[\Delta\xi_i]}^{\mu q} \Big\}^{1/q} \\ &\leqslant C_r a^{-(k-r+1/q-1/p)/\gamma} \Big\{ \sum_i \| f^{(k)}(t) \|_{p[\Delta\xi_i]}^{p} \Big\}^{1/q} \\ &\leqslant C_r a^{-(k-r+1/q-1/p)/\gamma} \| f^{(k)}(t) \|_{p(I)}^{p/q} \leqslant C_r a^{-(k-r+1/q+1/p)\gamma}. \end{aligned} \tag{38}$$

Substituting for γ and a their values in terms of l and α into this estimate, we get, finally, that, under the hypotheses of the theorem, for each $f \in \widetilde{W}_p^{(k)}$ there exists a spline $s(x) = s_l(x, f, \alpha, p, q, k)$ such that, for $r = 0, 1, \ldots, k-1$,

$$\| f^{(r)}(x) - s^{(r)}(x) \|_{q(I)} \leqslant C_r^{p,q} \left(\frac{\alpha}{l_1^{1/\alpha}} \right)^{-(k-r+1/q-1/p)}, \tag{39}$$

whence Theorem 1 follows for $I = [0, \infty)$. It is clear what changes must be made to construct a proof for the case where $I = (-\infty, \infty)$.

The hypothesis $\alpha \leqslant 1$ in Theorem 1 is essential, since for $\alpha > 1$ the classes $S_{N_\alpha(u)}(I)$ and the corresponding best approximation $E_{N_\alpha(u)}(f)_q$ are not defined. The case $\alpha = 0$ for $\alpha_k(p, q) = 0$ (i.e. for $k = p = 1$ and $q = \infty$) will be examined in §4. The condition $\alpha \geqslant \alpha_k(p, q)$ on the parameter α for $\alpha_k(p, q) \neq 0$ in Theorem 1 is the best possible. This follows from (6) for $n(u) = N(u) = N_\alpha(u)$ and the second part of the following theorem.

THEOREM 2. *Let $\alpha_k(p, q) \leqslant \alpha < \infty$, and suppose that the other hypotheses of Theorem 1 are satisfied. Then there exist constants $0 < D_r = D_r(\alpha, p, q, k, I) < \infty$ such that*

$$\sup_{f \in \widetilde{W}_p^k(I)} e_{N_\alpha(u)}(f)_{q(I)} \leqslant D_0, \tag{40}$$

$$\sup_{e \in \widetilde{W}_p^k(I)} \inf_{s \in s_{N_\alpha(u)}} \max_{0 \leqslant r \leqslant k-1} \{D_r^{-1} \| f^{(r)}(x) - s^{(r)}(x) \|_{q(I)}\} \leqslant 1. \tag{41}$$

If $0 < \alpha < \alpha_k(p, q)$, then there exists a function $f_\alpha(x) \in W_p^k(I)$ such that for any function $n_\alpha(u) = O(u^\alpha)$ $(u \longrightarrow \infty)$

$$e_{n_\alpha(u)}(f_\alpha)_{q(I)} = +\infty,$$

so that both the left-hand side of (40) *and the right-hand side of* (41) *become infinite for any finite D_r.*

PROOF. For $\alpha_k \leqslant \alpha \leqslant 1$ $(\alpha > 0)$, (40) and (41) follow from Theorem 1 and (5) and (6).

Let $\alpha > 1$. We put $n_1(u) = u + 1$ for $u \geqslant 1$, and $n_1(u) = 1$ for $0 < u \leqslant 1$. Since the distribution function $m_s(u)$ takes on only integral values, the condition $m_s(u) \leqslant n_1(u)$ is equivalent to $m_s(u) \leqslant u + 1$ $(u > 0)$. Consequently the classes $s_{u+1}(I)$ and $s_{n_1(u)}(I)$ coincide. Taking into account the fact that $u^\alpha + 1 \geqslant n_1(u)(u > 0)$, we get for $\alpha \geqslant 1$ the inclusions

$$\sigma_u^\alpha{+1}(I) \supset s_u^\alpha{+1}(I) \supset s_{n_1(u)}(I) = s_{u+1}(I) \supset S_{u+1}(I).$$

Hence, for $\alpha > 1$, (40) and (41) follow from Theorem 1.

We prove the second part of the theorem. We take any function $\varphi(x)$ with absolutely continuous derivative $\varphi^{(k-1)}(x)$ on $[0, 1]$ and for which $\psi(x) = \varphi^{(k)}(x) \in L_p[0, 1]$ and

$$E_l(\varphi, c, d)_q = \inf_{c_j} \left\| \varphi(x) - \sum_{j=0}^{l} c_j x^j \right\|_{q[c, d]} > 0 \quad ([c, d] = [0, {}^1/_2] \text{ and } [{}^1/_2, 1]).$$

Let $0 < \alpha < 1$, $\beta > 1/p\alpha$, $I = [0, \infty)$ and

$$x_0 = 0,\ x_i = x_i^{(\alpha)} = b \sum_{j=1}^{i} j^{\frac{1-\alpha}{\alpha}} \ (b > 0,\ i = 1, 2, \ldots).$$

We define the function $f_\alpha(x) = f_{\alpha,\beta,\varphi}(x)$ by putting

$$f_\alpha^{(k)}(x) = \begin{cases} Bi^{-\beta}\psi\left(\dfrac{x - x_{i-1}}{x_i - x_{i-1}}\right), & x_{i-1} \leqslant x < x_i, \\ i = 1, 2, \ldots \end{cases}$$

and

$$f_\alpha(x) = \frac{1}{(k-1)!}\int_0^x (x-t)^{k-1} f_\alpha^{(k)}(t)\,dt.$$

Since $x_i - x_{i-1} = bi^{(1-\alpha)/\alpha} \longrightarrow \infty$ as $i \longrightarrow \infty$, it follows that $f_\alpha(x)$ is defined on the whole semiaxis $I = [0, \infty)$ and $f_\alpha^{(k-1)}(x)$ is absolutely continuous. If $p < \infty$, then

$$\|f_\alpha^{(k)}(t)\|_{p[0,\infty)} = |B| \left\{ \sum_{i=1}^{\infty} i^{-\beta p} \int_{x_{i-1}}^{x_i} \left| \psi\left(\frac{x - x_{i-1}}{x_i - x_{i-1}}\right)\right|^p dx \right\}^{1/p}$$

$$= |B| \, \|\psi\|_{p[0,1]} \left\{ \sum_{i=1}^{\infty} i^{-\beta p} \Delta x_i \right\}^{1/p}$$

This last series converges $(-\beta p + (1-\alpha)/\alpha < -1)$; therefore, for a suitable choice of $B = B(\alpha, \beta, p, b)$, we will have $\|f_\alpha^{(k)}\|_{p(I)} \leqslant 1$. Similarly, if $p = \infty$,

$$\|f_\alpha^{(k)}\|_{\infty(I)} \leqslant |B| \, \|\psi\|_{\infty[0,1]} \leqslant 1.$$

If we begin instead with $\varphi(x)$ for which $\psi(x) = \varphi^{(k)}(x) \in C[0, 1]$, $\varphi^{(k)}_{(0)} = \varphi^{(k)}(1) = 0$, then we get an $f_\alpha(x)$ for which $f_\alpha^{(k)}(x)$ is continuous and $f_\alpha^{(k)}(x) \longrightarrow 0$ as $x \longrightarrow \infty$.

We show that, for $0 < \alpha < \alpha_k = \alpha_k(p, q)$ $(\leqslant 1)$ and for suitable b and β,

$$e_{n_\alpha(u)}(f_\alpha)_{q(I)} = \infty,$$

provided that $n_\alpha(u) = O(u^\alpha)$ for $u \longrightarrow \infty$.

Let $R = \{t_i\colon 0 = t_0 < t_1 < \cdots\}$ be an infinite set of points on I whose distribution function $m_R(u)$ does not exceed $n_\alpha(u)$, and let S_R be a collection of splines of order $l \geqslant k - 1$ whose knots belong to R. Then

$$e_{n_\alpha}(f_\alpha)_{q(I)} = \inf_R \inf_{s \in S_R} \|f_\alpha(x) - s(x)\|_{q(I)}. \tag{42}$$

We take any fixed R and consider those intervals of the system $\{(x_{j-1}, x_j)\}_{j=1}^m$ $(x_j = x_j^{(\alpha)})$ which contain no more than one point $t_i \in R$. Suppose there are $n = n(m, R)$ such intervals. Then the remaining intervals of the system $\{(x_{j-1}, s_j)\}_1^m$ contain at least $2(m - n)$ points $t_i \in R$. By the definition of R, for large m we have

$$2(m-n) \leqslant n_\alpha(x_m) \leqslant cx_m^\alpha < cb^\alpha \left(\int_0^{m+1} t^{(1-\alpha)/\alpha}\,dt \right)^\alpha = cb^\alpha \alpha^\alpha (m+1) \leqslant 2b^\alpha c\alpha^\alpha m,$$

where c does not depend on R. We choose the b in the definition of the x_i so that the last expression is less than m. Then $n > m/2$, and therefore for each R such that $m_R(u) \leqslant n_\alpha(u)$, there exist at least $n > m/2$ intervals of $\{(x_{j-1}, x_j)\}_{j=0}^n$ one of the halves of which is free of knots of any spline $s(x) \in S_R$. We denote the appropriate halves (the left one if there are two) of such intervals (x_{j-1}, x_j) by $(\overline{\Delta x_j})$. Since on each $(\overline{\Delta x_j})$ the spline $s(x) \in S_R$ is a polynomial $p_l(x)$ of degree l, it follows that for $q < \infty$

$$\inf_{s\in S_R} \|f_\alpha(x) - s(x)\|_{q(I)} \geqslant \inf_{s\in S_R} \Big\{ \sum_{(\overline{\Delta x_j})\subset(0,\, x_m)} \|f_\alpha - s\|^q_{q(\overline{\Delta x_j})} \Big\}^{1/q}$$

$$\geqslant \Big\{ \sum_{(\overline{\Delta x_j})\subset(0,\, x_m)} \inf_{s\in S_R} \|f_\alpha(x) - s(x)\|^q_{q(\overline{\Delta x_j})} \Big\}^{1/q}$$

$$\geqslant \Big\{ \sum_{(\overline{\Delta x_j})\subset(0,\, x_m)} \inf_{p_l} \int_{(\overline{\Delta x_j})} \frac{|B|^q}{j^{\beta q}} \Big| \int_{x_{j-1}}^{x_j} \frac{(x-t)_+^{k-1}}{(k-1)!} \psi\Big(\frac{t - x_{j-1}}{x_j - x_{j-1}}\Big)\, dt - p_l(x) \Big|^q dx \Big\}^{1/q}.$$

We make the change of variables $t = \Delta x_j \tau + x_{j-1}$, $x = \Delta x_j \xi + x_{j-1}$ in each of these integrals. The intervals $(\overline{\Delta x_j})$ become $(0, 1/2)$ or $(1/2, 1)$, the polynomials $p_l(x)$ become polynomials $Q_l(\xi)$, and the inside integrals become

$$\Delta x_j^k \int_0^\xi \frac{(\xi - \tau)^{k-1}}{(k-1)!} \psi^{(k)}(\tau)\, d\tau,$$

which coincides with $\Delta x_j^k \varphi(\xi)$ to within polynomials of degree $k - 1 \leqslant l$. As a result, we get

$$\inf_{s\in S_R} \|f_\alpha - s\|_{q(I)} \geqslant |B| \min\{E_l(\varphi, 0, {}^1/_2)_q,\ E_l(\varphi, {}^1/_2, 1)_q\} \times \Big\{ \sum_{j:(\overline{\Delta x_j})\subset(0,\, x_m)} j^{-\beta q} (b j^{(1-\alpha)/\alpha})^{kq+1} \Big\}^{1/q} \tag{43}$$

For $\beta q - (kq + 1)(1 - \alpha)/\alpha \leqslant 0$ this last sum is not less than

$$\sum_{1\leqslant j\leqslant m/2} j^{-\beta q + (kq+1)(1-\alpha)/\alpha},$$

and for $\beta q - (kq + 1)(1 - \alpha)/\alpha > 0$ not less than

$$\sum_{\frac{m}{2}\leqslant j\leqslant m} j^{-\beta q + (kq+1)(1-\alpha)/\alpha}.$$

Both of these expressions approach ∞ as $m \to \infty$ if $\beta q - (kq + 1)(1 - \alpha)/\alpha \leqslant 1$, i.e. if $\beta \leqslant 1/(q\alpha) + k(1 - \alpha)/\alpha$. This condition on β is consistent with the earlier restriction $\beta > 1/p\alpha$, since the inequality $1/p\alpha \leqslant 1/q\alpha + k(1/\alpha - 1)$ is equivalent to the assumption in the theorem that $\alpha < 1 + (1/k)(1/q - 1/p) = \alpha_k(p, q)$.

Similarly, if $q = \infty$,

$$\inf_{s\in S_R} \|f_\alpha(x) - s(x)\|_{\infty(I)} \geqslant \inf_{s\in S_R} \max_{(\overline{\Delta x_j})\subset(0,\, x_m)} \|f_\alpha - s\|_{\infty(\overline{\Delta x_j})}$$

$$\geqslant \max_{(\overline{\Delta x_j})\subset(0,\, x_m)} \inf_{p_l} \Big\| \int_{x_{j-1}}^{x} \frac{(x-t)^{k-1}}{(k-1)!} \frac{B}{j^\beta} \psi\Big(\frac{t - x_{j-1}}{x_j - x_{j-1}}\Big)\, dt - p_l(x) \Big\|_{\infty(\overline{\Delta x_j})} \tag{44}$$

$$\geqslant \max_{(\overline{\Delta x_j})\subset(0,\, x_m)} |B|\, j^{-\beta} b^k j^{k(1-\alpha)/\alpha} \min\{E_l(\varphi, 0, {}^1/_2)_\infty,\ E_l(\varphi, {}^1/_2, 1)_\infty\}.$$

Since the number of intervals $\overline{(\Delta x_j)} \subset (0, x_m)$ is not less than $m/2$ (for large m), it follows that we can extract a subsequence $j_m \longrightarrow \infty$ ($m \longrightarrow \infty$) such that $\overline{(\Delta x_{jm})} \subset (0, x_m)$. In the case under consideration we have $\alpha < \alpha_k(p, \infty) = 1 - 1/kp$, so it is possible to choose β so that $-\beta + k(1-\alpha)/\alpha > 0$ and $j_m^{-\beta+k(1-\alpha)/\alpha} \longrightarrow \infty$ ($m \longrightarrow \infty$), without violating the condition that $\beta > 1/p\alpha$.

The inequalities (43) and (44) hold for any R such that $m_k(u) \leqslant n_\alpha(u)$ and for all sufficiently large m. Therefore, by (42),

$$\inf_{s \in S_R} \| f_\alpha(x) - s(x) \|_{q(I)} = \infty, \quad e_{n_\alpha(u)}(f_\alpha)_{q(I)} = \infty.$$

Since $f_\alpha(x) \in \widetilde{W}_p^k(I)$, surely

$$\sup_{f \in \widetilde{W}_p^{(k)}(I)} e_{n_\alpha(u)}(f)_{q(I)} = +\infty \ (1 \leqslant q \leqslant p \leqslant \infty).$$

For $I = [0, \infty)$, the theorem is proved. If $\varphi(x)$ is chosen so that $\varphi(0) = \cdots = \varphi^{(k)}(0) = 0$, and if $f_\alpha(x) \equiv 0$ for $x < 0$, then clearly $f_\alpha(x) \in W_p^k(-\infty, \infty)$ and $e_{n_\alpha(u)}(f_\alpha)_{q(-\infty,\infty)} = \infty$. This completes the proof of the theorem.

REMARK. As is evident from the proof of Theorem 2, we may put

$$D_r = \begin{cases} C_r^{p,q}(k, l)\,(\alpha^{-\alpha} l_1)^{(k-r+1/q-1/p)/\alpha} & \text{for} \quad \alpha_k(p, q) \leqslant \alpha \leqslant 1,\ \alpha > 0, \\ C_r^{p,q}(k, l)\, l_1^{(k-r+1/q-1/p)} & \text{for} \quad 1 \leqslant \alpha < \infty \end{cases} \tag{45}$$

in (40) and (41). By virtue of (12) and (13) for $\alpha \geqslant 1$, and because the rest of the hypotheses of the theorem are satisfied, (40) and (41) remain valid if we replace $s_{N_\alpha(u)}(I)$ by $\sigma_{N_\alpha(u)}(I)$; that is,

$$\sup_{f \in \widetilde{W}_p^k(I)} \varepsilon_{N_\alpha(u)}(f)_{q(I)} \leqslant D_0, \tag{46}$$

$$\sup_{f \in \widetilde{W}_p^k(I)} \ \inf_{f \in \sigma_{N_\alpha(u)}(I)} \ \max_{0 \leqslant r \leqslant k-1} \{ D_r^{-1} \| f_{(x)}^{(r)} - s^{(r)}(x) \|_{q(I)} \} \leqslant 1. \tag{47}$$

For $\alpha < 1$, $\sigma_{N_\alpha(u)}(I)$ and $\epsilon_{N_\alpha(u)}(f)_{q(I)}$ are not defined.

§3. Finiteness of approximations in $L_q(I)$ of the classes $W_p^k(I)$ for $1 \leqslant q < p \leqslant \infty$

As in the preceding case the value of $\alpha_k = \alpha_k(p, q) = 1 - (1/p - 1/q)/k$ for $q < p$ is critical in the question of the finiteness of the best approximation in $L_q(I)$ of classes $W_p^k(I)$ by classes of splines which are defined by the majorants of the distribution densities of their knots $\nu(u) = n(u) = N(u) = N_\alpha(u) = u^\alpha + 1$. But now it belongs to the set of those α for which it is impossible to guarantee that the approximations of functions $f \in W_p^k(I)$ by the appropriate splines are finite. Besides, $\alpha_k(p, q) > 1$ for $q < p$, so the best approximation $E_{N_\alpha(u)}(f)_{q(I)}$ is not defined for $\alpha \geqslant \alpha_k(p, q)$.

THEOREM 3. *Let $I = [0, \infty)$ or $(-\infty, \infty)$; let $1 \leqslant q < p \leqslant \infty$; let $l \geqslant 0$ and $1 \leqslant k \leqslant l + 1$ be integers; let $\alpha_k = \alpha_k(p, q) = 1 + (1/q - 1/p)/k$; and let $N_\alpha(u) = u^\alpha + 1$. Then for $\alpha > \alpha_k(p, q)$*

$$\sup_{f\in \widetilde{W}_p^k(I)} e_{N_\alpha(u)}(f)_{q(I)} \leqslant C_0^{p,q}(k,l)\left(\frac{2\alpha}{\alpha-\alpha_k(p,q)}\right)^{1/q}(2l_1)^{k\alpha_k}, \tag{48}$$

where $l_1 = l$ *for* $l \geqslant 1$, *and* $l_1 = 1$ *for* $l = 0$. *Furthermore, if*

$$\alpha > \alpha_{k-v} = 1 + \frac{1}{k-v}\left(\frac{1}{q} - \frac{1}{p}\right) \quad (v = 1, 2, \dots, k-1),$$

then there exist constants $0 < D_r < \infty$ *such that*

$$\sup_{f\in \widetilde{W}_p^k(I)} \inf_{s\in s_{N_\alpha(u)}(I)} \max_{0\leqslant r\leqslant v} \{D_r^{-1}\| f^{(r)}(x) - s^{(r)}(x)\|_{q(I)}\} \leqslant 1, \tag{49}$$

and we may put

$$D_r = D_r(\alpha, k, l, p, q) = C_r^{p,q}(k,l)\left(\frac{2\alpha}{\alpha-\alpha_{k-r}}\right)^{1/q}(2l_1)^{(k-r)\alpha_{k-r}(p,q)}, \tag{50}$$

where $C_r^{p,q}(k, l)$ *are constants defined by* (22)–(24).

PROOF. We treat only the case $I = [0, \infty)$. Let $f(x) \in \widetilde{W}_p^k(I)$. Without loss of generality, we may assume that $\|f^{(k)}\|_{p(I)} = 1$. We change the algorithm used in the proof of Theorem 1 to define the ξ_j. Let $\alpha > 1$, and $\gamma = \alpha/(\alpha - 1)$, $a_1 = 2l_1$ and $a = a_1^{-\gamma}$. For the case $p = \infty$ we put

$$\xi_0 = 0, \quad \xi_j = \xi_{j-1} + a_1 j^{-1/\gamma} \quad (j = 1, 2, \dots).$$

For the case $1 \leqslant p < \infty$, we put $F(x) = \int_0^x |f^{(k)}(t)|^p\, dt$. Having defined $\xi_0, \xi_1, \dots, \xi_{j-1}$, we define ξ_j by the following conditions. Let x_j be the largest root of

$$F(x) = F(\xi_{j-1}) + a(x - \xi_{j-1})^\gamma. \tag{51}$$

Since $F(x) \uparrow 1$ and $(x - \xi_{j-1})^\gamma \uparrow +\infty$, it follows that x_j is a finite point in the interval $[\xi_{j-1}, \infty)$. In addition, $(x_j - \xi_{j-1}) \leqslant a^{-1/\gamma}[F(x_j) - F(\xi_{j-1})] \leqslant a^{-1/\gamma} = a_1$. If $x_j - \xi_{j-1} \leqslant a_1 j^{-1/\gamma}$, then we put $\xi_j = \xi_{j-1} + a_1 j^{-1/\gamma}$; if not, then we put $\xi_j = x_j$. Beginning with $\xi_0 = 0$ and continuing this process, we get a sequence $0 = \xi_0 < \xi_1 < \cdots$. Since $\Delta\xi_j = \xi_j - \xi_{j-1} \geqslant a_1 j^{-1/\gamma}$ and $\beta = 1/\gamma < 1$, it follows that

$$\sum_{j=1}^{\infty} \Delta\xi_j = +\infty \quad \text{and} \quad \bigcup_{j=1}^{\infty} [\xi_{j-1}, \xi_j) = I.$$

For what follows, we note that for $\Delta\xi_j > a_1 j^{-\beta}$

$$\int_{\xi_{j-1}}^{\xi_j} |f^{(k)}(t)|^p\, dt = (a\Delta\xi_j)^\gamma, \tag{52}$$

and for $\Delta\xi_j = a_1 j^{-\beta}$

$$\int_{\xi_{j-1}}^{\xi_j} |f^{(k)}(t)|^p\, dt \leqslant a(\Delta\xi_j)^\gamma \tag{53}$$

(otherwise x_j would not be the largest root of (51)).

We put $t_i = t_i^{(j)} = \xi_{j-1} + (i/l_1)\Delta\xi_j$ $(i = 0, 1, \ldots, l; j = 1, 2, \ldots)$. We estimate the distribution function $m_{\{t_i\}}(u)$ of this system of points. Let $\Delta\xi_0 = 0$. If

$$\sum_{j=0}^{n} \Delta\xi_j + \frac{i_0}{l_1}\Delta\xi_{n+1} < u \leqslant \sum_{j=0}^{n} \Delta\xi_j + \frac{i_0+1}{l_1}\Delta\xi_{n+1} \ (n \geqslant 0, 0 \leqslant i_0 < l_1),$$

then, by the definition of the distribution function, $m_{\{t_i\}}(u) - 1 = m_1 = l_1 n + i_0$. For $m_1 \geqslant 1$, taking into account the fact that $\Delta\xi_j \geqslant a_1 j^{-\beta} \geqslant a_1(j + i/l_1)^{-\beta}$ $(i = 0, 1, \ldots, l_1 - 1)$, we get

$$\sum_{j=0}^{n} \Delta\xi_j + \frac{i_0}{l_1}\Delta\xi_{n+1} = \sum_{j=0}^{n}\sum_{i=0}^{l_1-1} \frac{\Delta\xi_j}{l_1} + \sum_{i=0}^{i_0-1}{}' \frac{\Delta\xi_{n+1}}{l_1} \geqslant \sum_{j=1}^{n}{}'\sum_{i=0}^{l_1-1} \frac{1}{l_1}\frac{a_1}{j^\beta}$$

$$+ \sum_{i=0}^{i_0-1}{}' \frac{1}{l_1}\frac{a_1}{(n+1)^\beta} \geqslant \sum_{j=1}^{n}{}' \frac{a_1}{l_1}\sum_{i=0}^{l_1-1} \frac{1}{(j+i/l_1)^\beta} + \frac{a_1}{l_1}\sum_{i=0}^{i_0-1}{}' \frac{1}{(n+1+i/l_1)^\beta}$$

$$= \frac{a_1}{l_1^{1-\beta}} \sum_{k=l_1}^{l_1 n+i_0+l_1-1}{}' \frac{1}{k^\beta} = \frac{a_1}{l_1^{1-\beta}} \sum_{k=l_1}^{m_1+l_1-1}{}' \frac{1}{k^\beta} \geqslant \frac{a_1}{l_1^{1-\beta}} \int_{l_1}^{m_1+l_1} \frac{dx}{x^\beta}$$

$$= \frac{1}{1-\beta}\frac{a_1}{l_1^{1-\beta}} m_1^{1-\beta}\left[\left(1+\frac{l_1}{m_1}\right)^{1-\beta} - \left(\frac{l_1}{m_1}\right)^{1-\beta}\right]$$

$$\geqslant \frac{1}{1-\beta}\frac{a_1}{l_1^{1-\beta}} m_1^{1-\beta}[(1+l_1)^{1-\beta} - l_1^{1-\beta}] \geqslant \frac{1}{1-\beta}\frac{a_1 m_1^{1-\beta}}{l_1^{1-\beta}} l_1^{-\beta}(2^{1-\beta}-1)$$

$$= \frac{\alpha a_1(2^{1/\alpha}-1)}{l_1} m_1^{1/\alpha} \geqslant \frac{a_1}{2l_1} m_1^{1/\alpha} = m_1^{1/\alpha}.$$

Therefore

$$m_{\{t_i\}}(u) = m_1 + 1 \leqslant u^\alpha + 1 \quad (u > 0). \tag{54}$$

As before, we subject the spline $s(x) = s_l(x, f)$ with admissible knots at the points $t_i = t_i^{(j)}$ to the conditions 2)

$$s^{(r)}(t_0^{(j)}) = f^{(r)}(t_0^{(j)}) \ (r = 0, 1, \ldots, \min[k-1, l-1] = k_0; j = 1, 2, \ldots)$$

and

$$s^{(r)}(t_0^{(j)}) = 0 \ (\text{if } k_0 < l-1 \text{ and } k_0 < r \leqslant l-1,\ j = 1, 2, \ldots).$$

In the case $l = 0$, we assume, in addition, that $s(t_0^{(j)}) = s(t_0^{(j)} - 0)$.

Since $m_s(u) \leqslant m_{\{t_i\}}(u)$, it follows that

$$m_s(u) \leqslant u^\alpha + 1,\ s(x) \in s_{u^\alpha+1}(I).$$

Let $0 \leqslant r \leqslant v$ and $\alpha > \alpha_{k-v} = 1 + (1/q - 1/p)/(k - v)$ $(v = 0, 1, \ldots, k-1)$. We estimate the error $\|f^{(r)} - s^{(r)}\|_{q(I)}$.

According to (21), for $x \in [\xi_{j-1}, \xi_j] = [\Delta\xi_j]$

$$s^{(r)}(x) - f^{(r)}(x) = \frac{(-1)^{l-k}}{(l-r)!}\int_{\Delta\xi_j} \frac{d^{-k}}{dt^{l+1-k}}\left\{(x-t)_+^{l-r} - \sum_{i=0}^{l} \frac{(x-t_i)_+^{l-r}\,\omega_j(t)}{(t-t_i)\,\omega_j'(t_i)}\right\} f^{(k)}(t)\,dt,$$

from which we get, after introducing the notation of (24), estimating by means of the

Hölder and Minkowski inequalities, and making the change of variables $x = \Delta\xi_j z + \xi_{j-1}$, $t = \Delta\xi_j\tau + \xi_{j-1}$, as we did in (34),

$$\| f^{(r)}(x) - s^{(r)}(x) \|_{q[\Delta\xi_j]} \leqslant C_r^{p,q}(k, l)\, \Delta\xi_j^{k-r+1/q-1/p} \| f^{(k)} \|_{p[\Delta\xi_j]}. \tag{55}$$

Hence in the case $p = \infty$ ($q < \infty = p$) we get

$$\begin{aligned} \| f^{(r)}(x) - s^{(r)}(x) \|_{q(I)} &= \Big\{ \sum_{j=1}^{\infty} \| f^{(r)}(x) - s^{(r)}(x) \|^q_{l[\Delta\xi_j]} \Big\}^{1/q} \\ &\leqslant C_r^{\infty,q}(k, l) \Big\{ \sum_{j=1}^{\infty} (\Delta\xi_j)^{(k-r+1/q)q} \Big\}^{1/q} = C_r^{\infty,q}(k, l)\, a_1^{k-r+1/q} \Big\{ \sum_{j=1}^{\infty} j^{-q(k-r+1/q)/\gamma} \Big\}^{1/q}. \end{aligned} \tag{56}$$

Since $r \leqslant v$, $\alpha > \alpha_{k-v}(\infty, q) \geqslant \alpha_{k-r}(\infty, q)$, it follows that $\gamma = \alpha/(\alpha - 1) < 1 + q(k - v) < 1 + q(k - r)$, the last series (56) converges, and its sum does not exceed

$$\frac{1 + q(k - r)}{1 + q(k - r) - \gamma} = \frac{(\alpha - 1)\alpha_{k-r}}{\alpha - \alpha_{k-r}} < \Big(1 + \frac{\alpha}{\alpha - \alpha_{k-r}}\Big) < \frac{2\alpha}{\alpha - \alpha_{k-r}}.$$

Consequently, for such r and α, and for $p = \infty$ and $q < \infty$,

$$\| f^{(r)}(x) - s^{(r)}(x) \|_{q(I)} \leqslant C_r^{\infty,q}(k, l) \Big(\frac{2\alpha}{\alpha - \alpha_{k-r}} \Big)^{1/q} a_1^{k-r+1/q}. \tag{57}$$

Substituting the value of a_1 everywhere, we get the corresponding statement of the theorem for the case $p = \infty$.

In case $p < \infty$, we denote the set of all j for which $\Delta\xi_j > a_1 j^{-\beta}$ by A, and the set $\{j\}_1^\infty \setminus A$ by B. By virtue of the way the ξ_j were constructed, we have for $j \in B$ that $\Delta\xi_j = a_1 j^{-\beta}$. Applying (55), (52) and (53) in turn, we get

$$\begin{aligned} \| f^{(r)}(x) - s^{(r)}(x) \|_{q(I)} &\leqslant C_r^{p,q}(k, l) \Big\{ \Big(\sum_{j \in \mathscr{A}} + \sum_{j \in \mathscr{B}} \Big) \Delta\xi_j^{q(k-r+1/q-1/p)} \| f^{(k)}(t) \|^q_{p[\Delta\xi_j]} \Big\}^{1/q} \\ &\leqslant C_r^{p,q}(k, l) \Big\{ \sum_{j \in \mathscr{A}} a_1^{q(k-r+1/q-1/p}} \| f^{(k)}(t) \|^{qp(k-r+1/q-1/p)/\gamma}_{p[\Delta\xi_j]} \\ &\quad + \sum_{j \in \mathscr{B}} a_1^{-\gamma q/p} (\Delta\xi_j)^{(\gamma q/p)+q(k-r+1/q-1/p)} \Big\}^{1/q} \\ &= C_r^{p,q}(k, l)\, a_1^{(k-r+1/q-1/p)} \Big\{ \sum_{j \in \mathscr{A}} \| f^{(k)} \|^{p\mu}_{p[\Delta\xi_j]} + \sum_{j \in \mathscr{B}} j^{-\mu} \Big\}^{1/q}, \end{aligned}$$

where $\mu = q/p + (q/\gamma)(k - r + 1/q - 1/p)$. Since we have assumed that $\alpha > \alpha_{k-v}$ and $r \leqslant v$, it follows that $\mu > 1$, and therefore

$$\sum_{j \in \mathscr{B}} j^{-\mu} \leqslant \sum_{1}^{\infty} j^{-\mu} < \frac{\mu}{\mu - 1},$$

$$\sum_{j \in \mathscr{A}} \| f^{(k)}(t) \|^{p\mu}_{p[\Delta\xi_j]} \leqslant \sum_{j=1}^{\infty} \int_{[\Delta\xi_j]} | f^{(k)}(t) |^p \, dt = 1 < \frac{\mu}{\mu - 1}.$$

Consequently, for such r and α,

$$\| f^{(r)}(x) - s^{(r)}(x) \|_{q(I)} \leqslant C_r^{p,q}(k, l)\, a_1^{k-r+1/q-1/p} \Big(\frac{2\mu}{\mu - 1} \Big)^{1/q}, \tag{58}$$

which, together with the condition $a_1 = 2l$ and the estimate

$$\frac{\mu}{\mu-1} = 1 + \frac{\alpha}{q(k-r)(\alpha-\alpha_{k-r})} < \frac{2\alpha}{\alpha-\alpha_{k-r}}$$

is equivalent to the theorem for the case $p < \infty$. Theorem 3 is proved.

It is impossible to weaken the assumptions on α in Theorem 3. This follows from (12), (13) and the second part of the following theorem.

THEOREM 4. *Under the hypotheses of Theorem 3 and using the notation introduced there, for* $\alpha > \alpha_k(p, q)$ *we have*

$$\sup_{f\in\widetilde{W}_p^k(I)} \varepsilon_{N_\alpha(u)}(f)_{q(I)} \leqslant \mathcal{D}_0, \tag{59}$$

and, for $\alpha > \alpha_{k-v}(p, q)$, $v = 1, \ldots, k-1$,

$$\sup_{f\in\widetilde{W}_p^k(I)} \inf_{s\in\sigma_{N_\alpha(u)}(I)} \max_{0\leqslant r\leqslant v} \{\mathcal{D}_r^{-1}\|f^{(r)}(x) - s^{(r)}(x)\|_{q(I)}\} \leqslant 1. \tag{60}$$

If $1 \leqslant \alpha \leqslant \alpha_k(p, q)$, *then there exists a function* $f_\alpha(x) \in W_p^k(I)$ *such that*

$$\varepsilon_{N_\alpha(u)}(f_\alpha)_{q(I)} = \infty, \tag{61}$$

so that the left-hand side of (59) *for* $1 \leqslant \alpha \leqslant \alpha_k(p, q)$ *and the left-hand side of* (60) *for* $1 \leqslant \alpha \leqslant \alpha_{k-v}(p, q)$ *and for any finite* $\mathcal{D}_r$ *are infinite.*

PROOF. (59) and (60) follow from Theorem 3 and (12) and (13).

We prove the second part of the theorem, which says that the conditions on α are the best possible. We consider the case where $\alpha = \alpha_k(p, q)$. We put

$$x_0 = 0,\ x_i = x_{i-1} + \frac{1}{\alpha} i^{(1-\alpha)/\alpha}\ (i = 1, 2, \ldots),$$

$$f_\alpha(x) = K i^{((1-\alpha)/\alpha)(k-1/p)-1/p} (\log 3i)^{-1/q} \varphi\left(\frac{x - x_{i-1}}{x_i - x_{i-1}}\right) (x \in [\Delta x_i]),$$

where $\varphi(x) \in \widetilde{W}_p^k[0, 1]$, $\varphi^{(j)}(0) = \varphi^{(j)}(1) = 0\ (j = 0, 1, \ldots, k)$;

$$E_l(\varphi)_{q(\Delta)} = \inf_{p_l} \|\varphi - p_l\|_{q(\Delta)} > E > 0\ (\Delta = [0, {}^1/_4], \ldots, [{}^3/_4, 1]) \tag{62}$$

and

$$K = \alpha^{-k+1/p} \left(\sum_{i=1}^{\infty} \frac{1}{i(\log 3i)^{p/q}}\right)^{-1/p} \quad \text{for } p < \infty,$$
$$K = \alpha^{-k} \quad \text{for } p = \infty.$$

We have

$$\|f_\alpha^{(k)}\|_{p[\Delta x_i]} = K \Delta x_i^{-k+1/p} i^{\frac{1-\alpha}{\alpha}\left(k-\frac{1}{p}\right)-\frac{1}{p}} (\log 3i)^{-1/q} \|\varphi^{(k)}\|_{p[0,1]}$$
$$= \alpha^{+k-(1/p)} K i^{-1/p} (\log 3i)^{-1/q} \|\varphi^{(k)}(t)\|_{p[0,1]}, \text{ i.e. } \|f_\alpha^{(k)}\|_{p(I)} \leqslant 1.$$

Furthermore, if $s(x) \in \sigma_{u^\alpha+1}(I)$, then, by the definition, there exists a $c > 0$ such that for any $u > 0$ there exists $x_u \leqslant cu$ which satisfies $m(x_u, x_u + u) \leqslant u^\alpha + 1$. Since

$$\sum_{M+1}^{M+N} \Delta x_i \leqslant (M+N)^{1/\alpha} - M^{1/\alpha} \leqslant N^{1/\alpha},\ x_M = \sum_{1}^{M} \Delta x_i \geqslant (M/8)^{1/\alpha},$$

it follows, in particular, that for any $N > 1$ there exists $M < (8c)^\alpha N$ such that the interval $[x_M, x_{M+N})$ contains not more than $N+1$ knots t_j of $s(x)$. Therefore at least one-half of the interval $[\Delta x_i] \subset [x_{M+[N/2]}, x_{M+N}]$ contain not more than three points t_j. Consequently (cf. (62))

$$\|f_\alpha(x) - s(x)\|^q_{q[x_{M+[N/2]}, x_{M+N}]} \geqslant K^q \sum_{M+[N/4]+1}^{M+N} i^{q(k-1/p)(1-\alpha)/\alpha - q/p} (\log 3i)^{-1} \Delta x_i E^q$$

$$= \frac{(KE)^q}{\alpha} \sum_{M+[N/4]+1}^{M+N} i^{-1} (\log 3i)^{-1} \quad (\alpha = \alpha_k(p,q)).$$

$$\|f_\alpha(x) - s(x)\|^q_{q[x_{M+[N/2]}, x_{M+N}]} \geqslant \frac{(KE)^q}{\alpha} \frac{1}{\log(M+N)} \log \frac{M+N}{M+N/4}$$

$$\geqslant \frac{(KE)^q}{\alpha} \log \frac{(8c)^\alpha N + N}{(8c)^\alpha N + N/4} \frac{1}{\log[(8c)^\alpha N + N]}$$

$$= \frac{(KE)^q}{\alpha} \log \frac{4(8c)^\alpha + 4}{4(8c)^\alpha + 1} \frac{1}{\log[(8c)^\alpha + 1] N}.$$

We note that

$$x_{M+N} \leqslant (c+1) N^{1/\alpha},\ x_{M+N} - x_M \geqslant c_1 N^{1/\alpha} \ (c_1 > 0).$$

Now we define the sequence

$$N_0 > 1,\ N_1 = \left(\frac{2(c+1)}{c_1}\right)^\alpha N_0, \ldots, N_j = \left(\frac{2(c+1)}{c_1}\right)^{j\alpha} N_0, \ldots$$

Since, for suitable M_j ($j = 0, 1, \ldots$),

$$x_{M_{j-1}+N_{j-1}} \leqslant (c+1) N_{j-1}^{1/\alpha} = \left(\frac{2(c+1)}{c_1}\right)^j \frac{c_1}{2} N_0^{1/\alpha} = \frac{c_1}{2} N_j^{1/\alpha} \leqslant \frac{x_{M_j+N_j} - x_{M_j}}{2}$$

and, in addition,

$$x_{M_j+N_j} - x_{M_j+[N_j/2]} \leqslant \frac{1}{2}(x_{M_j+N_j} - x_{M_j}),$$

it follows that the intervals

$$[x_{M_j+[N_j/2]}, x_{M_j+N_j}] \quad (j = 1, 2, \ldots)$$

are disjoint. Consequently

$$\|f_\alpha(x) - s(x)\|^q_{q(I)} \geqslant \frac{(KE)^q}{\alpha} \lg \frac{4(8c)^\alpha + 4}{4(8c)^\alpha + 1} \sum_{j=1}^{\infty} \frac{1}{\lg\left\{((8c)^\alpha + 1) N_0 \left(\frac{2(c+1)}{c_1}\right)^{\alpha j}\right\}}$$

$$= K_1 \sum_{j=1}^{\infty} \frac{1}{j \lg\left(\frac{2c+2}{c_1}\right)^\alpha + \lg((8c)^\alpha + 1)} = \infty,$$

for any spline $s(x) \in \sigma_{u^\alpha+1}(I)$; that is,

$$\varepsilon_{u^{\alpha}+1}(f_{\alpha})_{q(I)} = +\infty \ (\alpha = \alpha_k(p, q)).$$

Since $f_{\alpha}(x) \in \widetilde{W}_p^k(I)$, it is surely true that

$$\sup_{f \in \widetilde{W}_p^k(I)} \varepsilon_{N_{\alpha}(u)}(f)_{q(I)} = +\infty \quad (\alpha = \alpha_k(p, q)). \tag{63}$$

If we put $n_{\alpha}(u) = 1$ for $0 < u < 1$ and $n_{\alpha}(u) = u^{\alpha} + 1$ for $u \geqslant 1$ and take into account the fact that the $\mu_s^{(c)}(u)$ in the definition of $\sigma_{\nu(u)}(I)$ are integer-valued, we get, for $\alpha < \alpha_k(p, q)$,

$$\sigma_{N_{\alpha}(u)}(I) = \sigma_{n_{\alpha}(u)}(I) \subset \sigma_{N_{\alpha_k}(u)}(I) \quad (\alpha_k = \alpha_k(p, q)).$$

Therefore (61) and (63) are certainly true for $\alpha \leqslant \alpha_k(p, q)$,

Let $1 \leqslant \alpha \leqslant \alpha_{k-v}(p, q)$. The set $\{g(x) = f^{(v)}(x) \colon f \in W_p^k(I)\}$ is clearly equal to $W_p^{k-v}(I)$. Furthermore, the collection $\{s_{l-v}(x) = s^{(v)}(x) \colon s(x) = s_l(x) \in \sigma_{N\alpha(u)}(I)\}$ is equal to the set $\sigma_{N_{\alpha}(u)}(I)$ of splines of order $l - v$. Therefore, applying the part of the theorem which has already been proved, with l and k replaced by $l - v$ and $k - v$, we get that for $1 \leqslant \alpha \leqslant \alpha_{k-v}(p, q)$

$$\sup_{f \in \widetilde{W}_p^k(I)} \inf_{s_l \in \sigma_{N_{\alpha}(u)}(I)} \| f^{(v)}(x) - s_l^{(v)}(x) \|_{L_q(I)} = \infty$$

and hence, for any finite $\mathcal{D}_r$ and for $1 \leqslant \alpha \leqslant \alpha_{k-v}(p, q)$

$$\sup_{f \in \widetilde{W}_p^k(I)} \inf_{s_l \in \sigma_{N_{\alpha}(u)}(I)} \max_{0 \leqslant r \leqslant v} \{\mathcal{D}_r^{-1} \| f^{(r)}(x) - s_l^{(r)}(x) \|_{q(I)}\} = \infty. \tag{64}$$

This proves Theorem 4 completely.

Remark. The condition $\alpha \geqslant 1$ occurs in Theorem 4 only because $\sigma_{N_{\alpha}(u)}(I)$ is not defined for $\alpha < 1$. From the inclusion relations among $\sigma_{\nu(u)}(I)$, $s_{n(u)}(I)$ and $S_{N(u)}(I)$, and from Theorem 4, it follows that for $0 \leqslant \alpha \leqslant \alpha_k(p, q)$

$$\sup_{f \in \widetilde{W}_p^k(I)} e_{N_{\alpha}(u)}(f)_{q(I)} = \infty \ (q < p), \tag{65}$$

and, for $0 \leqslant \alpha \leqslant 1$,

$$\sup_{f \in \widetilde{W}_p^k(I)} E_{N_{\alpha}(u)}(f)_{q(I)} = \infty \ (q < p). \tag{66}$$

Similarly (64) remains valid if in the left-hand side we replace $\sigma_{N_{\alpha}(u)}(I)$ by $s_{N_{\alpha}(u)}(I)$ for $0 \leqslant \alpha \leqslant \alpha_{k-v}(p, q)$, or by $S_{N_{\alpha}(u)}(I)$ for $0 \leqslant \alpha \leqslant 1$.

§4. Orders of the approximations

The results of the preceding section allow us to solve completely the problem of the orders of the best approximations of functions in the classes $W_p^k(I)$ $(1 \leqslant p \leqslant \infty$, $k = 1, 2, \ldots)$ by the classes $\sigma_{Au^{\alpha}+1}(I)$, $s_{Au^{\alpha}+1}(I)$ and $S_{Au^{\alpha}+1}(I)$ for $A \longrightarrow \infty$ and $\alpha \neq 0$, and to obtain bounds for the corresponding best approximations. This can be done on the basis of the following simple theorem.

THEOREM 5. *Let* $I = [0, \infty)$ *or* $(-\infty, \infty)$, *let* $1 \leqslant p, q \leqslant \infty$, $0 \leqslant r_1 \leqslant k \leqslant l + 1$, $f(x) \in W_p^k(I)$, $\lambda > 0$, *let* $n(u)$ *be a nondecreasing majorant which defines the class*

$s_{n(u)}(I)$ of splines $s(x)$ of order l, and let $s_{n(\lambda u)}(I)$ be the corresponding class defined by the majorant $n_1(u) = n(\lambda u)$. Then

$$e_{n(\lambda u)}(f(x))_{q(I)} = \lambda^{-1/q} e_{n(u)}\left(f\left(\frac{x}{\lambda}\right)\right)_{q(I)}, \tag{67}$$

$$\inf_{s(x)\in s_{n(\lambda u)}(I)} \max_{0\leqslant r\leqslant r_1} \left\{\frac{\| f^{(r)}(x) - s^{(r)}(x)\|_{q(I)}}{\lambda^{r-1/q}\mathcal{D}_r}\right\} = \inf_{s\in s_{n(u)}(I)} \max_{0\leqslant r\leqslant r_1} \left\{\mathcal{D}_r^{-1}\left\|\frac{d^r}{dx^r}\left[f\left(\frac{x}{\lambda}\right) - s(x)\right]\right\|_{q(I)}\right\}, \tag{68}$$

$$\sup_{f\in\widetilde{W}_p^k(I)} e_{n(\lambda u)}(f)_{q(I)} = \lambda^{-k+1/p-1/q} \sup_{f\in\widetilde{W}_p^k(I)} e_{n(u)}(f)_{q(I)}, \tag{69}$$

$$\sup_{f\in\widetilde{W}_p^k(I)} \inf_{s\in s_{n(\lambda u)}(I)} \max_{0\leqslant r\leqslant r_1} \left\{\frac{\| f^{(r)}(x) - s^{(r)}(x)\|_{q(I)}}{\lambda^{r-k+1/p-1/q}D_r}\right\} = \sup_{f\in\widetilde{W}_p^k(I)} \inf_{s\in s_{n(u)}(I)} \max_{0\leqslant r\leqslant r_1} \{\mathcal{D}_r^{-1}\| f^{(r)}(x) - s^{(r)}(x)\|_{q(I)}\}. \tag{70}$$

The equalities (67)–(70) continue to hold if we replace the classes $s_{n(u)}(I)$ and $s_{n(\lambda u)}(I)$ and the best approximations $e_{n(u)}(f)_{q(I)}$ and $e_{n(\lambda u)}(f)_{q(I)}$ by $\sigma_{n(u)}(I)$, $\sigma_{n(\lambda u)}(I)$ and $\epsilon_{n(u)}(f)_{q(I)}$, $\epsilon_{n(\lambda u)}(f)_{q(I)}$, respectively, or by $S_{n(u)}(I)$, $S_{n(\lambda u)}(I)$ and $E_{n(u)}(f)_{q(I)}$, $E_{n(\lambda u)}(f)_{q(I)}$.

We note that it is not assumed that the quantities in the theorem are finite; in particular, if one side of an equality is infinite, then so is the other.

PROOF. If we make the change of variable $x = \lambda\xi$, a spline $s(x)$ which satisfies the condition $m_s(u) \leqslant n(u)$ goes into the spline $s_1(\xi) = s(\lambda\xi)$ which has the same number of knots on the interval $[0, u]$ as $s(x)$ has on $[0, \lambda u]$, i.e. $m_{s_1}(u) = m_s(\lambda u) \leqslant n(\lambda u)$. The converse is also true: from $m_{s_1}(u) \leqslant n(\lambda u)$ it follows that $m_s(u) \leqslant n(u)$. Thus the relation $s_1(\xi) = s(\lambda\xi)$ establishes a one-to-one correspondence between $s_{n(u)}(I)$ and $s_{n(\lambda u)}(I)$. Hence (67) and (68) follow from

$$\| f^{(r)}(\xi) - s_1^{(r)}(\xi)\|_{q(I)} = \left\|\frac{d^r}{d\xi^r}(f(\xi) - s(\lambda\xi))\right\|_{q(I)} = \lambda^{r-1/q}\left\|\frac{d^r}{dx^r}\left(f\left(\frac{x}{\lambda}\right) - s(x)\right)\right\|_{q(I)} \quad (r = 0, 1, \ldots, r_1).$$

To prove the remaining relations, it suffices to note that

$$\| f^{(k)}(\xi)\|_{p(I)} = \lambda^{k-1/p}\left\|\frac{d^k}{dx^k} f\left(\frac{x}{\lambda}\right)\right\|_{q(I)},$$

and that consequently the class $\widetilde{W}_p^k(I)$ maps onto itself under the transformation

$$\xi \to x/\lambda, \quad f(\xi) \to F(x) = \lambda^{k-1/p} f\left(\frac{x}{\lambda}\right).$$

For the other classes of splines, the arguments are very similar to this one, so we may regard the theorem as proved.

By virtue of Theorem 5, the determination of the orders of the left-hand sides of (69) and (70) as $\lambda \longrightarrow \infty$ reduces to the question of whether their right-hand sides are finite. Their finiteness for $n(u) = u^\alpha + 1$ was studied in §§1 and 2. The following theorems are consequences of these results and Theorem 5.

THEOREM 6. *Let $I = [0, \infty)$ or $(-\infty, \infty)$, let $1 \leqslant p \leqslant q \leqslant \infty$, let $l \geqslant 0$ and k, $1 \leqslant k \leqslant l + 1$, be integers, let $l_1 = \max(1, l)$, let $\alpha_k = \alpha_k(p, q) = 1 - (1/p - 1/q)/k$, and let $C_r^{p,q}(k, l)$ be constants defined by (24). Then for $\alpha > 0$ and $\alpha_k = \alpha_k(p, q) \leqslant \alpha \leqslant 1$ we have*

$$\sup_{f \in \widetilde{W}_p^k(I)} E_{Au^\alpha+1}(f)_{q(I)} \leqslant C_0^{p,q}(k, l)\left(\frac{l_1}{A\alpha^\alpha}\right)^{k\alpha_k/\alpha}, \tag{71}$$

$$\sup_{f \in \widetilde{W}_p^k(I)} \inf_{s \in S_{Au^\alpha+1}(I)} \max_{0 \leqslant r \leqslant k-1} \left\{\frac{\| f^{(r)}(x) - s^{(r)}(x)\|_{q(I)}}{D_r(\alpha)\, A^{-(k-r-1/p+1/q)/\alpha}}\right\} \leqslant 1, \tag{72}$$

where we may put $D_r(\alpha) = C_r^{p,q}(k, l)(l_1\alpha^{-\alpha})^{(k-r-1/p+1/q)/\alpha}$.

The inequalities (71) and (72) remain valid for $\alpha > 0$, $\alpha \geqslant \alpha_k$, and $D_r(\alpha) = D_r(1)(\alpha \geqslant 1, r = 0, 1, \ldots, k, -1)$ if we replace $S_{Au^\alpha+1}(I)$ and $E_{Au^\alpha+1}(f)_{q(I)}$ by $s_{Au^\alpha+1}(I)$ and $e_{Au^\alpha+1}(f)_{q(I)}$. For $\alpha \geqslant 1$ $(D_r(\alpha) = D_r(1))$ they also remain valid if we replace $S_{Au^\alpha+1}(I)$ and $e_{Au^\alpha+1}(f)_{q(I)}$ by $\sigma_{Au^\alpha+1}(I)$ and $\epsilon_{Au^\alpha+1}(f)_{q(I)}$.

All of these estimates are precise with respect to order for each α and for $A \longrightarrow \infty$.

The second inequality in the theorem gives a serial solution of the problem of simultaneously approximating a function and its derivatives by splines. The fact that it is precise means that, if we replace any one of the $A^{(k-r-1/p+1/q)\alpha}$ by a function $\varphi_r(A)$ of more rapid growth, the left-hand side becomes an unbounded function of A as $A \longrightarrow \infty$.

PROOF. This theorem follows from Theorems 1, 2 and 5, and the remark following Theorem 2.

THEOREM 7. *Let $I = [0, \infty)$ or $(-\infty, \infty)$, let $1 \leqslant q < p \leqslant \infty$, let $l \geqslant 0$ and k, $1 \leqslant k \leqslant l + 1$, be integers, let $l_1 = \max(1, l)$, let $\alpha_k = \alpha_k(p, q) = 1 + (1/q - 1/p)/k$, and let D_r be constants defined by (50). Then for $\alpha > \alpha_k(p, q)$*

$$\sup_{f \in \widetilde{W}_p^k(I)} e_{Au^\alpha+1}(f)_{q(I)} \leqslant \overline{D}_0 A^{-k\alpha_k/\alpha}, \tag{73}$$

and for $\alpha > \alpha_{k-v} = 1 + (1/q - 1/p)/(k - v)$, $v = 1, \ldots, k - 1$,

$$\sup_{f \in \widetilde{W}_p^k(I)} \inf_{s \in s_{Au^\alpha+1}(I)} \max_{0 \leqslant r \leqslant v} \left\{\frac{\|f^{(r)}(x) - s^{(r)}(x)\|_{q(I)}}{D_r A^{-(k-r-1/p+1/q)/\alpha}}\right\} \leqslant 1. \tag{74}$$

The inequalities (73) and (74) remain valid for the same α if we replace $e_{Au^\alpha+1}(f)_{q(I)}$ and $s_{Au^\alpha+1}(I)$ by $\epsilon_{Au^\alpha+1}(f)_{q(I)}$ and $\sigma_{Au^\alpha+1}(I)$. All of these estimates are precise with respect to order for $A \longrightarrow \infty$.

PROOF. This theorem follows from Theorems 3, 4 and 5.

For all of the remaining values of $\alpha \geqslant 0$, except the case $\alpha = 0$, $k = p = 1$, q

$= \infty$ ($\alpha_k(p, q) = 0$), the corresponding upper bounds in (71)–(74) either are undefined or become infinite. Theorem 1 (for $A = l_1 \geqslant 1$) can be proved in this case with a different (finite) constant on the right-hand side of (25). However, it is impossible to deduce the corresponding estimate for $A \longrightarrow \infty$, as was done above for the case $\alpha > 0$. We consider this case directly. It is all the more important that, for $\alpha > 0$, $\alpha_k(p, q) = 0$, the upper bound in (71) not approach zero as $A \longrightarrow \infty$. In the case $\alpha = 0$, each of the conditions $m_s(u) \leqslant Au^\alpha + 1$, $M_s(u) \leqslant Au^\alpha + 1$ means that the spline $s(x)$ has no more than $A + 1$ knots on I; therefore

$$e_{A+1}(f)_{q(I)} = E_{A+1}(f)_{q(I)}$$

($\epsilon_{A+1}(f)_{q(I)}$ is undefined).

THEOREM 8. *Let* $A \geqslant l + 1$. *For any function* $f \in W_1^1(I)$,

$$E_{A+1}(f)_{\infty(I)} \leqslant \frac{l+1}{A} \|f'\|_{1(I)}. \tag{75}$$

PROOF. Let $I = (-\infty, \infty)$. We put $v = [A/l] + 1$ for $l = 1, 2, \ldots$, and $v = [A] + 1$ for $l = 0$; we define ξ_i by

$$\int_{-\infty}^{\xi_i} |f'(t)|\, dt = \frac{i}{v} \quad (i = 1, 2, \ldots, v-1,\ \|f'\|_{1(I)} = 1). \tag{76}$$

Since the left-hand side depends monotonically and continuously on ξ_i and approaches 0 (1) as $\xi_i \longrightarrow -\infty$ ($+\infty$), it follows that ξ_i exists for each $i = 1, \ldots, v-1$. If the solutions of (76) fill out the entire interval for some i, then, for definiteness, we choose ξ_i so that $|\xi_i|$ is a maximum. Putting $t_0 = t_0^{(i)} = \xi_{i-1}$ ($l \geqslant 0$), $t_j = t_j^{(i)} = \xi_{i-1} + (j/e)(\xi_i - \xi_{i-1})$ ($j = 1, 2, \ldots, l$; $l > 0$), we define the spline $s(x) = s_l(x, f)$ on $[\xi_{i-1}, \xi_i]$ ($i = 2, 3, \ldots, v-1$) by the following conditions: for $l > 0$,

$$s(\xi_{i-1}) = f(\xi_{i-1}),\quad s(\xi_i) = f(\xi_i),$$
$$s^{(r)}(\xi_{i-1}) = s^{(r)}(\xi_i) = 0 \quad (r = 1, 2, \ldots, l-1;\ l > 1);$$

the points t_j ($j = 0, 1, \ldots, l$) are admissible knots of $s(x)$. If $l = 0$, we put $s(x) \equiv f(\xi_i)$ for $x \in (\xi_{i-1}, \xi_i]$. For $x \leqslant \xi_1$ we put $s(x) \equiv f(\xi_1)$, and for $x \geqslant \xi_{v-1}$ we put $s(x) \equiv f(\xi_{v-1})$ ($l \geqslant 0$). It is clear that $s(x)$ will be a spline on the entire interval I, and, by virtue of the choice of v, $M_s(u) \leqslant A + 1$.

To estimate $f(x) - s(x)$ on $[\xi_{i-1}, \xi_i]$, we use (20), from which we get

$$|f(x) - s(x)| \leqslant \frac{1}{l!} \int_{\xi_{i-1}}^{\xi_i} \left| \frac{d^l}{dt^l} \left\{ (x-t)_+^l - \sum_{j=0}^{l} \frac{(x-t_j)_+^l \omega_i(t)}{(t-t_j)\omega_i'(t_j)} \right| |f'(t)| \right\} dt$$

$$= \int_{\xi_{i-1}}^{\xi_i} |\rho_{0,l,1}(z(x), \tau(t))| |f'(t)|\, dt,$$

where

$$z(x) = (x - \xi_{i-1})/(\xi_i - \xi_{i-1}),\quad \tau(t) = (t - \xi_{i-1})/(\xi_i - \xi_{i-1}),$$

and $\rho_{0,l,1}(z, \tau)$ is defined by (23). Hence we find that

$$|f(x) - s(x)| \leqslant \max_{0 \leqslant z, \tau \leqslant 1} |\rho_{0,l,1}(z, \tau)| \int_{\xi_{i-1}}^{\xi_i} |f'(t)|\, dt = \frac{C_0^{p,q}(1, l)}{v}.$$

If $x \leqslant \xi_1$ or $x \geqslant \xi_{v-1}$, then

$$|f(x) - s(x)| = |f(x) - f(\xi_1)| = \left| \int_x^{\xi_1} f'(t)\, dt \right| \leqslant \int_{-\infty}^{\xi_1} |f'(t)|\, dt = \frac{1}{v},$$

$$|f(x) - s(x)| = |f(x) - f(\xi_{v-1})| \leqslant \left| \int_{\xi_{v-1}}^{x} |f'(t)|\, dt \right| = - \int_{-\infty}^{\xi_{v-1}} |f'(t)|\, dt = \frac{1}{v}.$$

respectively. We compute $C_0^{p,q}(1, l)$. Writing $l_1 = \max(1, l)$, we have

$$\rho_{0,l,1}(z, \tau) = \frac{1}{l!} \frac{d^l}{d\tau^l} \left\{ (z - \tau)_+^l - \sum_{j=0}^{l} \frac{(z - j/l_1)_+ (\tau^l + \ldots)}{\prod_{i \neq j} [(j - i)/l_1]} \right\} = (-1)^l (z - \tau)_+^0$$

$$- \sum_{j=0}^{l} \frac{(z - j/l_1)_+^l\, l_1^l}{(-1)^{l-j} j!\,(l-j)!} = (-1)^l (z - \tau)_+^0 - \frac{l_1^l}{l!} \sum_{j=0}^{l} C_l^j (-1)^{l-j}$$

$$\times (z - j/l_1)_+^l = \left\{ (-1)^l (z - \tau)_+^0 - \frac{l_1^l}{l!} \Delta_{1/l_1}^l (z - x)_+^l \right\}_{x=0}$$

where $\{\Delta_{1/l_1}^l (z - x)_+^l\}_{x=0}$ is the lth difference with increment $1/l_1$ with respect to x for $x = 0$. Therefore

$$\rho_{0,l,1}(z, \tau) = (-1)^l (z - \tau)_+^0 - \frac{1}{l!} \Delta_{1/l_1}^{v} \frac{d^{l_1-1}}{dx^{l_1-1}} (z - x)_+^l \big|_{x=x_z} = (-1)^l [(z - \tau)_+^0$$

$$- l_1 \Delta_{1/l_1}^{v} (z - x_z)_+^0],$$

where $v = l/l_1$ and $0 \leqslant x_z \leqslant 1$. Hence $C_0^{1,\infty}(1, l) = 1$.

Thus $|f(x) - s(x)| \leqslant 1/v$ for all $x \in I$. It remains to notice that $v \geqslant A/l$ for $l > 0$, and $v \geqslant A$ for $l = 0$.

In case $I = [0, \infty)$, we must replace (76) by

$$\int_0^{\xi_i} |f'(t)|\, dt = \frac{i}{v} \quad (i = 0, 1, \ldots, v - 1)$$

and ξ_1 by ξ_0 in all of the arguments. This completes the proof of Theorem 8.

BIBLIOGRAPHY

1. J. H. Ahlberg, E. N. Nilson and J. L. Walsh, *The theory of splines and their application*, Academic Press, New York, 1967. MR **39** #684.

2. Ju. N. Subbotin and N. I. Černyh, *The order of the best spline approximations of certain classes of functions*, Mat. Zametki 7 (1970), 31–42 = Math. Notes 7 (1970), 20–26. MR **41** #4077.

3. Géza Freud and V. A. Popov, *Certain questions connected with approximation by spline-functions and polynomials*, Studia Sci. Math. Hungar. **5** (1970), 161–171. (Russian) MR **42** #2225.

Translated by H. T. JONES